Telecommunications Cabling Installation

Library of Congress Cataloging-in-Publication Data

Telecommunications cabling installation / BICSI
   p. cm.
  ISBN 0-07-137205-9
   1. Telecommunication wiring.   I. BICSI.

TK5103.12.T46 2001
621.382—dc21

00-053377

McGraw-Hill

A Division of The McGraw·Hill Companies

Copyright © 2001 by BICSI. All rights reserved. Printed in the United States of America. Except as permitted under the United States Copyright Act of 1976, no part of this publication may be reproduced or distributed in any form or by any means, or stored in a database or retrieval system, without the prior written permission of the publisher.

1 2 3 4 5 6 7 8 9 0   DOC/DOC   9 0 9 8 7 6 5 4 3 2 1 0 9

ISBN 0-07-137205-9

The sponsoring editor for this book was Zoe G. Foundotos, the editing supervisor was Sally Glover, and the production supervisor was Sherri Souffrance. It was set in Century Schoolbook by North Market Street Graphics.

Printed and bound by R. R. Donnelley & Sons Company.

 This book is printed on recycled, acid-free paper containing a minimum of 50% recycled, de-inked fiber.

McGraw-Hill books are available at special quantity discounts to use as premiums and sales promotions, or for use in corporate training programs. For more information, please write to the Director of Special Sales, McGraw-Hill, Two Penn Plaza, New York, NY 10121-2298. Or contact your local bookstore.

Information contained in this work has been obtained by The McGraw-Hill Companies, Inc. ("McGraw-Hill") from sources believed to be reliable. However, neither McGraw-Hill nor its authors guarantees the accuracy or completeness of any information published herein and neither McGraw-Hill nor its authors shall be responsible for any errors, omissions, or damages arising out of use of this information. This work is published with the understanding that McGraw-Hill and its authors are supplying information but are not attempting to render engineering or other professional services. If such services are required, the assistance of an appropriate professional should be sought.

Acknowledgments

The BICSI officers and membership wish to thank the following members of the BICSI Engineering and Methods Steering Committee Panel 200 who contributed to the development of the *Telecommunications Cabling Installation Manual,* 2nd edition, and provided important feedback through their reviews of this manual.

Chair: J. Ray Craig, RCDD/LAN Specialist, *Craig Communications Service*

Committee: J. Allen Byrne, *Advance Power Systems*
Richard Dunfee, RCDD, *BICSI*
Robert Faber, Jr., RCDD/LAN Specialist, *The Siemon Company*
Joan Hersh, *BICSI*
Nelda Hills, *BICSI*
Phil Klingensmith, RCDD, *Ohio Valley Consulting*
Chuck Lohrmann, RCDD, *Lohrmann & Associates*
Bryan Moffitt, *Lucent Technologies*
Joe O'Brien, *Nelson Firestop Products*
Frank L. Perniciaro, *BellSouth Telecommunications Inc.*
Vic Phillips, RCDD, *Information Transport System Designers*
David E. Rittenhouse, RCDD, *Siecor Corporation*
John Siemon, P.E., *The Siemon Company*
Scott Smith, RCDD, *TeleTech Communications, Inc.*
Mike St. Angelo, RCDD, *Holmes & Narver, Inc.*
John Brazee, RCDD, ex officio, *Consultant*

About BICSI

BICSI, a not-for-profit telecommunications association, was founded in 1974 to serve and support the telephone company building industry consultants (BICs). BICs are responsible for the design and distribution of telecommunications wiring in commercial and multifamily buildings.

BICSI has grown dramatically since those early days and is expected to serve 22,000 members from every state and 85 countries around the world by the end of 2001. Our programs and interests cover the broad spectrum of voice, data, and video technologies. BICSI offers courses, conferences, publications, and registration programs for telecommunications cabling distribution designers and installers.

BICSI Member Benefits and Opportunities

BICSI members receive substantial discounts on quality education—design courses, conferences, and manuals. Especially for cabling installers, BICSI offers a Telecommunications Cabling Installation Training and Registration Program (details follow) as well as a Cabling Workshop and Cabling Installation Expo every year in the fall.

Members also gain access to valuable telecommunications information—with the *BICSI News, Region News, District News,* standards and regulatory updates, and BICSI's Web site.

BICSI members may also pursue and attain prestigious credentials—RCDD®; RCDD/LAN Specialty; Registered Installer, Level 1; Installer, Level 2; and Technician; and more.

For Further Information

BICSI looks forward to continuing growth in the future, both in size and in quality of service offered. For a complete packet of information, please contact:

BICSI World Headquarters
8610 Hidden River Pkwy.
Tampa, FL 33637-1000 USA
800-242-7405 or 813-979-1991
fax: 813-971-4311
e-mail: bicsi@bicsi.org
Web site: http://www.bicsi.org

BICSI Telecommunications Cabling Installation Training and Registration Program

As the telecommunications industry evolves and changes, cabling installation practices take on an important role in overall system performance. Even the most carefully designed system cannot function optimally unless it is properly installed.

BICSI has designed a comprehensive training and testing program for the cabling installation community within our industry. The reasons for this undertaking are pressing and compelling.

- High-performance cabling systems require proper design and installation.
- A multitiered career path is needed for the professional development of industry cabling installers.
- The industry needs vendor-independent criteria to identify competent cabling installers.
- Voluntary monitoring of the industry's cabling installation quality through registration and training is essential.
- The program allows prospective customers a method of selecting firms that employ BICSI registered cabling installers who adhere to high quality standards.
- Employment of BICSI registered cabling installers provides employers with a means of measuring the competency of job applicants.

The BICSI Program

The goal of BICSI's Telecommunications Cabling Installation Training and Registration Program is to produce highly competent cabling installers in a minimal amount of time and at a reasonable cost. Upon completion of training, program participants have been taught to conduct site surveys, pull wire/cable, and terminate and test copper and fiber optic cabling to the highest level of specification (currently Category 5e).

BICSI's program provides three levels of increased knowledge and experience: **Installer, Level 1; Installer, Level 2;** and **Technician.** The program offers core skills training, registration examinations, and structured on-the-job training (OJT) to meet the diverse needs of the telecommunications cabling industry.

The multiple levels of competency provide, for the first time, a career path for cabling installers. BICSI's cabling installation training provides the opportunity for continued professional and career development.

BICSI Policy for Numeric Representation of Units of Measure

BICSI technical manuals primarily follow the modern metric system, known as the International System of Units (SI). The SI is intended as a basis for worldwide standardization of measurement units. All units of measure in this manual are expressed in SI terms, followed by an equivalent empirical (U.S. customary) unit of measure in parentheses (see exceptions listed below).

Style Guidelines

- In general, SI units of measure are converted to an empirical unit of measure and placed in parentheses. Exception: When the reference material from which the value is pulled is provided in empirical units only, the empirical unit is the benchmark.
- In general, soft (approximate) conversions are used in this manual. Soft conversions are considered reasonable and practicable; they are not precise equivalents. In a few instances, precise equivalents (hard conversions) may be used when it is a:
  - Manufacturer requirement for a product.
  - Standard or code requirement.
  - Safety factor.
- For metric conversion practices, refer to ANSI/IEEE/ASTM SI 10-1997 *Standard for Use of the International System of Units (SI): The Modern Metric System.*
- Trade size is approximated for both SI and empirical purposes. Example: 100 mm (4 in) trade size. Although the units of millimeters and inches are indicated on trade size dimensions, it is important to note that the values do not necessarily correspond to the physical size of the conduit.

- American wire gauge (AWG) and plywood are not assigned dual designation SI units.
- When Celsius temperatures are used, an equivalent Fahrenheit temperature is placed in parentheses.

Contents

Chapter 1 Background Information   1

Industry Orientation   1

Overview   1
    Before and after divestiture   1
    Emergence of entrepreneurs and competition   2
    Merging of computers and telephone wiring systems   3
    Need for high performance standards and installers 4
BICSI® History   6

Structured Cabling Systems   6

Overview   6
    Standardization of cabling installation   7
    Generic structured cabling installation   8
Entrance Facilities   9
    Underground entrances   10
    Buried entrances   12
    Aerial entrances   13
Backbone Cables   14
Horizontal Cables   16
Work Areas   17
Equipment Rooms   18
Telecommunications Closets   19
Cross-Connects   20
Topologies   21
    Star topology   22
    Bus topology   23
    Ring topology   24
    Hybrid topologies   25

Standards and Codes   28

Introduction   28
Purpose of Codes and Standards   29
Recognized Regulatory Bodies   30

Codes Affecting Telecommunications — 31

National Fire Protection Association (NFPA) — 31
The National Electrical Code® (NEC) — 31
The Canadian Electrical Code, Part 1 — 33
    Applicable Telecommunications Sections — 33
Standards Affecting Telecommunications — 34
    Overview — 34
ANSI/TIA/EIA-568-A — 34
    Commercial Building Telecommunications Cabling Standard — 34
    ANSI/TIA/EIA-568-A-1—Propagation Delay and Delay Skew Specifications for 100 Ω 4-pair Cable — 36
    ANSI/TIA/EIA Telecommunications Systems Bulletin (TSB-67)—Transmission Performance Specifications for Field Testing of Unshielded Twisted-Pair Cabling Systems — 36
    ANSI/TIA/EIA Telecommunications Systems Bulletin (TSB-72)—Centralized Optical Fiber Cabling Guidelines — 37
    ANSI/TIA/EIA Telecommunications Systems Bulletin (TSB-75)—Additional Horizontal Cabling Practices for Open Offices — 37
ANSI/TIA/EIA-569-A — 37
    Commercial Building Standard for Telecommunications Pathways and Spaces — 37
ANSI/EIA/TIA-570 — 38
    Residential and Light Commercial Telecommunications Wiring Standard — 38
ANSI/TIA/EIA-606 — 39
    Administration Standard for the Telecommunications Infrastructure of Commercial Buildings — 39
ANSI/TIA/EIA-607 — 40
    Commercial Building Grounding and Bonding Requirements for Telecommunications — 40
Related Standards — 41
    Institute of Electrical and Electronics Engineers — 41
    International Standards Organization (ISO)/International Electrotechnical Commission (IEC) — 42

Plans and Specifications — 42

Overview — 42
Drawings — 43
Construction Specifications — 49
Request for Quote — 51
Summary — 51

Media — 52

Overview — 52
Unshielded Twisted-Pair (UTP) Copper Cable — 53
Screened Twisted-Pair (ScTP) Cable — 58
    Category 5 screened twisted-pair copper cable — 58
Shielded Twisted-Pair (STP) Copper Cable — 59
Coaxial Cable — 60
RG-8, IEEE 802.3 Coaxial Cable (Thicknet) — 62
RG-58A/U, IEEE 802.3 Coaxial Cable (Thinnet) — 63
Optical Fiber Cable — 63

Contents xiii

Connectorization — 69

Overview — 69
Types of Outlet Interfaces — 69
    Unshielded twisted-pair (UTP) connectors — 70
    Screened twisted-pair (ScTP) connectors — 77
    Shielded twisted-pair (STP) cable data connectors — 77
    Enhanced shielded twisted-pair (STP-A) cable data connectors — 78
    Coaxial connectors — 78
    "F" series coaxial connectors — 78
    "N" series coaxial connectors — 79
    BNC coaxial connectors — 80
    Optical fiber connectors — 80
Matching Connectors to Media Type — 82

Transmission — 83

Overview — 83
Power, Current, and Voltage — 84
AC and DC — 84
Frequency — 85
Analog — 85
Digital — 86
Megahertz and Megabits — 86
Decibel — 86

Copper Cable Media — 87

Overview — 87
Resistance — 87
Inductance — 89
Capacitance — 89
Characteristic Impedance — 89
Attenuation — 90
Return Loss — 90
Insulation — 91
Pair Twist — 92
Shielding — 92
Digital Transmission Speed — 93
Crosstalk — 93
Hardware — 94

Optical Fiber Cable Media — 95

Overview — 95
Bandwidth — 95
Attenuation — 96

Grounding and Bonding — 97

Overview — 97
Grounding System Components — 97

Safety — 98
Avoiding Electric Shock — 98
Preventing Electric Shock — 99
References — 99
Electrical Exposure (Telecommunications) — 101
Lightning Exposure — 101
Electrical Exposure (Building Structure) — 102
Ground — 103
Bonding — 103
American Wire Gauge — 103

Protection Systems — 104

Lightning Protection System — 104
Electrical Power Systems — 105
Grounding Electrode System — 106
Electrical Bonding and Grounding — 107
Electrical Power Protection — 107
Telecommunications Bonding and Grounding — 108
Telecommunications Bonding Principles — 109

Telecommunications Grounding Practices — 110

Overview — 110
Grounding Choices — 110
Using the Electrical Service Ground — 110
Installing a Grounding Electrode — 111
Physical Protection — 112
CATV — 112
Water Pipes — 113

Telecommunications Bonding Practices — 113

Bonding Connections — 113
Bonding Conductors — 116
Inspection — 116
Exposed Cables — 117

System Practices — 117

Small Systems — 117
Large Systems — 117
Telecommunications Closets — 119
Backbone Cable Protection — 119
Telecommunications Bonding Backbone (TBB) — 121
Coupled Bonding Conductor (CBC) — 121
Unshielded Backbone Cable — 122
Backbone Cable Shields — 122
Shielded Cabling Systems — 123
Exposed Cable Sheath Terminations — 123

Equipment Grounding — 124
Single-Point Equipment Grounding — 124
Receptacle Outlet Grounding — 124

Telecommunications Circuit Protectors — 124
Overview — 124

Protector Technology — 126
Primary Protectors — 126
    Carbon blocks — 126
    Gas tubes — 126
    Solid state — 126
Fuses and Fuse Links — 126
Secondary Protectors — 127
    Heat coil — 127
    Sneak current fuse — 127
    PTC resistors — 127
Enhanced Protection — 127

Primary Protector Installation Practices — 128
Overview — 128

Common Safety Practices — 128
Overview — 128
OSHA — 129
First Aid — 129
    First Aid, CPR, and the Law — 130
    First Aid Kits — 130
Emergency Rescue — 131
Communication — 132
Designating Work Areas — 133
Tools and Equipment — 134
Ladder Safety — 135
Personnel Lifts — 138

Personal Protective Equipment — 139
Overview — 139
Headgear — 139
Eye Protection — 140
Breathing Protection — 141
Lifting Belt — 141
Protective Footwear — 142
Gloves — 142
Detection Badges/Exposure Monitors — 142
Hearing Protection — 143
Safety Harness — 144

Contents

Clothing	144
Grooming	145

Hazardous Environments—Indoor 145

Overview	145
Electrical Hazards	145
Lightning Hazards	148
Access Floors	148
Catwalk Hazards	148
Crawl Space Hazards	149
Confined Spaces	149
Optical Fiber Hazards	151
Battery Hazards	152
Asbestos Hazards	152
Chemical Hazards	153

Hazardous Environments—Outdoors 154

Outside Plant	154

Professionalism 154

Overview	154
Customer Relations	155
Project Team	156
Courteous Communication Skills	157
Professional Appearance	158

Chapter 2 Planning 161

Overview	**161**
Customer's Drawings	**162**
Designer's Drawings	**163**
Materials List	**164**
Scope of Work	**164**
Contract	**165**
Project Schedule	**165**
Project Log	**166**
Site Survey	**166**
Initial Construction Meeting	**169**
Materials Ordered	**170**
Materials Received	**170**

Storage of Project Materials — 171
Job site — 171
Company location — 172
Distributor — 172

Distribution of Materials on Site — 173
Development of a Project Schedule — 173
Project Log — 174
Preinstallation Meeting — 174
Meetings — 174
Project Safety Plan — 175
Job Change Orders — 175
Strategy — 176
Attachment A Designer's Drawings — 177
Attachment B Material List — 178
Attachment C Labor List — 179
Attachment D Project Time Line — 180
Attachment E Project Schedule — 181
Attachment F Site Survey Checklist — 182
Attachment G Job Change Orders — 183
Attachment H Scope of Work (Example) — 184

Chapter 3 Installing Supporting Structures — 189

Overview — 189
Setting Up Telecommunications Closets — 190
Overview — 190
Plywood Backboards — 192
Installing Plywood — 193
Cable Trays — 195
Tubular Construction — 195
Rod Stock — 197
Installation — 200
D-Ring Installation — 202
Conduits — 203
Ground Wires — 204
Relay Racks — 205
Floor-Mounted Cabinets — 207
Wall-Mounted Racks and Cabinets — 208
Installation Checklist — 209

Installing Backbone Pathways — 209

Overview — 209
Electrical Metallic Tubing (EMT) — 210
IMC Conduit — 211
Galvanized Rigid Conduit (GRC) — 213
Cutting and Threading IMC and GRC — 214
Joint Make-Up — 215
Conduit Elbows and Bends — 216
Hangers — 217
Conduit Terminations — 218
Securing Conduit Formations — 220
Pathway Preparation — 220
Anchors — 221
Fasteners — 223

Installing Horizontal Pathways — 228

Stub-Up/Stub-Out Conduits — 228
Surface-Mounted Raceway — 228

Installing Grounding Infrastructure — 231

Overview — 231
Local Code Requirements — 235
Standards and Codes — 235
Ground Source — 236
Grounding Hardware — 236
Telecommunications Main Grounding Busbar (TMGB) — 237
Steps—Installing TMGB — 238
Telecommunications Bonding Backbone (TBB) — 240
Telecommunications Grounding Busbar (TGB) — 240
Telecommunications Equipment Bonding Conductor (TEBC) — 240
Telecommunications Bonding Backbone Interconnecting Bonding Conductor (TBBIBC) — 240
Alternating Current Equipment Ground (ACEG) — 241
Steps—Ground Test — 241

Installing Cable Support Systems — 242

Overview — 242
Steps—Install Cable Support Systems — 242

Chapter 4 Pulling Cable — 251

Overview — 251

Cable Pulling Setup — 254

Overview — 254
Steps—Pulling Setup — 255

Pulling Horizontal Cable in Conduit with Fishtape	**259**
Overview	259
Steps—Conduit Pull	260
Pulling Horizontal Cable in Open Ceiling	**266**
Overview	266
Steps—Open Ceiling	267
Pulling Backbone in Vertical Pathway—from Top Down	**269**
Overview	269
Steps—Vertical (Top Down)	272
Pulling Backbone in Vertical Pathway—from Bottom Up	**276**
Overview	276
Steps—Vertical (Bottom Up)	279
Pulling Backbone—Horizontal	**283**
Overview	283
Steps—Horizontal Backbone	285
Pulling Optical Fiber Cable	**287**
Overview	287
Steps—Optical Fiber	288

Chapter 5 Firestopping — 293

Overview	**293**
Firestopping	**296**
Role of Firestopping in Fire Protection	**297**
Secondary Functions of Firestop Seals	**298**
Firestopping Systems	**298**
Introduction	298
Appropriate Systems	299
Selecting Firestop Materials/Systems	299
Qualified Electrical Apparatus	299
Testing and Guidelines for Firestops	**300**
Fire Rating Classifications—United States	300
Fire Rating Classifications—Canada	301

Evaluation of Firestop Systems — 302

Introduction — 302
Qualification Testing for Openings — 302
Other Qualification Information — 302
Selecting a Firestop Assembly — 302
Firestop Systems — 304

Categories of Firestop Systems — 305

Introduction — 305
Mechanical Systems — 305
Nonmechanical Systems — 307

Nonmechanical Firestop Systems — 307

Introduction — 307
Types of Putty — 308
Collars — 308
Types of Caulk — 309
Cementitious Materials — 310
Intumescent Sheets — 310
Intumescent Wrap Strips — 310
Silicone Foams — 311
Premanufactured Pillows — 311
Sprays — 312
Firestop Blankets — 312

Firestopping for Brick, Concrete Block, and Concrete Walls — 312

Pipes, Cables, Conduits, Cable Trays, and Innerducts — 312
Pipes, Cables, Conduits, and Innerducts in Cored or Sleeved Openings — 313
Cable Trays — 315

Firestopping for Gypsum Board Walls — 315

Pipes, Conduits, and Innerducts — 315
Communications Cable — 315
PVC Innerduct — 316
Firestopping a Shaft Condition — 316
Cable Trays — 317

Firestopping for Floor Assemblies — 317

Making Penetrations — 317
Pipes, Cables, Conduits, Ducts, Innerduct, and Cable Trays — 317

Firestopping for Floor/Ceiling Assemblies — 317

Introduction — 317
Effects of Fire on Ceilings — 318

Pipe, Conduit, Innerduct, Cable Trays, and Cable Penetrations (in Ceilings) — 318
Pipe, Conduit, Innerduct, Cable Trays, and Cable Penetrations (in Floor/Ceilings) — 318
Choosing a Seal System — 318

Firestopping Considerations (General) — 319

Match Existing Conditions — 319
Selection Criteria — 319

Typical Installations — 319

Sealing a Floor Penetration with Putty — 319
Sealing an Outlet Box with Putty — 322
Steps—Restore Penetrations (General) — 323

Chapter 6 Cable Termination Practices — 327

Overview — 327

Pretermination Functions — 329

Overview — 329
Steps—Pretermination — 329

Copper IDC Termination — 333

Overview — 333
Steps—IDC Termination — 334
66-Block Termination — 336
110-Style Hardware — 338
BIX Hardware — 338
LSA Hardware — 340
Patch Panels — 341
Work Area Outlets — 343
Direct Connection — 344
Field-Constructed Patch Cords — 344
Screened Twisted-Pair (ScTP) — 345
Shielded Twisted-Pair (STP) — 346

Coaxial Cable Terminations — 348

Overview — 348
Steps—Coaxial Cable — 350

Fiber Termination — 351

Overview — 351
Steps—Heat-Cured Termination — 355
Steps—UV-Cured Termination — 362
Steps—Crimp-Style Termination — 368
Steps—Anaerobic-Style Termination — 373

Chapter 7 Splicing Cable — 375

Copper Cable — 375

Overview — 375
Safety — 381
Steps—Copper Splicing — 381

Optical Fiber Cable — 389

Overview — 389
Safety — 391
Steps—Length Determination — 391
Steps—Stripping — 392
Steps—Single-Fiber Mechanical — 393
Steps—Single-Fiber Automatic Fusion Splicers — 398
Steps—Single-Fiber Semi-Automatic Fusion Splicers — 404
Summary — 408

Chapter 8 Testing Cable — 409

Copper Cable — 409

Overview — 409
Introduction — 410
Field Testers — 411
  Volt-ohm-milliampere multimeter — 411
  Induction amplifier/tone generator — 411
  Wire map testers — 411
  Cable-end locator kit — 412
  Certification field testers — 412
  Certification test sets — 412
  Time domain reflectometer (TDR) — 413
  Optical fiber flashlight — 413
  Infrared conversion card — 413
  Low intensity laser — 414
  Strand identifier — 414
  Optical light source and power meter — 414
  Optical time-domain reflectometer (OTDR) — 414
  Telephone test set (butt set) — 414
  Test adapters, leads, and cables — 414

Horizontal Cabling—Unshielded Twisted-Pair (UTP), Screened Twisted-Pair (SCTP), and STP-A Cables — 414

Introduction — 414
Continuity Testing — 414
Cable Testers — 415
  Setup — 416
  Calibration — 416
  NVP calibration — 417
  Sanity checks — 418
  Testing — 418

TIA/EIA TSB-67 and ANSI/TIA/EIA-568-A and Addenda Requirements   418
Additional Tests   420
Tester Performance   422
    Interface adapters   423
    Durability   423
    Downloads   423
    Backup   424
    Delay skew—ANSI/TIA/EIA-568-A and Addenda   424
    Noise   424
    Maintenance and upgrades   424
    Batteries   425
Measurement Problems   425

Backbone Cable   426

Introduction   426
Continuity Testing   427
Length   427

Coaxial Cable—Data   428

Introduction   428

50-Ohm Coaxial Cable   429

Continuity Test   429
Length Determination   429

75-Ohm Coaxial Cable   431

Continuity Test   431
Length Determination   431

Optical Fiber Cable   432

Overview   432
    Preinstallation testing   432
    Acceptance testing   433
    Preventive maintenance testing   434

Light Source and Power Meter Testing—Channel or Link   435

Overview   435
Tools and Equipment   435
    Light source, multimode   435
    Light source, singlemode   435
    Power meter   436
    Reference adapter   437
    Test jumpers   437
    Two-way radios   438
Steps—Administrative   438
Steps—Connector Cleaning   439

Steps—Adapter Cleaning 439
Steps—Link Attenuation 440
Safety 441
Steps—Calibration 442
Steps—Reference Reading 446
Steps—Link Measurements 446
Test Measurement Documentation 448
    Optical fiber link attenuation record 448

Light Source and Power Meter Testing—Patch Cables   449

Overview 449
Reference Components 449
Mode Filter 449
Test Configuration 449

Optical Time Domain Reflectometer   450

Overview 450
Tools and Equipment 450
Applications 450
Theory of Operation 451
    Rayleigh scattering 451
    Fresnel reflection 452
    Backscatter level vs. transmission loss 452
The OTDR 452
    Laser light source 452
    Coupler/Splitter 453
    Optical sensor section 453
    Controller section 453
    Display section 455
OTDR Specifications 455
    Dynamic range 455
    Dead zone 456
    Importance of dead zone 457
    Resolution 458
    Accuracy and linearity 458
Wavelength 460
Operation 461
    Configuration 461
    Mainframe and optical card 461
    Fiber type 461
    Wavelength 461
    Connector 462
    Measurement parameters 462
    Distance range 462
    Resolution 463
    Pulse width 463
    Averaging 463
    Fault location 464
    Distance measurements 465
    Loss measurements 465

Quality factor (dB/km)	467
Reflectance	467
Measurement Problems	**467**
Nonreflective break	468
Gain splice	468
Ghost reflection	470
Testing Procedures	**471**
Considerations	471
Automatic Measurements	**472**
Autorange feature	472
Fiber analysis software	473
Field Test Equipment Selection	**474**
Copper media	475
Fiber media	475
Training	476
Documentation	476
Test data analysis	476
Manufacturer/Vendor support	477
Compatibility	477

Appendix A Fiber Performance Calculations Worksheet — 478

Appendix B Fiber Link Attenuation Record — 479

Chapter 9 Troubleshooting — 481

General Reference — 481

Overview	481
Test equipment	481
Wire map tester	482
Handheld cable tester	482
Certification test set	482
Cable tracer	483
Cable-end locator kit	483
Optical light source and power meter	483
Optical time domain reflectometer (OTDR)	483
Optical fiber flashlight	483
Infrared conversion card	483
Low-intensity laser	483
Multimeter	484
Electromagnetic field strength meter	484
Toner/Wand	484
Telephone test set	484
Test adapters, leads, and cables	484
Other equipment	484
Ladder	484
Two-way radio	484
Alcohol and wipes	485
Compressed air	485
Notebook	485
Inspection microscope	485
Documentation	485

Communication skills	485
Industry and standards knowledge	486
Safety standards and procedures	486
Troubleshooting tips for Isolation	486

Copper Cable — 489

Overview	489
Using Test Equipment for Troubleshooting	491
Interpreting test results	491
High attenuation (copper)	491
Attenuation-to-crosstalk ratio (ACR)	492
Incorrect capacitance	492
Excessive Near-End Crosstalk (NEXT)	493
Incorrect cable length	493
Excessive loop resistance	494
Incorrect connections	494
Incorrect impedance	495
Noise	495
Locating cables	496
Cable tracer	496
Cable-end locator kit	496
Copper Media	496
Isolation	497
Steps—UTP Cable	502
Steps—Coaxial Cable	504

Optical Fiber Cable — 505

Overview	505
Verify	506
Isolate	506
Repair	509
Safety	509
Test	510
Test Equipment	510
Steps—Fault Isolation	513

Appendix A Typical OTDR Fault Presentations — 516

Chapter 10 Retrofit Installations — 519

Retrofit Installations — 519

Overview	519
Planning for a Retrofit Installation	523
Scope of Work	526
Contract	526
Project Schedule	527
Project Log	527

Site Survey	527
Verify Existing Infrastructure	532
Steps—Verify Existing Infrastructure	534
Evaluating the Existing Infrastructure	537
Steps—Evaluating the Existing Infrastructure	540
Job Plan	541
Initial Construction Meeting	542
Materials	542
Development of a Project Schedule	543
Preinstallation Meeting	544
Meetings	544
Project Safety Plan	544
Strategy	545

System Cutover 545

Overview	545
Steps—Cutover	548
Remove Abandoned Cable and Equipment	549
Steps—Post Cutover Removal	550

Glossary 551

Bibliography 591

Index 593

Figures

Chapter 1 Background Information 1

Structured Cabling Systems 6

Figure 1.1	Example of structured cabling installation	8
Figure 1.2	Example of underground entrance	11
Figure 1.3	Example of buried entrance	12
Figure 1.4	Example of aerial entrance	14
Figure 1.5	Example of structured backbone cabling installation	15
Figure 1.6	Example of horizontal cable alternatives	17
Figure 1.7	Example of typical equipment room	19
Figure 1.8	Cross-connection and interconnection	20
Figure 1.9	Star topology	23
Figure 1.10	Bus topology	24
Figure 1.11	Ring topology	25
Figure 1.12	Tree topology	26
Figure 1.13	Star-wired ring topology	27
Figure 1.14	Clustered star topology	27
Figure 1.15	Hierarchical star topology	28

Plans and Specifications 42

Figure 1.16	A typical plot plan	44
Figure 1.17	A typical floor plan	45
Figure 1.18	A typical cross-section	46
Figure 1.19	Typical telecommunications plan	48
Figure 1.20	Another typical telecommunications plan	49
Figure 1.21	Representative symbols and abbreviations	50

Media 52

Figure 1.22	Typical UTP cable	53
Figure 1.23	Switchboard plug	54
Figure 1.24	Undercarpet cable	57
Figure 1.25	Screened twisted-pair (ScTP) cable	59
Figure 1.26	Shielded twisted-pair (STP) cable	60
Figure 1.27	RG-6 coaxial cable	61
Figure 1.28	RG-11U coaxial cable	61

Figure 1.29	RG-8 Thicknet coaxial cable	62
Figure 1.30	RG-58A/U Thinnet coaxial cable	63
Figure 1.31	Multimode fiber	64
Figure 1.32	Singlemode fiber	65
Figure 1.33	Tight-buffered optical fiber cable	66
Figure 1.34	Single stranded tight-buffered fiber cable	66
Figure 1.35	Loose-tube optical fiber cable	66
Figure 1.36	Side view of loose-tube optical fiber cable	67
Figure 1.37	Loose-tube furcating harness	67

Connectorization 69

Figure 1.38	8P8C modular plugs non-keyed	71
Figure 1.39	8P8C modular jack	72
Figure 1.40	Telecommunications outlet and connectors	73
Figure 1.41	Eight-position jack pin/pair assignments T568A	74
Figure 1.42	Optional pinout T568B	74
Figure 1.43	Optional pinout USOC	75
Figure 1.44	Telco 50-pin connector	76
Figure 1.45	DB connector	76
Figure 1.46	ScTP modular jack	77
Figure 1.47	Two 150 Ω STP-A outlet connectors	79
Figure 1.48	F-type connector	79
Figure 1.49	N-type connector	80
Figure 1.50	BNC connector	80
Figure 1.51	Connector ST compatible	81
Figure 1.52	Connector 568SC compatible	82

Transmission 83

Figure 1.53	Analog sine wave	85
Figure 1.54	Ohm's law	88
Figure 1.55	Crosstalk paths	94

Grounding and Bonding 97

Figure 1.56	Electrical hazards	100
Figure 1.57	Zone of protection	102
Figure 1.58	Cone of protection	103
Figure 1.59	Lightning protection system	105
Figure 1.60	Typical (small) electrical power system	106
Figure 1.61	Typical telecommunications ground	111
Figure 1.62	Compression connectors	114
Figure 1.63	Exothermic welding	114
Figure 1.64	Exothermic welding mold	115
Figure 1.65	Typical small system arrangement	118
Figure 1.66	Busbar	118
Figure 1.67	Typical building grounding infrastructure	120
Figure 1.68	Bullet bond clamp kit	122
Figure 1.69	Installation of a bond clamp	123
Figure 1.70	Fuse protection	127

Common Safety Practices 128

| Figure 1.71 | First aid kit | 130 |
| Figure 1.72 | Communications headset | 132 |

Figure 1.73	Designated work area	133
Figure 1.74	Tag broken tools	134
Figure 1.75	Step ladder	136
Figure 1.76	Extension ladder	137
Figure 1.77	Hardhat	140
Figure 1.78	Eye protection	140
Figure 1.79	Gloves	142
Figure 1.80	Hearing protection	143
Figure 1.81	Fire extinguisher	147
Figure 1.82	Breathing respirator	150
Figure 1.83	Rubber apron and gloves	152

Professionalism 154

Figure 1.84	Communications perceptions	155

Chapter 3 Installing Supporting Structures 188

Figure 3.1	Corner installation of plywood backboards	192
Figure 3.2	Installation using toggle bolts in drywall construction	193
Figure 3.3	Plywood installed using toggle bolts	193
Figure 3.4	Tubular cable tray	195
Figure 3.5	Suspended cable tray	195
Figure 3.6	Wall bracket (tubular)	196
Figure 3.7	Multilevel cable tray	197
Figure 3.8	Cable retaining posts	197
Figure 3.9	Rod-stock cable tray	198
Figure 3.10	Directional transition	199
Figure 3.11	Plan view of a typical telecommunications closet with cable tray installed on two walls	199
Figure 3.12	Elevation view of cable tray installed on the rear wall of a telecommunications closet	200
Figure 3.13	Typical backboard layout for D-ring installation	201
Figure 3.14	Mushroom	202
Figure 3.15	Vertical cable tray	203
Figure 3.16	Conduits on channel stock	204
Figure 3.17	Grounding bushing	204
Figure 3.18	Relay rack	205
Figure 3.19	Wall-mounted rack with hinge	207
Figure 3.20	EMT couplings	210
Figure 3.21	Intermediate metallic conduit	211
Figure 3.22	Intermediate metallic conduit coupling	211
Figure 3.23	Galvanized rigid conduit	212
Figure 3.24	Galvanized rigid conduit coupling	213
Figure 3.25	Cross-section of conduit OD vs ID	215
Figure 3.26	Pipe hanger	217
Figure 3.27	Screw heads	223
Figure 3.28	Example of H and C connectors	231
Figure 3.29	Typical building grounding infrastructure	233
Figure 3.30	TMGB	236
Figure 3.31	TGB	237
Figure 3.32	J-hook	243
Figure 3.33	Compression coupling	245
Figure 3.34	Set-screw coupling	245
Figure 3.35	Conduit hangers	246

Chapter 4 Pulling Cable 249

Figure 4.1	Secured area with safety cones and caution tape	254
Figure 4.2	Large reel and adjustable jackstands	254
Figure 4.3	Cable tree	255
Figure 4.4	Sheave and pulley hangers	256
Figure 4.5	Rolling hitch knot	260
Figure 4.6	Example of marked job floor plans with common symbols	262
Figure 4.7	Vacuum blowing a ball or a bag	263
Figure 4.8	Vacuuming a ball	264
Figure 4.9	Beam clamps, support rings, and J-hooks	265
Figure 4.10	A reel brake attached to a cable reel	270
Figure 4.11	A pull rope attached to cable lead	271
Figure 4.12	Bullwheel	271
Figure 4.13	Channel with straps	272
Figure 4.14	Cable on tray from vertical pathway	273
Figure 4.15	Backboard layout with D-rings	273
Figure 4.16	Wire mesh grips	278
Figure 4.17	Tugger in position and properly secured to a concrete slab	279
Figure 4.18	A properly secured pulley	279
Figure 4.19	Swivel to prevent cable twisting	283
Figure 4.20	Innerduct	285
Figure 4.21	Four-inch conduit with four one-inch innerducts	287
Figure 4.22	Connecting aramid yarn	287
Figure 4.23	Multiweave wire mesh grip with swivel pulling eye	287

Chapter 5 Firestopping 293

Figure 5.1	Elastomeric modules (within frames)	306
Figure 5.2	Mechanical firestop systems	307
Figure 5.3	Conduit penetration in concrete floor	313
Figure 5.4	Cable penetration in concrete wall	313
Figure 5.5	PVC innerduct penetration in concrete wall	314
Figure 5.6	PVC innerduct penetration in concrete floor	314
Figure 5.7	Qualified cable tray seal system in concrete wall	315
Figure 5.8	A qualified steel pipes system in gypsum board wall	316
Figure 5.9	Communications cable seal system for gypsum board wall	316
Figure 5.10	PVC innerduct penetration of gypsum board wall	317
Figure 5.11	Tear off putty	320
Figure 5.12	Building bottom of penetration seal	320
Figure 5.13	Filling the penetration	321
Figure 5.14	Building top on penetration	321
Figure 5.15	Overlap the pad on outlet box	322
Figure 5.16	Second pad of putty on outlet box	322
Figure 5.17	Joining the pads on outlet box	323
Figure 5.18	Conduit penetration through masonry, wall, or floor	324
Figure 5.19	Fire seal of drywall	324
Figure 5.20	Fire seal of wall or floor sleeve	325
Figure 5.21	Fire seal cable tray penetration	325

Chapter 6 Cable Termination Practices 327

Figure 6.1	Wiring schemes	330
Figure 6.2	Work area outlet	330

Figure 6.3	Cable management products	332
Figure 6.4	66-type, 110-type, BIX, and LSA tools	333
Figure 6.5	Sheath removal tools	334
Figure 6.6	66-block and 89 brackets	337
Figure 6.7	66-block distribution frame	337
Figure 6.8	110-style hardware	339
Figure 6.9	BIX mount with connector	339
Figure 6.10	LSA 8-pair and 10-pair block	340
Figure 6.11	Patch panels	341
Figure 6.12	Cables terminated on rear of panel with cable management bar	342
Figure 6.13	4-port faceplate	343
Figure 6.14	Modular furniture faceplate	343
Figure 6.15	Type A connector	347
Figure 6.16	Type B connector	348
Figure 6.17	Typical coaxial cable construction	349
Figure 6.18	Coaxial cable termination tools	349
Figure 6.19	Captive pin BNC connector	350
Figure 6.20	Cross section of an optical fiber cable	352
Figure 6.21	Fiber connectors	353
Figure 6.22	Multimode fiber	354
Figure 6.23	Fiber stripping tool	354
Figure 6.24	Exploded view of ST connector	355
Figure 6.25	Mark the buffer	355
Figure 6.26	Buffer removal	356
Figure 6.27	Mark syringe	357
Figure 6.28	Apply epoxy	357
Figure 6.29	Install connector	358
Figure 6.30	Curing oven	358
Figure 6.31	Scribe fiber	359
Figure 6.32	Remove excess fiber	359
Figure 6.33	Nub removal	359
Figure 6.34	Install boot	360
Figure 6.35	Polish jigs on disks	360
Figure 6.36	100x microscope	361
Figure 6.37	Fiber end examples	362
Figure 6.38	Mark the buffer	362
Figure 6.39	Buffer removal	363
Figure 6.40	Clean fiber	363
Figure 6.41	Mark syringe	363
Figure 6.42	Inject adhesive	364
Figure 6.43	Place connector on fiber	364
Figure 6.44	UV curing lamp	365
Figure 6.45	Scribe fiber	365
Figure 6.46	Remove excess fiber	365
Figure 6.47	Install strain-relief collar	366
Figure 6.48	Install boot	366
Figure 6.49	Moisten disks	366
Figure 6.50	Polish fiber	367
Figure 6.51	Compressed air cleaning	368
Figure 6.52	Preinstall boot	368
Figure 6.53	ST components	369
Figure 6.54	SC components	369
Figure 6.55	Mark the buffer	369
Figure 6.56	Cleave fiber	369
Figure 6.57	Cam tool	370
Figure 6.58	Insert fiber	370
Figure 6.59	Cam fiber	371

Figure 6.60	Crimp tube	371
Figure 6.61	Install boot	372
Figure 6.62	Install SC housing	372

Chapter 7 Splicing — 375

Copper Cable — 375

Figure 7.1	Horizontal, in-line vault splice supported in a rack	377
Figure 7.2	Splice opening with sheath bonds and endplate installed	378
Figure 7.3	In-line splice	378
Figure 7.4	Foldback splice	379
Figure 7.5	Modular connector	379
Figure 7.6	2-Bank splice	380
Figure 7.7	Wrapped splice with endcaps	380
Figure 7.8	Completed splice in cable rack	380
Figure 7.9	Scuffed sheath with markings on the sheath	383
Figure 7.10	Sheath opening with core wrapper in place and vinyl tape protection from shield	384
Figure 7.11	Sheath opening with shield connector, braid, and insulating sleeve	384
Figure 7.12	Support tube assembly	385
Figure 7.13	Double splicing head setup	386
Figure 7.14	Position of the splicing head to the cable groups	386
Figure 7.15	Cable pairs being placed into splice head	387
Figure 7.16	Module being checked for correct color code	387
Figure 7.17	Crimper being placed on splicing head	387
Figure 7.18	Spliced cable after being tied down	388
Figure 7.19	A splice wrapped with polyethylene and endcaps	388

Optical Fiber Cable — 389

Figure 7.20	Routing required splicing	389
Figure 7.21	Sample mechanical splice	393
Figure 7.22	Sample mechanical splice tool	393
Figure 7.23	Loading mechanical splice	394
Figure 7.24	Fiber stripper tool	394
Figure 7.25	Cleave length	395
Figure 7.26	Isopropyl alcohol (99 percent pure) wipe	395
Figure 7.27	Cleaving fiber	396
Figure 7.28	Close mechanical splice	397
Figure 7.29	Fusion splicer	398
Figure 7.30	Fiber alignment	399
Figure 7.31	No-Nik tool	400
Figure 7.32	Wipe fiber	400
Figure 7.33	Cleave fiber	401
Figure 7.34	Inserting fiber	402
Figure 7.35	Common occurrences	403

Chapter 8 Testing — 409

Figure 8.1	Basic link test configuration	416
Figure 8.2	Channel test configuration	417
Figure 8.3	Calibration of NVP	418

Figure 8.4	Typical attenuation graphic display	419
Figure 8.5	Typical NEXT graphic display	420
Figure 8.6	Typical ACR graphic display	422
Figure 8.7	Uncertainty zone for certification field testers	423
Figure 8.8	Ethernet segment—typical	428
Figure 8.9	Multimode light source	435
Figure 8.10	Singlemode light source	435
Figure 8.11	Power meter	436
Figure 8.12	Sample power meter reading	436
Figure 8.13	Reference adapter	437
Figure 8.14	Test jumpers	437
Figure 8.15	Calibration of meters	440
Figure 8.16	Field tester	442
Figure 8.17	Dust cap removal	442
Figure 8.18	Attach connector A	443
Figure 8.19	Prepare test jumper 2	443
Figure 8.20	Dust cap removal—light source	444
Figure 8.21	Attach connector C	444
Figure 8.22	Reference adapter	445
Figure 8.23	Attach connector B	445
Figure 8.24	Attach connector D (SC)	445
Figure 8.25	Connect test jumper 1	447
Figure 8.26	Meter reading	447
Figure 8.27	Test setup for the testing of an optical fiber jumper	450
Figure 8.28	OTDR trace	454
Figure 8.29	Dead zone measurements	457
Figure 8.30	Interpret fiber trace	464
Figure 8.31	LSA splice loss	466
Figure 8.32	Gain splice	469
Figure 8.33	Ghost reflection	471
Figure 8.34	OTDR test configuration—end to end attenuation	471
Figure 8.35	OTDR trace with automatic event selection	473
Figure 8.36	Event table linked to figure 8.35	474

Chapter 9 Troubleshooting   481

Copper Cable   481

Figure 9.1	Twisted-pair wire	497
Figure 9.2	Schematic of a transmission line	497
Figure 9.3	Troubleshooting flow chart	498

Optical Fiber Cable   505

Figure 9.4	Connector loss C2 to C1	508
Figure 9.5	Connector loss C1 to C2	508
Figure 9.6	Reflective loss	508
Figure 9.7	Reflective loss	509
Figure 9.8	Reflective break	516
Figure 9.9	Nonreflective break	516
Figure 9.10	Defective fusion splice	517
Figure 9.11	Defective mechanical splice	518

Chapter 10 Retrofit Installations — 519

Figure 10.1	Circuit layout records (CLRs)	521
Figure 10.2	Pre- and post-stressed concrete	530
Figure 10.3	Typical "as builts"	533
Figure 10.4	Signal generator and inductive amplifier	534
Figure 10.5	Conversion card	536
Figure 10.6	Float a backboard	538

Tables

Chapter 1 Background Information — 1

Industry Orientation — 1
Table 1.1 Telecommunications categories — 3

Standards and Codes — 28
Table 1.2 NEC codes affecting telecommunications — 32
Table 1.3 CEC sections — 33

Plans and Specifications — 42
Table 1.4 Specification book — 50

Media — 52
Table 1.5 High-pair-count cable, color-coded pairs — 55
Table 1.6 Color codes for 100 Ω UTP patch cables — 56
Table 1.7 High-pair-count cable, color-coded pairs, and binder groups — 57
Table 1.8 Optical fiber color code chart — 68
Table 1.9 Horizontal and backbone 62.5/125 μm optical fiber cable transmission performance parameters — 68
Table 1.10 Backbone singlemode optical fiber cable transmission performance parameters — 68

Connectorization — 69
Table 1.11 Cable connectors — 82

Grounding and Bonding — 97
Table 1.12 American Wire Gauge (AWG) — 104

Professionalism — 154
Table 1.13 Project communications — 158
Table 1.14 Appearance — 158

Chapter 3 Installing Supporting Structures — 188

Table 3.1	Checklist for a telecommunications closet installation	208
Table 3.2	Bend radius	215
Table 3.3	Adapting designs	218

Chapter 5 Firestopping — 293

Table 5.1	United States' fire rating classifications	301
Table 5.2	Canada's fire rating classifications	301

Chapter 6 Cable Termination Practices — 327

Table 6.1	Binder color-code chart	335
Table 6.2	Data patch cord pin wiring	345
Table 6.3	Voice patch cord pin wiring	345
Table 6.4	Coaxial cable characteristics	349

Chapter 7 Splicing Cable — 375

Copper Cable — 375

Table 7.1	Telecommunications cable color code	376

Chapter 8 Testing Cable — 409

Table 8.1	Troubleshooting	426

Chapter 9 Troubleshooting — 481

Table 9.1	Test set selection	499

Chapter 10 Retrofit Installations — 519

Table 10.1	Detailed cutover plan	520

Chapter 1

Background Information

Industry Orientation	1
Structured Cabling Systems	6
Standards and Codes	28
Plans and Specifications	42
Media	52
Connectorization	69
Transmission	83
Grounding and Bonding	97
Common Safety Practices	128
Professionalism	154

Industry Orientation

Overview

Advances in technology, along with the high levels of technical expertise demanded in all aspects of the telecommunications field, make training a greater requirement than ever before.

Before and after divestiture. In the early 1900s, telecommunications was defined as the provisioning of telephone service to residential and business subscribers. Many telephone companies provided a local dial tone within the same geographic area. The technology of the times used open wire on poles and cross arms with glass insulators. It was commonplace to see 10 to 15 cross arms with 10 pins and insulators on poles on both sides of the street.

People living across the street from each other had to call "long distance" to communicate via telephone. This created so much confusion that Congress enacted the Communications Act of 1934. This legislation regulated the telecommunications industry and provided for defined monopolies and geographic areas of operation. It allowed the Federal Communications Commission (FCC) to regulate all telecommunications on a nationwide basis.

Regulation provided order within the industry and allowed subscribers to deal with a single telephone company. This method of operation was maintained until Congress enacted legislation providing for the creation of the Rural Electrification Administration (REA).

Under the guidance of REA, rural areas of the United States were provided telephone service. Many local "Telephone Cooperatives" were formed and operated in a rural environment. These REA Cooperatives, as they were known, continue to operate today under the Rural Utilities Service (RUS).

A defining moment in the telephone business in the United States occurred in the 1984 antitrust court decision to divest AT&T of its monopoly control of long distance service. This breakup of the Bell System was called "divestiture."

Before divestiture, AT&T:

- Held a nationwide long-distance monopoly.
- Operated individual local telephone companies.
- Rented telephone equipment and systems to customers.

AT&T was responsible for the installation of telephone equipment and cable within Bell System operating areas. Bell System installation practices developed by AT&T were used by most local telephone companies.

From the 1930s until the 1950s,

- There were almost 5000 independent (non-AT&T) telephone companies.
- The total revenues earned by the independents amounted to less than 10 percent of all telephone company revenues.
- The state and federal government provided regulation.
- All nationwide long distance service was provided by AT&T.
- Local companies received rebates from AT&T from long distance revenues, which helped subsidize local service.

Emergence of entrepreneurs and competition. By 1950, competition started to appear for the Bell System. Challenges from new companies began to emerge in the form of lawsuits, and business for AT&T changed.

Tom Carter of Texas became a pivotal figure in the eventual breakup of the Bell System. Carter developed a cradle and telephone receiver that Texas oil men could take with them and get phone calls on the job without having to go 16 km (10 mi) to answer a call. AT&T sued Mr. Carter to deny him the right to connect privately owned telephone equipment to the Bell System network, and lost.

In St. Louis, AT&T cut off its service to a hospital rather than allow the hospital's doctors to use Carter's telephone, the "Carterfone."

In 1984, a decision by Judge Harold Greene resulted in the:

- Breakup of the Bell System into seven Regional Bell Operating Companies (RBOCs).
- Deregulation of the telecommunications industry.
- Start of the competitive environment.

Since divestiture, customers may own wiring inside of buildings and have the responsibility for:

- New installations.
- Retrofits on old work.
- Maintenance.

Merging of computers and telephone wiring systems. The means for transporting voice, data, and video communications are merging. At this time, each of these services utilize similar design and cabling methods.

In the past, voice, data, and video were either analog or digitally based. There is a common misconception that local area networks (LANs) involve only data systems; this is untrue. A LAN may also distribute voice and video.

Telecommunications usually fall into one or more of the following categories; however, the distinctions are blurring.

TABLE 1.1 Telecommunications Categories

Category	Examples
Voice	Telephone, modem, fax
Data	LAN, metropolitan area network (MAN), wide area network (WAN), mainframe computers, personal computers (PCs)
Video	Cable Television (CATV), Closed Circuit Television (CCTV), interactive video

Networks have been growing in size, dimension, and importance since the early 1980s. By 1990, over 58 percent of PCs used in business, industry, and government were connected to networks. Some of these networks extend for miles.

Earlier, small networks may have been casually wired over desks and along baseboards by the customer. As networks grew in size, complexity, and importance, the cabling installer's job became a critical part of new and retrofitted installations. The growth in the types of networks, adapters, software, and applications required that the cabling support sophisticated LAN technologies.

The increased size, complexity, and types of network management technologies require that the cabling installer be:

- Knowledgeable of cable, connectors, and systems.
- Capable of understanding standards-based designs.
- Able to efficiently install a vast variety of equipment.

Time, money, and performance become determining factors in the quality of the installation. To this end, new technologies such as optical fiber cable should be considered for new LAN systems. Optical fiber transmits audio, video, and data as coded light pulses, with many advantages over older methods, which utilized copper cables. Some advantages of optical fiber include:

- Expanded bandwidth.
- Immunity to external interference (electromagnetic interference/radio frequency interference [EMI/RFI]).
- High-speed transmission.
- Low transmission loss.
- Lightweight materials.
- Small size.

Still, the most commonly used type of cable in telecommunications installations in North America is copper unshielded twisted-pair (UTP). In Europe, it is screened twisted-pair (ScTP). The wires are in pairs that are twisted together to minimize the effect of electrical interference.

The unwanted transfer of information from one circuit to another (crosstalk) is dependent upon such factors as the number of different pair twists per foot and the relationship of these twists.

Need for high performance standards and installers. When standards were resident in telephone companies, AT&T controlled the quality of the installations. Before 1984, the Bell System and other organizations prepared and utilized extensive documentation that provided detailed instructions on every task associated with telephone equipment and cabling installation. The REA, now RUS, prepared and disseminated the documentation used by the telephone cooperatives.

After divestiture, national and industry-sponsored standards bodies such as the American National Standards Institute (ANSI), Telecommunications Industry Association (TIA), and Electronic Industries Alliance (EIA) assumed leadership and began to provide standards for acceptable performance and installation.

With divestiture, the open, competitive market created a condition in which:

- Any contractor was eligible to compete for installation work.
- The required level of expertise was not always available.
- Customers became responsible for evaluating the qualifications of the installers and the quality of the installation.

In February of 1996, the Telecommunications Reform Act of 1996 was signed into law. One of the basic features of the Act is extensive deregulation, as seen by provisions that:

- Permit telephone companies to engage in telephone equipment manufacturing.
- Deregulate cable rates.
- Enable local phone companies to offer long distance telephone service.
- Allow competition in local telephone service markets.
- Eliminate prohibitions on cable and telephone company cross-ownership.

This Act provides an important stepping stone for companies to compete in the business by providing local dial tone to both business and residential customers.

The CATV industry has taken advantage of this opportunity to begin providing local telephone service over their existing facilities in several locations. In addition, some of the local telephone companies are entering the CATV business.

In compliance with this legislation, the FCC is reviewing several applications for local telephone companies to enter the long distance market. Access to this market has been denied to the RBOCs since divestiture.

The telecommunications market has become deregulated to the point that now, with enough financial resources, any company can enter any aspect of telecommunications.

BICSI® history

The Architects and Builders Service was created in the late 1920s, when telephone companies realized the need to work with architects and engineering firms. This service assisted the architects and engineers in developing specifications for installing telephone service inside of commercial and residential buildings. One of the first publications issued was the *Planning for Home Telephone Conveniences*. Published in 1929 and distributed without charge to architectural and engineering firms, the document established methods for installing inside telephone plants and was a forerunner of this and other BICSI manuals available today.

The Architects and Builders Service continued through the 1960s until AT&T determined that a better way was needed to provide the service. Woody Bradley, with AT&T, and Ross Cotton, with Bell Canada, teamed up to develop a program called the Building Industry Consulting Service. This program was implemented in the United States and Canada, almost simultaneously, in the late 1960s.

At about the same time, engineers who worked for regulated telephone companies began to hold informal meetings at the University of Kentucky. They exchanged information on how to better design telephone structures in commercial buildings. These informal gatherings progressed to a formal gathering and an annual conference in Kentucky and Florida.

In 1997, BICSI was incorporated as the Building Industry Consulting Service International, Inc., in the state of Kentucky. In 1980, BICSI established its corporate offices in Tampa, Florida. In the early 1990s, as the industry changed and deregulation occurred, the term Building Industry Consulting Service was not widely recognized as a telecommunications term. Therefore, BICSI now refers to itself simply as "BICSI, a Telecommunications Association."

The first formal manual, BICSI *Engineering and Standards Handbook,* was published in the early 1980s. The *Telecommunications Distribution Methods Manual,* now in its eighth edition, is considered a valuable reference tool and is the basis for the Registered Communications Distribution Designer (RCDD®) program. BICSI currently produces a variety of technical publications, including the *Telecommunications Cabling Installation Manual,* which is now in its second edition.

Structured Cabling Systems

Overview

A structured cabling system is defined as the complete collective configuration of cabling and associated hardware at a given site that has been installed to provide a comprehensive telecommunications infrastructure. This infrastructure is intended to serve a wide range of usage, such as to provide telephone service or computer networks, and should not be device dependent.

In this manual, a structured cabling system is further defined in terms of ownership. The structured cabling system begins at the point at which the service provider (SP) terminates. This point is also known as the point of demarcation (demarc) or Network Interface Device (NID).

For example, in a telephone system installation, the SP furnishes one or more service lines (per customer requirements). The SP connects the service lines at the point of demarcation.

Every structured cabling system is unique. This is due to variations in:

- The architectural structure of the building which houses the cabling installation.
- Cable and connection products.
- The function of the cabling installation.
- The types of equipment that the cabling installation will support, both present and future.
- The configuration of an already installed cabling installation (in the cases of upgrades and retrofits).
- Customer requirements.
- Warranties offered by manufacturers.

Standardization of cabling installation. Although the specifics of an installation might be unique, the overall components of a structured cabling system and the methods used to complete and maintain the installation are relatively standard. The standardization of cabling installations is necessary because of the need to ensure acceptable system performance from increasingly complex arrangements. The cabling industry in the United States accepts ANSI/TIA/EIA as the responsible organization for providing and maintaining standards and practices within the profession.

ANSI, in conjunction with TIA/EIA, has published a series of standards to design, install, and maintain cabling installations of all types. These standards, which are described in some detail later in this chapter, help to ensure a proper cabling installation.

Benefits of standards include the following:

- Consistency of design and installation
- Conformance to physical and transmission line requirements
- Provides a basis for examining a proposed system expansion and other changes
- Provides for uniform documentation

Generic structured cabling installation. The following illustration shows a generic structured cabling installation. The industry standard term for a network installation that serves a relatively small area (such as a structured cabling installation serving a building) is a LAN.

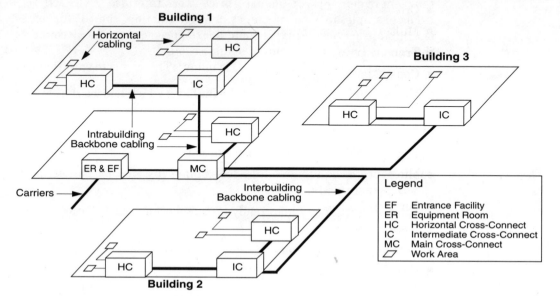

Figure 1.1 Example of structured cabling installation.

There are other types of network configurations, including MAN and WAN. Other sections in this chapter provide details about LANs, MANs, WANs, and other types of networks which involve cabling installations.

Keep in mind that the only type of installation covered by the tasks in this document is an installation within a single building. Outside plant systems are covered in the BICSI *Customer-Owned Outside Plant Manual*.

Structured cabling installations include most or all of the following components:

- Entrance facilities
- Vertical and horizontal backbone pathways
- Vertical and horizontal backbone cables
- Horizontal pathways

- Horizontal cables
- Work area outlets
- Equipment rooms
- Telecommunications closets
- Cross-connect facilities
- Multiuser Telecommunications Outlet Assemblies (MUTOA)
- Transition points
- Consolidation points

Entrance facilities

The entrance facility includes the cabling components needed to provide a means to connect the outside service facilities to the premises cabling. This can include the following:

- Service entrance pathways
- Cables
- Connecting hardware
- Circuit protection devices
- Transition hardware
- Other equipment

The service provider is responsible for the installation of the above services to a specified point of demarcation, which is the interface between the service provider's facility and the customer. For purposes of this manual, the cabling installer is responsible for extending services from the demarc to the structured cabling system of the tenant(s).

Entrance facility pathways and spaces are to be completed according to the requirements set forth in ANSI/TIA/EIA-569-A. Refer to the section on standards later in this chapter for more information.

An entrance facility is meant to provide the following:

- The network demarcation point between the service providers and customer premises cabling (if required).
- Placement of the electrical protection governed by the applicable electrical codes.
- House the transition between cabling used in the outside plant to cabling approved for intrabuilding construction. This usually involves

transition to fire-rated cable. (Refer to Chapter 1, Section 3 of this manual and ANSI/NFPA-70 *National Electrical Code®*, 1999 edition, Section 800-50 and Section 770-53, for details.)

An entrance facility must enter and terminate within the building at the most suitable location needed to serve the occupants of the building. The service entrance facility includes the:

- Route these facilities follow on private property.
- Entrance point to the building.
- Termination point within the building.

The service entrance facility depends on the:

- Type of facility being used (e.g., underground, tunnel, buried, or aerial).
- Route for the facility.
- Building architecture.
- Aesthetic considerations.

Although optical fiber cables are specified in many situations, the most common medium for providing connections to the service providers is still copper cable.

There are four principal types of entrance facilities:

- Underground
- Tunnel
- Buried
- Aerial

Underground entrances. Underground entrances use conduit or other types of mechanical pathways to provide out-of-sight service to a building. The pathway for an underground service:

- Is usually provided by the building owner from the building to the property line (or to another customer-owned building on the same property).
- Runs from the building entrance location to a pole, pedestal, or maintenance hole provided and maintained by the service provider.

The advantages of underground entrances are that they:

- Preserve the aesthetic appearance of the building.
- Are adaptable for future facility placement or removal.
- Are economical over a long life.
- Provide the security of additional physical cable protection.
- Minimize the need for possible subsequent repairs to the property when growth is required for existing facilities.

The disadvantages of underground conduit entrances are that they:

- Have a high initial installation cost.
- Require careful route planning.
- Provide a possible path for water or gas to enter buildings if improperly sealed.
- Usually take more time to install where facility provisioning must be expedited.

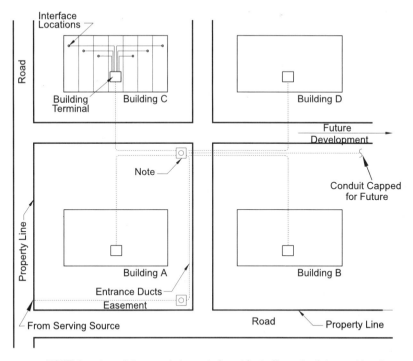

NOTE: Locate maintenance holes out of road for traffic and safety considerations.

Figure 1.2 Example of an underground entrance.

Buried entrances. Buried entrances (trenched or plowed) are a means of providing out-of-sight service to a building without conduit. The trench:

- Is usually provided by the building owner from the building to the property line (or to another customer-owned building on the same property).

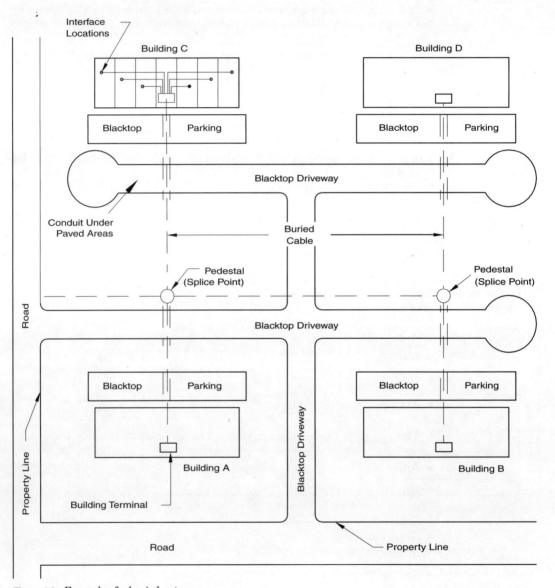

Figure 1.3 Example of a buried entrance.

- Runs from the building entrance location to a pole, pedestal, or maintenance hole. From here, the service provider takes it to a pole or maintenance hole in the provider's system.
- Requires a sleeve or entrance conduit through the perimeter wall.

The advantages of buried entrances are that they:

- Preserve the aesthetic appearance of the building.
- Usually have a lower initial installation cost than an underground installation.
- Can easily bypass obstructions, compared to underground installations.

The disadvantages of buried entrances are that they:

- Are inflexible for future service reinforcements or changes.
- Do not provide the same physical protection for the cable sheath as underground systems.
- May be difficult to locate, unless metallic warning tape or other means are colocated, especially in the case of optical fiber with no metallic member.
- Discourage accurate route planning and record keeping.

Aerial entrances. Aerial entrances are another means of providing service to a building. *Aerial* refers to the cables placed overhead.

The advantages of aerial entrances are that they:

- Usually provide the lowest installation cost.
- Are readily accessible for maintenance.

The disadvantages of aerial entrances are that they:

- Affect the aesthetic appearance of the building.
- Are subject to traffic and pedestrian clearances.
- Can damage building exterior.
- Are susceptible to environmental conditions, such as falling tree limbs and wind and ice loading.
- Are usually joint-use installations with the power company, CATV company, and telephone or data service providers.

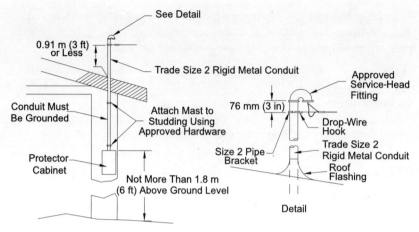

NOTES: 1. Iron pipe must be permanently and effectively grounded.
2. This arrangement is limited to conduit attachments of up to four lines.
3. Service mast must be sufficiently high to provide proper drop-wire clearance over sidewalks, streets, or roadways in compliance with codes and regulations.

Figure 1.4 Example of an aerial entrance.

Backbone cables

The term *backbone* is used to describe cables which handle the major network traffic. The ANSI/TIA/EIA-568-A standard defines backbone cabling as follows:

> The function of the backbone cabling is to provide interconnections between telecommunications closets, equipment rooms, and entrance facilities in the telecommunications cabling system structure. Backbone cabling consists of the backbone cables, intermediate and main cross-connects, mechanical terminations, and patch cords or jumpers used for backbone-to-backbone cross-connection. Backbone cabling also includes cabling between buildings.

There are two types of backbone cables: interbuilding and intrabuilding. Interbuilding backbone cable is defined as a cable that handles traffic between buildings. Intrabuilding backbone cable is defined as a cable that handles traffic between closets in a single building.

The standard also identifies two administrative levels of backbone cabling: first-level and second-level backbone. A first-level backbone is a cable that is installed between a main cross-connect (MC) and either an intermediate cross-connect (IC) or a horizontal cross-connect (HC). A second-level backbone is a cable that is installed between an IC and an HC.

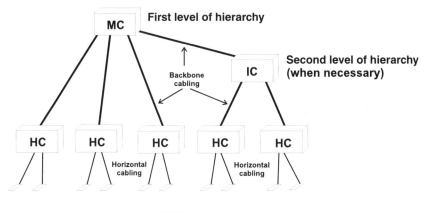

Figure 1.5 Example of a structured backbone cabling installation.

The main components of backbone cabling are as follows:

- Cable pathways—shafts, conduits, raceways, and floor penetrations (such as sleeves or slots) that provide routing space for the cables
- The actual cables—optical fiber, twisted-pair copper, coaxial copper, or some combination of these cables
- Connecting hardware—connecting blocks, patch panels, interconnections, cross-connections, or some combination of these components
- Miscellaneous support facilities—cable support hardware, firestopping, and grounding hardware

Certain considerations should be addressed when designing a backbone cabling system:

- The useful life of the backbone cabling system is expected to consist of several planned growth periods (typically three to ten years). This is shorter than the overall life of the premises cabling system.
- Prior to the start of a planning period, the maximum amount of backbone cable for the planning period should be projected; growth and changes during this period should be accommodated without installing additional cabling.
- Planning the routing and support structure for copper cabling should avoid areas where potential sources of EMI may exist.

> **Note.** The terms *horizontal* and *backbone* (previously called riser) have evolved from the orientations that are typical for functional cables of these types. However, the physical orientation of the cabling has no bearing on classifying the cable as horizontal or backbone.

Horizontal cables

The term *horizontal* is used since this part of the cabling system typically runs horizontally along the floor(s) and ceiling(s) of a building. The ANSI/TIA/EIA-568-A standard defines horizontal cabling as follows:

> The horizontal cabling is the portion of the telecommunications cabling system that extends from the work area telecommunications outlet/connector to the horizontal cross-connect in the telecommunications closet. The horizontal cabling includes the horizontal cables, the telecommunications outlet/connector in the work area, the mechanical termination, and patch cords or jumpers located in the telecommunications closet.

Horizontal cabling systems consist of two basic elements:

- Horizontal cable and its connecting hardware provides the means of transporting signals between the work area outlets and the horizontal cross-connect located in the telecommunications closet. This type of cabling and its connecting hardware is referred to as a basic link.
- Horizontal pathways and spaces are used to distribute and support horizontal cable and connecting hardware between the work area outlet and the telecommunications closet.

Certain considerations should be addressed when selecting horizontal cabling:

- Maximum horizontal cable length is 90 m (295 ft).
- A minimum of two telecommunications outlets per work area is required to meet ANSI/TIA/EIA-568-A standards.
- The horizontal cabling is usually not readily accessible. The time, effort, and skills required can make future changes costly.
- It should accommodate a variety of user applications (i.e., voice, data, video, and other low-voltage services) to minimize required changes as needs evolve.
- The routing and support structure for copper horizontal cabling shall avoid areas where potential sources of EMI may exist.

ANSI/TIA/EIA-568-A requires the selection of two media to service the telecommunications in a work area:

Background Information

- A 100 Ω, unshielded twisted-pair cable (Category 3, 4, or 5).
- A 100 Ω, screened twisted-pair cable (Category 3, 4, or 5).
- A STP-A, enhanced, shielded twisted-pair cable.
- A pair of 62.5/125 μm strands of optical fiber.

Any combination of the above media can be used, as long as one of the media is a 100 Ω, unshielded twisted-pair cable.

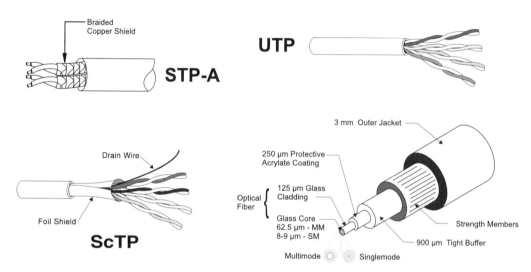

Figure 1.6 Example of horizontal cable alternatives.

Work areas

A typical work area is usually an area of approximately 10 square meters (100 square feet). Work areas include the components that extend from the work area outlet to the equipment. They are outside of the scope of ANSI/TIA/EIA-568-A. These components can include equipment such as telephones, data terminals, video equipment, and computers. Also included within the work area is the cabling that routes from this equipment to the work area outlet.

When planning for work area cabling, keep the following items in mind:

- Patch cords are designed to provide easy routing changes.
- The maximum horizontal cable length has been specified with the assumption that a maximum length of 3 m (10 ft) of patch cord has been used in the work area.
- A patch cord with identical connectors on both ends is most commonly used. Patch cords should be factory manufactured.

- Copper patch cordage is required to be constructed with stranded cable. This allows for flexibility. Patch cords induce 20 percent higher attenuation.
- When application-specific adapters (i.e., baluns, modular adapters, etc.) are needed at the work area, ANSI/TIA/EIA-568-A requires that they be external to the work area outlet.

> **Note.** Cabling adapters in the work area may have detrimental effects on the transmission performance of the cabling system. It is important, therefore, that adapter compatibility with premises cabling and equipment be considered before they are connected to the telecommunications network.

Equipment rooms

An equipment room is a special-purpose room that provides space and maintains a suitable operating environment for large telecommunications equipment. Equipment rooms are generally considered to serve an entire building (or even a campus), whereas a telecommunications closet serves one floor of a building or a portion of a floor.

Equipment rooms are sometimes referred to as main closets. Equipment rooms:

- Terminate and cross-connect backbone and horizontal cables.
- Provide work space for service personnel.
- Are designed according to specific requirements associated with the cost, size, growth, and complexity of the equipment.
- Can also serve as a portion of an entrance facility or as a telecommunications closet.
- House large pieces of common control equipment such as voice, data, video, fire alarm, energy management equipment, or intrusion detection equipment.

Although an equipment room usually serves an entire building, occasionally buildings use more than a single equipment room in order to provide one or more of the following:

- Separate facilities for different types of equipment and services
- Redundant facilities and disaster avoidance
- A separate facility for each tenant in a multitenant building

Certain considerations should be kept in mind for an equipment room:

- It must be versatile. An equipment room must be designed to accommodate both current and future applications. It must have provisions for growth and the ability to go through numerous equipment replacements and upgrades during its life, with minimal service disruption and cost.
- It must meet lighting, air conditioning, floor loading, electrical, and minimum space requirements.

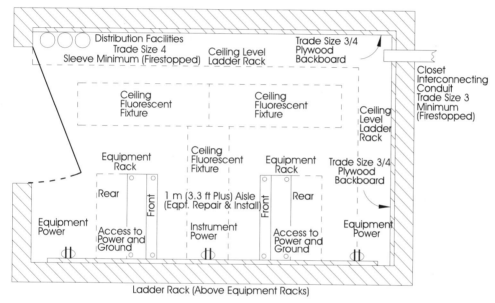

Figure 1.7 Example of a typical equipment room.

Telecommunications closets

Telecommunications closets differ from equipment rooms and entrance facilities in that they are generally considered to be floor-serving as opposed to building-serving facilities.

Note. The word *closets* is a misnomer since most of the spaces used as closets are actually rooms. The ANSI/TIA/EIA-568-A standard requires that dedicated, floor-serving telecommunications closets be provided to service flexible, high density horizontal distribution systems in floors and ceilings to facility work areas.

These closets:

- Serve as a point of termination for horizontal and backbone cables on compatible connecting hardware.

- House the horizontal cross-connect—a collective reference for the connecting hardware, jumpers, and patch cords used for completing cross-connection of horizontal and backbone cable terminations.
- May contain intermediate cross-connect points for different parts of the backbone cabling system.
- Provide a controlled environment for the telecommunications equipment, connecting hardware, and splice closures.

Consider the following in planning telecommunications closets:

- The size of the building, the usable floor space served, the occupant needs, and the telecommunications service to be used.
- Optimize the ability of the telecommunications closet to accommodate change.
- Meet lighting, air conditioning, floor loading, electrical, and spacing requirements.

Observe all applicable codes and rules regarding the use of telecommunications closets during construction.

Cross-connects

One of the primary functions of the telecommunications closet is to house cross-connections and interconnections. The following figure shows a comparison of the two:

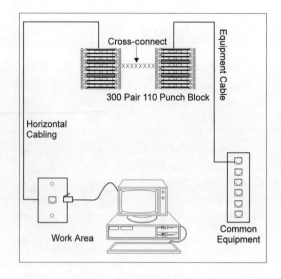
Cross-Connection

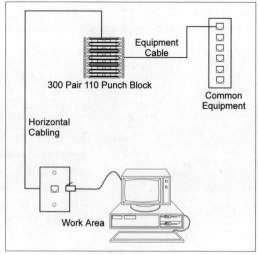
Interconnection

Figure 1.8 Cross-connection and interconnection.

Cross-connections are required between horizontal and backbone cable and when multiport equipment connections (e.g., 25-pair connectors) attach to horizontal or backbone cabling. Interconnections may be made when single-port equipment connections (e.g., 4-pair and optical fiber connectors) attach to horizontal or backbone cabling.

There are three types of cross-connects:

- Main cross-connect—the cross-connect in the equipment room for connecting entrance cables, backbone cables, and equipment cables
- Intermediate cross-connect—the cross-connect points located between the main cross-connect and horizontal cross-connects in backbone cabling
- Horizontal cross-connect—a location for the cross-connect of horizontal cabling to other cabling and equipment

Cross-connections are available in two types:

- Patch cords (UTP, ScTP, STP-A, optical fiber)
- Copper jumpers

ANSI/TIA/EIA-568-A makes the following recommendations pertaining to all cross-connects:

- Horizontal and backbone cabling are to be permanently terminated on connecting hardware that meets the requirements of ANSI/TIA/EIA-568-A.
- All connections between backbone and horizontal cables are to be cross-connections.
- Equipment cables that consolidate several ports on a single connector are to be terminated on dedicated connecting hardware.
- Equipment cables may be either cross-connected or interconnected to horizontal or backbone terminations (refer to Figure 1.8).
- Direct interconnections reduce the number of connections required to configure a link; however, this may reduce flexibility.
- When used as cross-connects, patch cords shall be made of stranded cable.

Topologies

The definition of the word *topology* states that an item's topology defines its physical appearance. For example, a topological map represents the physical appearance of the area shown. Topologies can have a physical, electrical, or logical configuration. In many ways, a standard cabling plan's topology is the same—it is representative of the standard cabling plan's physical appearance.

LAN topology is related to the topology of the physical medium that supports it. LAN topology is primarily determined by how transmission channels are used to connect network devices. Typically, it refers to how the LAN is physically set up, its electrical and logical configuration, and the cabling strategy being used. It is acknowledged that topology is the foundation of a LAN.

It should be pointed out that within the context of LANs, the word *topology* takes on a dual meaning. Both aspects are important to how the LAN will function.

- First, topology refers to the physical appearance of the LAN. This is known as the physical topology.
- The second aspect refers to how the LAN functions. This logical topology is determined by how the messages are transmitted from device to device.

There are many instances where a LAN has a certain physical appearance but logically transmits its messages in a different manner. For this reason, it is necessary to make the distinction between the physical topology and the logical topology of a LAN. The purpose of this section is to illustrate the physical appearance of a LAN.

There are three fundamental topologies—star, bus, and ring. From these three, a number of hybrid topologies have developed, including tree, star-wired ring, clustered star, and hierarchical star.

In structured cabling systems, the required physical topology is the star topology. Because it can be configured to accommodate each of the logical topologies (star, bus, ring, and hybrids), the star topology offers the greatest flexibility.

Star topology. In a star topology, the hub or switch is placed in the physical, as well as logical, center of the network. The remaining network devices are connected to this central hub like the points on a star.

Each device has its own direct, dedicated line to the hub or switch. Any network device wanting to send a message to another network device does so through the central hub. The station sending the message sends it to the hub. The hub then routes the message to the specified destination station—this is known as switching. Since the hub handles all the message switching, the stations on the network do not require any extra technology to route signals over the transmission channel.

Star topology advantages:

- Cabling is easier to install and maintain.
- If one device is disabled or isolated from the central hub, it is the only device affected.

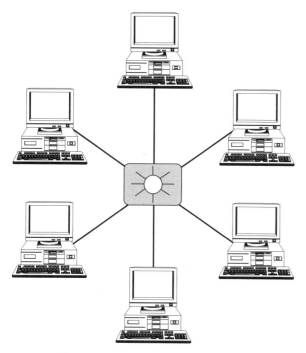

Figure 1.9 Star topology

- Faults are easier to locate and isolate.
- It provides a central location for managing the network.

Star topology disadvantage:

- It may be vulnerable to breakdown, as the network is essentially controlled by one device at a central location.

Bus topology. A bus topology is a linear configuration. It places all of the network devices on one length of cable, similar to stops on a city bus route. The hubs, server, stations, and peripheral devices all use the same continuous length of transmission channel.

The ends of the transmission channel, in this arrangement, are not connected to network devices. Ordinarily, problems would occur when the transmitted signal was sent along the cable and it reached either of the ends. For this reason, each end of the cable is connected to a terminator.

When a message is sent utilizing this topology, the following event takes place: The transmission signal leaves the sending device and travels along the cable in both directions. The device for which the message is intended will recognize the transmission and read the message.

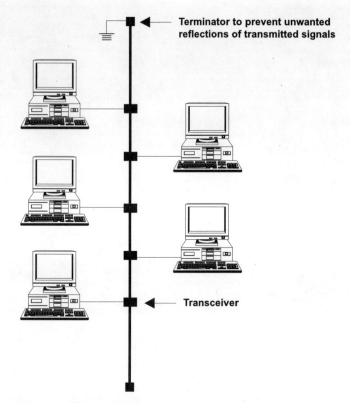

Figure 1.10 Bus topology.

Bus topology advantages:

- Easily adaptable to many environments—it can be configured to suit most situations.
- Easily expandable by adding devices at various points along the cable.

Bus topology disadvantages:

- Lacks central control—finding a fault is difficult.
- If the cable is damaged or if either end of the cable loses its termination, the entire network will fail.

Ring topology. A ring topology places all of the network devices in a circle. It uses one transmission channel to connect all devices. Each device is connected to the next one. The last device is connected to the first—closing the circle.

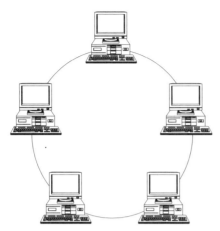

Figure 1.11 Ring topology.

When a message is sent, it travels from device to device around the circle. The sending device sends its message towards the destination device. Each device between the sender and the receiver listens to the message as it passes by. If the message is not intended for a particular device, it resends the message and the next device in the ring repeats the procedure. This continues until the intended destination receives the message.

Ring topology advantage:

- There is no reliance on a central device—all messages pass through all devices.

Ring topology disadvantages:

- Additional network devices can only be connected while the network is inoperative, since breaking the ring would cause network failure.
- If any device fails, the entire network is affected.

Although the ring topology is considered one of the three fundamental network topologies, it has never been popular in its basic form. The more popular dual-ring topology provides two paths between stations—a primary path and a backup path. In the event of a failure in the primary path, the signal can be diverted to the backup path by stations on either side of the point of failure—preventing total network failure.

Hybrid topologies. Hybrid topologies resulted from a need to meet specific requirements or industry technological advancements. While there are many variations of the three topologies discussed above, certain hybrids are more popular than others.

- *Tree topology.* The tree topology is an extension of the bus topology. By adding cable extensions to the basic bus topology, a larger network can be achieved using less cable. Each additional cable extends from the underlying bus structure and supports multiple network devices along its length.

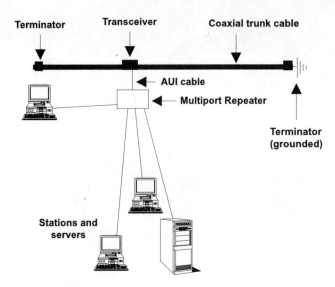

Figure 1.12 Tree topology.

- *Star-wired ring topology.* A star-wired ring is also referred to as a collapsed ring. In this configuration, the network devices are connected to each other as they are in a ring topology. The difference is that they are connected through a central unit which acts as a wire center. The transmission method is the same as with the ring topology, except now dual rings are present, one primary and the other backup. All messages must first pass through the wire center before moving to the next device. The main improvement over the ring topology is that the failure of a single device will no longer cause the whole network to fail, due to active monitoring by the wire center.

- *Clustered star topology.* A clustered star is much like the tree topology, except that there are clusters of devices at the end of each branch. There exists an underlying bus configuration that supports cable extensions. Each of these cable extensions has a cluster of network devices at its end. This is also known as a "bus star" configuration.

- *Hierarchical star topology.* A hierarchical star topology is an extension of the star topology. In this configuration, departmental network

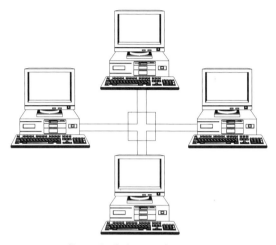

Figure 1.13 Star-wired ring topology.

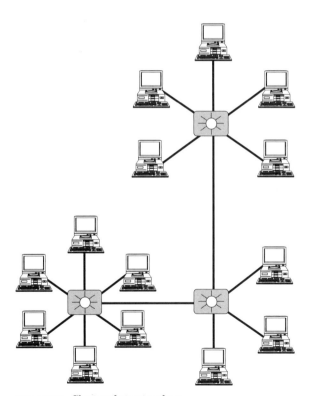

Figure 1.14 Clustered star topology.

devices are connected to a hub or switch, as in a star topology. These hubs or switches are then connected to each other via a central hub, also following a star configuration. This is the recommended topology for structured cabling backbone systems in buildings and in campus environments.

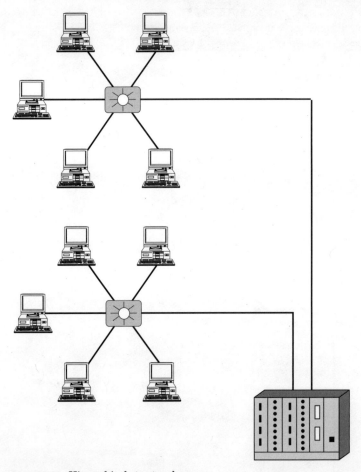

Figure 1.15 Hierarchical star topology.

Standards and Codes

Introduction

Because telecommunications cabling installers work with other disciplines, such as data processing and the building industry, they must follow a myriad of rules in everyday business. While codes address minimum safety requirements that must be adhered to, standards are intended to ensure system performance by providing requirements and guidelines for proper installation.

Construction in virtually all areas of North America is regulated by building codes and standards. These are normally enforced by a local jurisdictional agency. Codes and standards encompass most, if not all, aspects of the construction industry. Installation methods, materials, and electrical products must conform to local code requirements. Both codes and standards contain two words, *shall* and *should*, which can have a major impact in how tasks are accomplished. These are defined as:

- Shall—A mandatory requirement.
- Should—A recommendation.

Purpose of codes and standards

Building codes and standards govern installation practices and materials used when constructing telecommunications facilities. The purpose of codes and standards is to:

- Protect life, health, and property.
- Ensure construction quality.

Codes are not intended to cover measures that may be required to protect telecommunications and equipment from:

- Intrusion.
- Induced noise.
- Events that can disrupt the flow of information.

In general, standards are established as a basis to compare, measure, or judge:

- Capacity.
- Quantity.
- Content.
- Extent.
- Value.
- Quality.
- Performance.
- Limits.

Independent organizations exist that specialize in establishing, certifying, and maintaining these standards.

Recognized regulatory bodies

Various organizations publish codes, standards, and methods for materials and testing. Much of this information is adopted by manufacturers to ensure standardization. Some local enforcement agencies adopt or adhere to such standards as evidence of quality in installation. Some of the major recognized agencies/organizations are:

American National Standards
  Institute (ANSI)
430 Broadway
New York, NY 10018
212-642-4900; fax: 212-398-0023
http://www.ansi.org

American Society for Testing and
  Materials (ASTM)
100 Barr Harbor Drive
West Conshohoken, PA 19428-2959
610-832-9585; fax: 610-832-9555
e-mail: service@astm.org

Electronic Industries Alliance (EIA)
2500 Wilson Boulevard
Arlington, VA 22201-3834
703-907-7500; fax: 703-907-7501
http://www.eia.org

Federal Communications
  Commission (FCC)
1919 M Street NW, Room 702
Washington, DC 20554
202-418-0200; fax: 202-418-0232
e-mail: fccinfo@fcc.gov
http://www.fcc.gov

Institute of Electrical and
  Electronics Engineers, Inc. (IEEE)
IEEE Service Center
445 Hoes Lane, PO Box 1331
Piscataway, NJ 08855-1331
732-981-0060; fax: 732-981-9667
e-mail: customer.service@iee.org
http://www.ieee.org

International Electrotechnical
  Commission (IEC)
3, Rue de Varembé
CH-1211 Genéve 20
Switzerland
+41-22-919-02-11; fax: +41-22-919-03-00
e-mail: inmail@iec.ch
http://www.iec.ch

International Organization for
  Standardization (ISO)
1, Rue de Varembé
Case Postale 56
CH-1211 Genéve 20
Switzerland
+41-22-749-01-11; fax: +41-22-733-34-30
e-mail: central@iso.ch
http://www.iso.ch

National Electrical Manufacturers
  Association (NEMA)
1300 N. 17th Street
Suite 1847
Rosslyn, VA 22209
703-841-3200; fax: 703-841-3300
http://www.nema.org

National Fire Protection Association
  (NFPA)
1 Batterymarch Park
PO Box 9101
Quincy, MA 02269-9101
617-770-3000; fax: 617-770-0700
e-mail: custserv@nfpa.org
http://www.nfpa.org

Occupational Safety and Health
  Administration (OSHA)
200 Constitution Avenue NW
Washington, DC 20210
202-219-8151; fax: 202-219-5986
http://www.osha.gov

Telecommunications Industry
  Association (TIA)
2500 Wilson Boulevard
Suite 300
Arlington, VA 22201-3836
703-907-7700; fax: 703-907-7727
http://www.tiaonline.org

Underwriters Laboratories, Inc. (UL)
333 Pfingsten Road
Northbrook, IL 60062
847-272-8800; fax: 847-272-8129
http://www.ul.com

To order copies of any of these standards, contact:

Global Engineering Documents
15 Inverness Way
Englewood, CO 80112
800-854-7179; fax: 314-726-6418
e-mail: global@ihs.com
http://www.global.ihs.com

Codes Affecting Telecommunications

National Fire Protection Association (NFPA)

The National Fire Protection Association (NFPA) develops and produces the following fire codes relating to telecommunications:

- National Electrical Code (NEC): ANSI/NFPA-70.
- Central Station Signaling Systems: ANSI/NFPA-71.
- National Fire Alarm Code: ANSI/NFPA-72.
- Protection of Electronic Computer Data Processing Equipment: ANSI/NFPA-75.
- Standard for Installation of Lightning Protection Systems: ANSI/NFPA-780.

The *National Electrical Code®* (NEC)

The NFPA sponsors, controls, and publishes the NEC within the United States' jurisdictional area. The NEC is intended to protect people and property from electrical hazards. The NEC specifies the minimum provisions necessary to safeguard persons and property from electrical hazards.

Most federal, state, and local municipalities have adopted the NEC, in whole or in part, as their legal electrical code. Some states or localities adopt the NEC and add more stringent requirements.

The NEC is used by:
- Lawyers and insurance companies to determine liability.
- Fire marshals and inspectors (electrical) in loss prevention and safety enforcement.
- Installation designers to ensure a compliant installation.

The code sets the minimum standards that must be met to protect people and property from electrical hazards. It is revised every three years. ANSI/NFPA-70 (NEC) is arranged by chapter, article, and section (i.e., Chapter 8, Article 800, Section 800-52). Portions within the code having significant importance to the telecommunications cabling installer are shown in Table 1.2.

TABLE 1.2 NEC Codes Affecting Telecommunications

NEC Reference	Title	Description
Section 90-2	Scope	The Scope provides information about what is covered in the NEC. This section offers reference to the NESC for industrial or multibuilding complexes.
Section 90-3	Code Arrangement	This section explains how the NEC chapters are positioned. Specifically, Chapter 8, Communications Systems, is an independent chapter, except where reference is made to other chapters.
Article 100-Part A	Definitions	Definitions are those not commonly defined in English dictionaries. Some terms of interest include *accessible, bonding, explosion-proof apparatus, ground, premises wiring,* and *signaling circuit*.
Section 110-16	Working Space About Electric Equipment (600 Volts, Nominal, or Less)	This section explains the space for working clearances around electrical equipment. This information is useful when placing a terminal in an electrical closet or electronic components on a communications rack.
Article 250	Grounding	This article is referenced from Article 800. It contains specific requirements for the communications grounding and bonding network.
Article 500	Hazardous (Classified) Locations	All of Article 500 is referenced in Article 800 (Section 800-7). These include hazardous locations such as gasoline stations and industrial complexes. Additionally, health care facilities (Section 517-80) are of particular importance. Theaters and marinas are included in this article.
Article 725	Class 1, Class 2, and Class 3 Remote-Control, Signaling, and Power-Limited Systems	Article 725 specifies circuits other than those used specifically for electrical light and power.
Article 760	Fire Alarm Systems	Article 760 contains requirements of wiring and equipment of fire alarm systems.
Article 770	Optical Fiber Cables and Raceways	Article 770 pertains to optical fiber cables and Raceways. Within this section are the requirements for listing of cable, marking, and installation.

TABLE 1.2 NEC Codes Affecting Telecommunications (*Continued*)

NEC Reference	Title	Description
Article 800	Communications Systems	Article 800 contains the requirements for communications. This chapter should be known by the telecommunications installer.
Article 810	Radio And Television Equipment	Article 810 contains the requirements for radio and television.
Article 820	Community Antenna Television and Radio Distribution Systems	Article 820 contains the requirements for community antenna television and radio distribution systems.

It is important to remember that the NEC standards specify only a safe environment, not an environment in which telecommunications systems are guaranteed to operate free of any interference or errors.

The Canadian Electrical Code, Part 1

The Canadian Standards Association sponsors, controls, and publishes the Canadian Electrical Code, Part 1 (CEC, Part 1). The intent of this code is to establish safety standards for the installation and maintenance of electrical equipment, including telecommunications. As with the NEC, the CEC, Part 1 is a voluntary code that may be adopted and enforced by the provincial and territorial regulatory authorities.

Applicable telecommunications sections

Telecommunications installers must be familiar with all sections of the CEC, Part 1. The following chart lists sections of the CEC, Part 1 that are of primary interest to the telecommunications installer.

TABLE 1.3 CEC Sections

CEC Reference	Title	Description
2	General Rules	Provides information on: • Permits. • Marking of cables. • Flame-spread requirements for electrical wiring and cables.
10	Grounding and Bonding	Contains detailed grounding and bonding information and requirements for using and identifying grounding and bonding conductors.
12	Wiring Methods	Involves the requirements for installing wiring systems. It outlines: • Raceway systems. • Boxes. • Other system elements.
56	Optical Fiber Cables	Contains the requirements for installing optical fiber cables.
60	Electrical Communication Systems	Contains the requirements for installing communications circuits.

Standards affecting telecommunications

Overview. ANSI/TIA/EIA publishes standards for the manufacturing, installation, and performance of electronic and telecommunications equipment and systems. Five of these ANSI/TIA/EIA standards govern telecommunications cabling in buildings. Each standard covers a specific part of building cabling. They address the required cable, hardware, equipment, design, and installation practices. In addition, each ANSI/TIA/EIA standard lists related standards and other reference materials that deal with the same topics.

Most of the standards include sections that define important terms, acronyms, and symbols. The ANSI/TIA/EIA standards which govern telecommunications cabling in buildings are:

- ANSI/TIA/EIA-568-A, *Commercial Building Telecommunications Cabling Standard*
    - ANSI/TIA/EIA-568-A-1, *Propagation Delay and Delay Skew Specifications for 100 Ω 4-pair Cable*
    - ANSI/TIA/EIA Telecommunications Systems Bulletin (TSB-67), *Transmission Performance Specifications for Field Testing of Unshielded Twisted-Pair Cabling Systems*
    - ANSI/TIA/EIA Telecommunications Systems Bulletin (TSB-72), *Centralized Optical Fiber Cabling Guidelines*
    - ANSI/TIA/EIA Telecommunications Systems Bulletin (TSB-75), *Additional Horizontal Cabling Practices for Open Offices*
- ANSI/TIA/EIA-569-A, *Commercial Building Standard for Telecommunications Pathways and Spaces*
- ANSI/EIA/TIA-570, *Residential and Light Commercial Telecommunications Wiring Standard*
- ANSI/TIA/EIA-606, *Administration Standard for the Telecommunications Infrastructure of Commercial Buildings*
- ANSI/TIA/EIA-607, *Commercial Building Grounding and Bonding Requirements for Telecommunications*

In addition, the *National Electrical Code*® (NEC) ANSI/NFPA-70, published by the National Fire Protection Association (NFPA), provides electrical safety standards that protect people and property from fires and electrical hazards.

The following briefly describes these five ANSI/TIA/EIA standards.

ANSI/TIA/EIA-568-A

Commercial Building Telecommunications Cabling Standard. ANSI/TIA/EIA-568-A covers telecommunications cabling in commercial buildings. The standard provides specifications for a generic building-cabling system that can be created and used with a variety of products from many different manufacturers. In addition to design specifications, the standard includes performance specifications for the cables and components used in commercial building cabling.

Five sections of ANSI/TIA/EIA-568-A contain general requirements for specific cabling applications. These applications are:

- *Horizontal cabling (Section 4).* Recognizes three types of cable for use in horizontal cabling:
  - Four-pair 100 Ω UTP (unshielded twisted-pair)
  - Four-pair 100 Ω ScTP (screened twisted-pair)
  - Two-pair 150 Ω STP-A (shielded twisted-pair)
  - Two-fiber 62.5/125 μm optical fiber

  For existing wiring only, 50 Ω coaxial cable is also recognized. Screened 100 Ω cabling is also allowed.

  This section also specifies star topology for horizontal cabling, the maximum length for horizontal cables [90 m (295 ft), with an additional 10 m (30 ft) per channel for patch cords and jumpers], and the installation of two telecommunications outlets/connectors for each work area.

- *Backbone cabling (Section 5).* Recognizes four types of cable for use in backbone cabling:
  - 100 Ω UTP (unshielded twisted-pair)
  - 100 Ω ScTP (screened twisted-pair)
  - 150 Ω STP-A (shielded twisted-pair)
  - 62.5/125 μm multimode optical fiber
  - Singlemode optical fiber

  For existing wiring only, 50 Ω coaxial cable is also recognized. Screened 100 Ω cabling is also allowed.

  This section also specifies the topology for backbone cabling (star), the maximum length for backbone cables (intrabuilding and interbuilding), the maximum length for cross-connect jumpers, and grounding considerations.

- *Telecommunications closet cabling (Section 7).* Describes the functions of telecommunications closets and describes the cabling practices that should be followed.

- *Equipment room cabling (Section 8).* Describes the functions of equipment rooms.

- *Entrance facility cabling (Section 9).* Describes the functions of entrance facilities and discusses the network demarcation point.

Four sections of ANSI/TIA/EIA-568-A contain specifications for the different types of cabling which may be used in commercial buildings, including:

- *100 Ω unshielded twisted-pair (UTP) cabling (Section 10).* Defines three categories of UTP cabling and connectors:
  - Category 3, with transmission characteristics specified up to 16 MHz
  - Category 4, with transmission characteristics specified up to 20 MHz
  - Category 5, with transmission characteristics specified up to 100 MHz

  This section also specifies the physical characteristics, transmission characteristics, and color coding for both horizontal and backbone UTP cables. Section 10 also specifies the connecting hardware.

  Installation practices for UTP cabling are covered in subsection 10.6 of the standard. The precautions against untwisting pairs are especially important to cabling installers.

- *150 Ω shielded twisted-pair (STP-A) cabling (Section 11).* Specifies the physical characteristics, transmission characteristics, and color coding for both horizontal and backbone STP-A cables. Section 11 also specifies the connecting hardware.

  Installation practices for STP-A cabling are covered in subsection 11.6 of the standard. The grounding of the braided shield to the telecommunications grounding busbar in the telecommunications closet is especially important.

- *Optical fiber cabling (Section 12).* Recognizes:
  - 62.5/125 µm multimode optical fiber cable for horizontal and backbone cabling.
  - Singlemode optical fiber cable for backbone cabling only.

  The physical characteristics of optical fiber cables and connectors are noted. Section 12 also specifies connectors and their color coding.

  Installation practices for optical fiber cable are covered in subsection 12.7 of the standard.

- *Hybrid and undercarpet cabling (Section 13).* Installation considerations are covered in subsection 13.2.2 of the standard.

ANSI/TIA/EIA-568-A does not include specifications for work area cabling.

ANSI/TIA/EIA-568-A-1—Propagation Delay and Delay Skew Specifications for 100 Ω 4-pair Cable. This addendum to ANSI/TIA/EIA-568-A specifies the propagation delay for all recognized categories of 100 Ω 4-pair cabling. In addition, it also specifies the delay skew for 4-pair cables for the recognized cabling categories.

ANSI/TIA/EIA Telecommunications Systems Bulletin (TSB-67)—Transmission Performance Specifications for Field Testing of Unshielded Twisted-Pair Cabling Systems. This bulletin provides specifications for the electrical characteristics of field testers, test methods, and minimum transmission requirements for UTP cabling. It defines the test link, test channel, tests to be completed, and the pass/fail criteria for each type.

ANSI/TIA/EIA Telecommunications Systems Bulletin (TSB-72)—Centralized Optical Fiber Cabling Guidelines. This bulletin provides implementation guidelines and connecting hardware requirements for the implementation of a centralized optical fiber cabling network. These guidelines are in addition to those found in Section 4, Horizontal Cable, and Section 12.4.6, Connecting Hardware, of the *Commercial Building Telecommunications Cabling Standard* ANSI/TIA/EIA-568-A.

ANSI/TIA/EIA Telecommunications Systems Bulletin (TSB-75)—Additional Horizontal Cabling Practices for Open Offices. This bulletin provides optional practices for open office environment installation of horizontal cables recognized in ANSI/TIA/EIA-568-A. It provides optional topologies and connecting schemes to support that portion of the modular or movable office furniture that is easily and frequently reconfigured. The specification and installation procedures for the multiuser telecommunications outlet assemblies (MUTOA) and the consolidation point are contained in this bulletin.

ANSI/TIA/EIA-569-A

Commercial Building Standard for Telecommunications Pathways and Spaces. This standard provides specifications for designing and constructing the pathways and spaces for telecommunications cabling in commercial buildings. The standard is especially useful for writing construction bids and contracts.

The standard's specifications are based on the belief that the building, the telecommunications needs of its occupants, and the telecommunications technology available may change several times during the life of a building. In this standard, telecommunications includes all low voltage systems which carry information inside a building.

In addition to setting specifications, the standard offers a good working definition of each type of space or pathway discussed. The standard also contains many illustrations of pathways and equipment.

The section on horizontal pathways (Section 4) describes specifications for:

- Underfloor duct systems.
- Conduit. This subsection includes the length and placement of conduit runs, the minimum radius for conduit bends, and the proper use of pull boxes and splice boxes.

- Access floors.
- Cable trays and wireways. This subsection defines the types of trays and wireways and explains their proper support and installation.
- Ceiling pathways, including information on utility columns.
- Perimeter pathways.

The section on backbone pathways (Section 5) describes specifications for:

- Intrabuilding backbones.
- Interbuilding backbones.

The remaining sections describe specifications for:

- Workstation pathways (Section 6).
- Telecommunications closets (Section 7).
- Equipment rooms (Section 8).
- Entrance facilities (Section 9), including:
  - Underground.
  - Tunnel.
  - Aerial.
  - Direct buried.

The Miscellaneous Items (Section 10) includes information about clearance from sources of electromagnetic energy.

Firestopping information can be found in ANSI/TIA/EIA-569-A Appendix A and is a part of the standard. Interbuilding backbone pathways and related spaces have been consolidated into Appendix C and are now a part of the standard.

ANSI/EIA/TIA-570

Residential and Light Commercial Telecommunications Wiring Standard. ANSI/EIA/TIA-570 provides specifications for premises cabling systems which connect one to four exchange access lines to customer premises equipment.

The standard gives an overview of premises cabling systems and defines important concepts and components. It also provides example illustrations of a typical:

- Premises cabling system.
- Residential cabling layout.
- Multioccupant cabling layout.

Section 4 establishes a cable grading system based on the type of service to be provided.

The section on installation requirements (Section 5) warns that the telecommunications cabling must be disconnected at the demarcation point (or the auxiliary disconnect) during any cabling operations. Any other power sources which might endanger the cabling installer should also be disconnected. The section also specifies color coding and jack contact assignments.

The installation requirements section includes a subsection on cabling rearrangements and how they differ from original cabling.

Specific procedures for premises cabling are included in the subsection on installation (Section 5.3). Included is information about installing:

- An auxiliary disconnect outlet.
- The distribution device, including its input and output connections.
- Station and distribution cable, including the requirements for maximum pulling tension of 110 N (25 lbf), minimum slack for making connections (457 mm [18 in]), and maximum sheath removal for making connections (76–152 mm [3 to 6 in]).
- Undercarpet cable.
- Telecommunications outlets.

The technical requirements for premises cabling components are also included in the standard under Section 6.

Though it is not a part of the standard, Appendix A covers line assignments in selected network interface jacks. Appendix B (Multitenant Residential Building) includes useful information about premises cabling that explains:

- Planning.
- Rough-in cabling.
- Finish cabling.
- Systems testing.
- Rearrangements.

ANSI/TIA/EIA-606

Administration Standard for the Telecommunications Infrastructure of Commercial Buildings. ANSI/TIA/EIA-606 describes the requirements for the keeping of records and the information that must be available in order to properly administer the telecommunications system in a commercial building. To ensure that administration records are accurate and up-to-date, cabling installers must understand:

- What information they must record.
- How the information should be recorded.
- What devices and structures must be labeled.
- How the devices and structures should be labeled.

The standard requires that an administration system track all aspects of the telecommunications system—including cables, terminations, termination hardware, patching and cross-connect facilities, splices, conduits and other pathways, bonding and grounding, telecommunications closets, equipment rooms, and other telecommunications spaces.

Each component of the telecommunications system must have a unique identification code. This code is used for the component's record, for the record of connected or related components, and for any drawings in which the components appear. If the component requires labeling, the code must be printed on the component's label or on the component itself. Details about identification codes appear in the Identifiers subsection (Section 4.2) of the standard.

The Records subsection (Section 4.3) explains what information must be recorded. Examples of records appear throughout the standard. These examples show both the required information and optional information, which would make the record more complete and provide linkage between components.

Although not part of the standard, Appendix C contains a list of common telecommunications symbols used in representing elements of the telecommunications infrastructure.

ANSI/TIA/EIA-607

Commercial Building Grounding and Bonding Requirements for Telecommunications. ANSI/TIA/EIA-607 specifies the grounding and bonding requirements for the telecommunications system in a commercial building. It also specifies the interconnections between telecommunications grounding and other grounding. This standard specifies the:

- Ground reference for telecommunications systems in entrance facilities, telecommunications closets, and equipment rooms.
- Bonding and connecting of conductors, cable shields, pathways, and hardware in entrance facilities, telecommunications closets, and equipment rooms.

ANSI/TIA/EIA-607 also describes in detail the physical requirements for the five major components of a telecommunications grounding and bonding system:

- Telecommunications bonding backbone (TBB)
- Telecommunications main grounding busbar (TMGB)
- Telecommunications grounding busbar (TGB)
- Telecommunications bonding backbone interconnecting bonding conductor (TBBIBC)
- Bonding conductor for telecommunications

The interconnections between these components and their relationship to other building grounds are explained in detail in the standard. The explanations include installation considerations.

It is specified that all bonding conductors must be a 6 AWG (or larger) insulated-copper wire marked by a distinctive green color. The main bonding conductor must also be labeled with a warning, which asks that any disconnection of this bonding conductor be reported immediately to the building's telecommunications manager. The Canadian Standards Association (CSA) publishes standards for the manufacturing, installation, and performance of electronic and telecommunications equipment and systems within Canada. Among the most important are:

- CAN/CSA-T524, *Residential Wiring for Telecommunications* (harmonizes with ANSI/EIA/TIA-570)
- CAN/CSA-T527, *Bonding and Grounding for Telecommunications in Commercial Buildings* (harmonizes with ANSI/TIA/EIA-607)
- CAN/CSA-T528, *Telecommunication Administration Standard for Commercial Buildings* (harmonizes with ANSI/TIA/EIA-606)
- CAN/CSA-T529, *Design Guidelines for Telecommunications Wiring Systems in Commercial Buildings* (harmonizes with ANSI/TIA/EIA-568-A)
- CAN/CSA-T530, *Building Facilities, Design Guidelines for Telecommunications* (harmonizes with ANSI/TIA/EIA-569-A)

Related standards

Institute of Electrical and Electronics Engineers. IEEE is the world's largest professional engineering society. IEEE provides:

- Standards for rating the performance of equipment and materials.
- Standards for installation and maintenance of equipment.

- Courses to allow engineers to keep abreast of developments in the electrical and electronic engineering fields.

IEEE standards with which cabling installers interact include IEEE 802.3 Ethernet and IEEE 802.5, Token Ring.

International Standards Organization (ISO)/International Electrotechnical Commission (IEC). As we live and deal in an international community, there also exists the requirement for the interoperability of electronic networks connecting this community. One of the functions of the ISO/IEC is to define the international cabling standards that make this interoperability possible. Of primary importance today is the ISO/IEC Standard 11801 *Information Technology—Generic Cabling for Customer Premises*. This standard accepts both 100 and 120 Ω unshielded twisted-pair (UTP) cabling and also allows for shielding of both of these types as an option. Cabling standards for the future, such as Class E (Category 6) up to 200 MHz and Class F (Category 7) up to 600 MHz, are being developed in conjunction with the Joint Technical Committee (JTC)/Sub Committee (SC) 25/Working Group (WG) 3, to which the U.S. Technical Advisory Group (TAG) is a contributor.

Plans and Specifications

Overview

Plans are sometimes referred to as blueprints, building plans, or drawings. They provide the pictorial representation of a project. Specifications provide a detailed written description of a project. Architects and engineers design the building and lay out the drawing and specifications.

Using symbols and abbreviations provides graphical guidelines for architectural, mechanical, electrical, plumbing, and telecommunications installations.

To understand the blueprints, it becomes necessary to understand the design used in creating them. The overall design of a structure is normally the responsibility of the architect, who will prepare the drawings and specifications for the complete project. The architect will hire engineers and consultants to prepare the structural, mechanical, electrical, plumbing, and communications drawings.

The drawing set shows the dimensions and the relationship between components but does not provide sufficient detail to complete the project. Specifications provide the additional detail required to complete the project. Specifications are a written description of the work to be performed and the responsibilities and duties required of the architect, engineer, and owner. Together with the drawings, these specifications form the basis of the contractual requirements.

Drawings

The drawing set, sometimes referred to as AME, (architectural, mechanical, electrical) or AMES (architectural, mechanical, electrical, structural) is normally composed of distinct segments. They are:

- Architectural—Architectural and structural (engineering) drawings. In some instances, the structural drawings may be in a separate section.
- Mechanical—The installation of the plumbing, heating, ventilating, and air conditioning (HVAC) systems within and into the building.
- Electrical—Electrical systems for lighting, power, alarms and communications, special electrical systems, and related electrical systems.
- Structural—Engineered drawings detailing the construction of all walls, floors, ceilings, load-bearing structures as well as detailed drawings providing cross-sectional details of these structures.

The architect is normally responsible for portions of the drawing set, including the:

- Title sheet.
- Index sheet(s)—These provide general information, including the site address, owner's name, and architect's name.
- Legend sheet(s)—These provide a list of standard symbols and abbreviations used throughout the particular drawing set.

Site drawings, also known as the plot plan, are prepared by a civil engineer and:

- Indicate the relationship of the various buildings on the project.
- Depict:
  - The location of the building on the site.
  - External features, such as parking areas.
  - The entry point of the utilities, including communications.

Floor plans show the layout of each floor.

Elevations of all the exterior faces of the structure are shown.

Vertical cross sections provide a detailed view of the floor levels as well as a detail of the footings, floors, walls, and roof.

Detail drawings are large-scale drawings with detailed construction as required.

With the exception of the detail drawings, all of the drawings will be to scale. Detail drawings intended to show relationships may or may not be to scale.

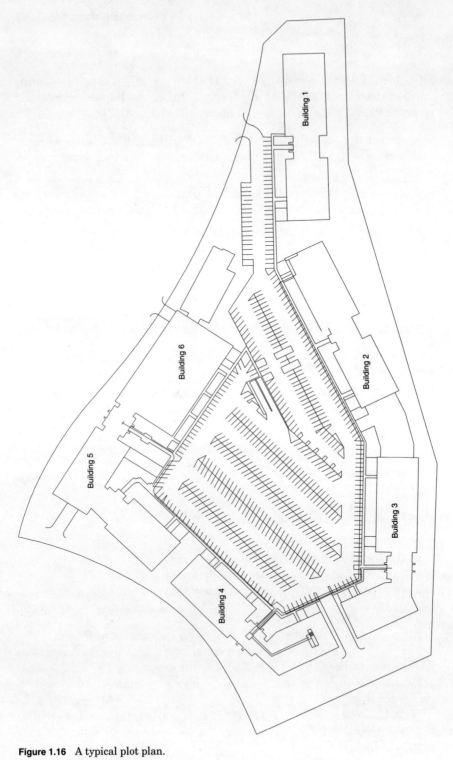

Figure 1.16 A typical plot plan.

Background Information   45

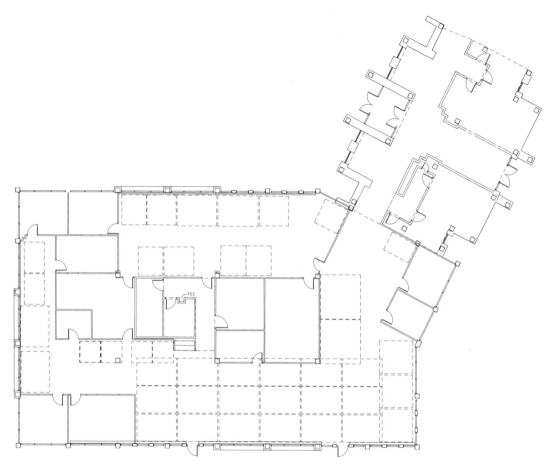

Figure 1.17 A typical floor plan.

Structural drawings provide the detail necessary to produce the structural support components of the building. These drawings are based on engineering guidelines set up for vertical loads, lateral stress, etc. Structural drawings may be included with the architectural drawings or in a separate section within the drawing set.

These drawings indicate the essential information required for cabling installation, including:

- Spaces and pathways.
- Telecommunications closets.
- Equipment rooms.
- Entrance facilities.

46 Chapter One

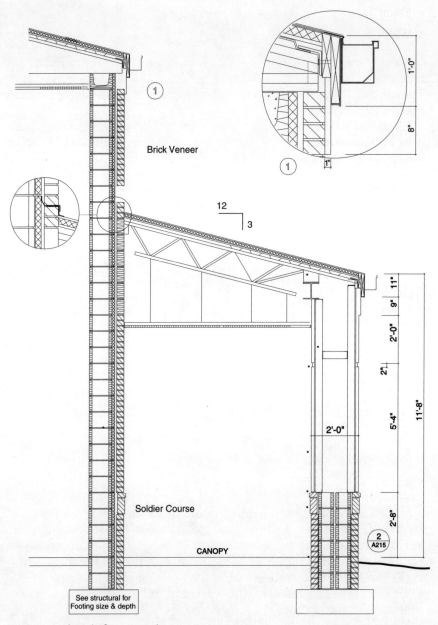

Figure 1.18 A typical cross section.

- Floor loading.
- Wall loading.

Normally completed by the mechanical engineer, mechanical drawings show the complete layout and design of the mechanical systems. They include floor plan layouts, cross sections of the building, and all necessary detail drawings.

Electrical drawings are normally completed by the electrical engineer and, sometimes, in the case of the telecommunications portion, by a Registered Communications Distribution Designer (RCDD®). These drawings may include a plot or site plan showing the entry points for all electrical and communications services. Floor plans that show the location of all outlets, fixtures, panels, backboards, etc., are a part of this set.

In addition, the electrical drawings include:

- Power and communications riser diagrams.
- Symbols list (legend).
- Schematic diagrams.
- Large-scale details (where necessary).

Blueprints for new construction are generally an accurate representation of the actual site. However, during the construction process, changes may occur in the location of:

- Rooms and closets.
- Telecommunications equipment.
- Work area outlets.
- Electrical equipment.
- HVAC equipment and ducting.
- Pathways.
- Utilities.

As explained in Chapter 2, Planning, a site survey is a critical part of planning. The convergence of the information on the blueprint, the elements of the job plan, and the physical design of the building should meld into a cohesive concept for the total project.

Projects which require new construction or retrofitting use plans as a locator for new or planned pathways.

Plans available for retrofits may not reflect the site "as-built" and should be carefully checked for accuracy. Again, the site survey becomes

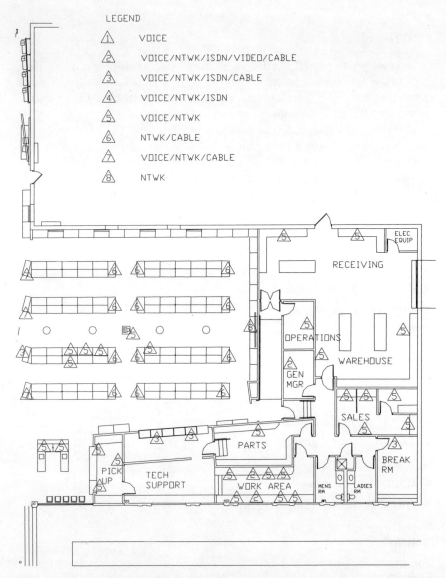

Figure 1.19 Typical telecommunications plan.

a critical evaluation method, producing an opportunity to match the information on the blueprint with observed:

- Spaces and pathways.
- Work area outlets.
- Telecommunications closets.

Background Information 49

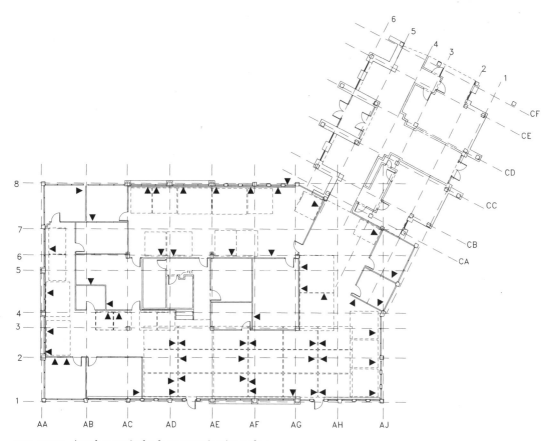

Figure 1.20 Another typical telecommunications plan.

- Equipment rooms.
- Entrance facilities.

Construction specifications

For new construction or when the telecommunications systems are a part of the general contract, construction specifications will be included in the overall project specifications. This specification book, prepared by the Construction Specifications Institute (CSI), is used by the American Institute of Architects (AIA) to define the requirements of a specific part of the construction work on the project. The following chart shows each division, its title, and the specific area it covers.

Note. Division 17 is not in the current (AIA) standard. Telecommunications is usually included under Section 16.

	Floor-mounted Telecom Outlet
	Voice Only
	Data Only
	Combination Voice/Data Outlet
TBB	Telephone Backboard
TC	Telecommunications Closet
MC	Main Cross-connect
W	Wall Phone
P	Payphone

Figure 1.21 Representative symbols and abbreviations.

TABLE 1.4 Specification Book

Division	Title	Description
1	General Requirements	A summary of the work, alternatives, schedule of the project, project meetings, submittals, quality control, temporary facilities, products, and the project closeout.
2	Site Work	Foundations, fill, drains, underground utilities, and other areas outside the building.
3	Concrete	All concrete work. Maintenance holes are covered in this section.
4	Masonry	All external and internal masonry.
5	Metals	Structural, metal joists, decking, framing, fabrication, etc.
6	Carpentry	All wood products and their usage within the project. Also includes plastics if a written specification is provided.
7	Thermal and Moisture Protection	Covers such items as waterproofing, insulation, shingles and roofing tiles, preformed roofing and siding, sheet-metal work, wall flashing, roof accessories, and sealants.
8	Doors and Windows	All doors and windows, hardware and specialties, glazing, and window/curtain walls.
9	Finishes	All finishes for floors, walls, and ceilings.
10	Specialties	Special items such as chalkboards and tackboards, modular furniture, louvers and vents for the HVAC system, access flooring, protective covers, etc. are part of this division.
11	Equipment	Vaults, darkrooms, food service, industrial, laboratory equipment, etc.

TABLE 1.4 Specification Book (*Continued*)

Division	Title	Description
12	Furnishing	Furniture, fabrics, rugs and mats, seating, etc.
13	Special Construction	Air-supported structures, incinerators, and other special items.
14	Conveying Systems	Any and all types of conveying systems.
15	Mechanical	All work of the HVAC and plumbing contractor.
16	Electrical	All of the interior and exterior electrical work. **Includes all communications unless Division 17 is used.**
(17)	Communications	All communications components for the project.

In the case of a retrofit, or if the telecommunications systems are part of a separate contract, the Request for Quote (RFQ) will usually contain the specifications.

Request for quote

An RFQ is a document commonly used to solicit quotes for telecommunications projects or equipment. Companies or individuals may use an RFQ to solicit quotes for:

- Private Branch Exchanges (PBX).
- Key telephone systems.
- Central office-based technologies.
- Distribution system design or installation.
- Preparation of an RFQ (as a consultant).

An RFQ may include any or all of the items listed above. For example, an RFQ might request a quote for the installation of a new telephone switch that could be premises based or central office based. Such an RFQ might include only the terminal equipment (PBX and station instruments) or the installation of associated wire and cable.

For a detailed description of the RFQ, refer to the BICSI *Telecommunications Distribution Methods Manual* (*TDMM*), 8th Edition, Chapter 24: Request for Quote (RFQ).

Summary

In either case, the construction specifications and the request for quote for a building or project are a written description of the work to be performed and of the various duties of the installation contractor. The drawings along with the specifications of the project form the basis of the contract for all work to be completed.

Media

Overview

Telecommunications signals can travel in many different ways, including:

- As radio waves through the air.
- Over copper cables as electrical signals.
- As sonar waves through the water.
- As beams of light, through the air.
- As beams of light, over an optical fiber cable.

In this document, we are concerned with the various types of telecommunications cable commonly in use. The major telecommunications cables are:

- UTP—Unshielded twisted-pair copper cable
  - Category 5
  - Category 4
  - Category 3
- ScTP—Screened twisted-pair copper cable
  - Category 5
  - Category 4
  - Category 3
- STP—Shielded twisted-pair copper cable
  - STP
  - STP-A
- Coaxial cable
  - Thin Ethernet IEEE 802.3
  - Thick Ethernet IEEE 802.3
  - Video—RG-11 backbone and RG-6 horizontal drop
  - RG-59
  - RG-62
- Fiber—Optical fiber cable
  - Multimode (50/125 μm)
  - Multimode (62.5/125 μm)
  - Singlemode (8.3/125 μm)

Unshielded twisted-pair (UTP) copper cable

Unshielded twisted-pair (UTP) cable has been used for a long time for both voice and data cabling. It has the following characteristics:

- Composed of pairs of wires twisted together
- Commonly available in various pair counts (from 2 to 1800 pairs)
- Is not normally shielded up to 600 pairs and has an overall aluminum–steel shield up to 1800 pairs
- Electrical interference protection by twisting of wires
- Has a characteristic impedance of 100 Ω
- Recommended conductor size is 22–24 AWG
- Solid conductors

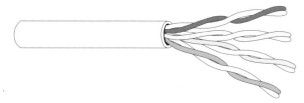

Figure 1.22 Typical UTP cable.

To improve information throughput, significant performance improvements have been made to UTP. In ANSI/TIA/EIA-568-A, specifications for several performance levels of UTP and associated connecting hardware are established as follows:

- Category 3—UTP cables and associated connecting hardware whose transmission characteristics are specified up to 16 MHz
- Category 4—UTP cables and associated connecting hardware whose transmission characteristics are specified up to 20 MHz
- Category 5—UTP cables and associated connecting hardware whose transmission characteristics are specified up to 100 MHz

Category 1 and Category 2 UTP are not recognized for new installations. They may still be in service in telephone (voice) applications.

Category 3 UTP is the minimum grade recommended in ANSI/TIA/EIA-568-A for new installations.

Category 5 UTP copper cable:

- Is the recommended grade, if UTP is selected as the second cable to the work area outlet. (ANSI/TIA/EIA-568-A permits the first cable to be Category 3.)

- Will support 10BASE-T Ethernet, 100BASE-T Fast Ethernet, 1000BASE-T Gigabit Ethernet, ATM 155 Mb/s, TP-PMD 100 Mb/s, and other LAN configurations.
- Can provide bandwidth up to 100 MHz.

When working with twisted-pair cables, the cabling installer should be able to identify individual pairs and conductors within the cable.

Standard color codes for copper and optical fiber cables have been developed. These color codes enable the cabling installer to quickly identify any pair within a cable.

Cabling installers must be able to identify individual conductors within each pair. In telephony, the individual conductors are referred to as tip and ring conductors. Each pair has a tip and a ring conductor, or a positive and negative conductor respectively.

The terms *tip* and *ring* originated from the earliest types of telephone systems, in which the operator had to physically use patch cords to route the calls. The operator's switchboard plug had three conductors: tip, ring, and sleeve.

Figure 1.23 Switchboard plug.

The tip conductor was connected to the very tip of the plug and had a positive voltage. The ring conductor had a negative voltage and was connected to a small collar or ring, just back and isolated from the tip. Located between the ring, the sleeve or ground conductor insulated the shaft of the plug.

It takes two conductors to make a pair. Each pair has a tip conductor and a ring conductor. The colors used to identify tip conductors are different from the colors used to identify ring conductors. There are five colors associated with tip conductors, and five different colors associated with ring conductors.

When the two conductors are paired, the two different colors identify the pair number. There are 25 possible color combinations when the five tip and five ring color codes are mixed, as can be seen in the color code table below. However, two tips or two rings are never used to make a pair.

The tip colors are white, red, black, yellow, and violet. There are five pairs within a tip color. Tip colors indicate in which group a pair is located. For example, white indicates that the pair group is 1, 2, 3, 4, or 5. Black indicates that it is pair group 11, 12, 13, 14, or 15.

TABLE 1.5 High-Pair-Count Cable, Color-Coded Pairs

	Ring conductors				
Tip conductors	Blue 1st Pair	Orange 2nd Pair	Green 3rd Pair	Brown 4th Pair	Slate 5th Pair
White (1–5)	1	2	3	4	5
Red (6–10)	6	7	8	9	10
Black (11–15)	11	12	13	14	15
Yellow (16–20)	16	17	18	19	20
Violet (21–25)	21	22	23	24	25

Note. The tip color does not identify a specific pair number until the combination of the tip and ring colors are matched.

The ring colors are blue, orange, green, brown, and slate. The ring color identifies the position of the pair within the group of five tip possibilities (i.e., blue is the first pair and green is the third pair within the tip group).

To identify the pair number, the tip and ring colors must be viewed together (i.e., the white/blue pair is actually a white tip and a blue ring). White indicates the pair is between pairs 1 through 5, while blue indicates it is the first pair. The first pair of 1 through 5 is "#1" pair.

Note. Another example: the black/brown pair. Black indicates pairs 11 through 15, and brown indicates fourth pair. The fourth pair is the fourteenth pair in the binder group.

The standard color code can be used to identify up to 25 pairs without duplicating any pair color combinations.

For cables over 25 pairs, the first group of 25 pairs is formed into a bundle and has a colored binder (tape or thread) placed around the bundle. Additional bundles (25 pairs) have their own unique colored binder. The colored binder wraps follow the same color code as the individual-pair color code. They are referred to as binder groups or binders.

The white/blue binder wrap is first and contains cable pairs 1–25, the white/orange binder wrap is second and contains cable pairs 26–50, and the red/blue binder wrap is the sixth and contains cable pairs 126–150. This system of 25-pair binder groups works for a cable containing up to 625 pairs of cable (25 binders × 25 pairs per binder = 625 total pairs); however, the system stops at 600 pairs. The violet/slate binder wrap is not used. On some smaller cables (100 pairs or less), the manufacturer may indicate binders with only the ring color, since it can be assumed that its tip color is white.

Once a cable exceeds 600 pairs each, 600 pairs (24 binders) are wrapped in a colored super binder. The colored super-group binder denotes 600 pairs. The color code changes from the normal tip/ring identifiers to a simple tip color. For example, white indicates 1–600 pairs, red indicates 601–1200, and black indicates 1201–1800 pairs.

UTP patch cords:

- Use stranded conductors for flexibility.
- Are allowed to exhibit up to 20 percent more attenuation than solid conductors.
- Should meet the same standards as the horizontal cabling in use.
- Usually have 8-position, 8-contact (8P8C) connectors on the ends.
- Must be twisted-pair construction.

If using stranded conductor cable, make sure that connectors are designed for these conductors. The wrong kind of connector may damage the wire and make a faulty connection.

TABLE 1.6 Color Codes for 100 Ω UTP Patch Cables

Pair	Identification	Conductor color Code-option 1	Conductor color Code-option 2
Pair 1	Tip	White-Blue (W-BL)	Green (G)
	Ring	Blue-White (BL-W)	Red (R)
Pair 2	Tip	White-Orange (W-O)	Black (BK)
	Ring	Orange-White (O-W)	Yellow (Y)
Pair 3	Tip	White-Green (W-G)	Blue (BL)
	Ring	Green-White (G-W)	Orange (O)
Pair 4	Tip	White-Brown (W-BR)	Brown (BR)
	Ring	Brown-White (BR-W)	Slate (S)

Another form of UTP cable is undercarpet cable. Undercarpet cable is available in performance levels up through Category 5. Undercarpet cable should not be the cabling installer's first choice. It is not recommended due to its:

- Susceptibility to damage.
- Limited flexibility for moves, adds, and changes.

When using undercarpet cable, a cabling installer should avoid:

- High traffic areas.
- Heavy office furniture locations.
- Undercarpet power cable.

Most of the cable terminations will involve 4-pair cable from a telecommunications closet to a work area. There is currently an

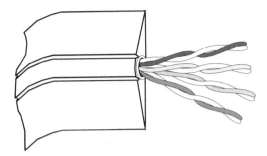

Figure 1.24 Undercarpet cable.

increased use of multimode fiber to work area outlets utilizing undercarpet cable for the support of high-bandwidth services, such as graphic design and high-performance workstations.

Twisted-pair cables are color coded and must be terminated in a specific order to maintain a universal standard of installation. Pay close attention to the color coding when terminating and splicing cables.

TABLE 1.7 High-Pair-Count Cable, Color-Coded Pairs, and Binder Groups

Pair Number	Binder group			
	Tip	Ring	Color	Pair Count
1	White	Blue	White-Blue	001–025
2	White	Orange	White-Orange	026–050
3	White	Green	White-Green	051–075
4	White	Brown	White-Brown	076–100
5	White	Slate	White-Slate	101–125
6	Red	Blue	Red-Blue	126–150
7	Red	Orange	Red-Orange	151–175
8	Red	Green	Red-Green	176–200
9	Red	Brown	Red-Brown	201–225
10	Red	Slate	Red-Slate	226–250
11	Black	Blue	Black-Blue	251–275
12	Black	Orange	Black-Orange	276–300
13	Black	Green	Black-Green	301–325
14	Black	Brown	Black-Brown	326–350
15	Black	Slate	Black-Slate	351–375
16	Yellow	Blue	Yellow-Blue	376–400
17	Yellow	Orange	Yellow-Orange	401–425
18	Yellow	Green	Yellow-Green	426–450
19	Yellow	Brown	Yellow-Brown	451–475
20	Yellow	Slate	Yellow-Slate	476–500
21	Violet	Blue	Violet-Blue	501–525
22	Violet	Orange	Violet-Orange	526–550
23	Violet	Green	Violet-Green	551–575
24	Violet	Brown	Violet-Brown	576–600
25	Violet	Slate	No Binder	—

High-pair-count cable (more than 4 pairs) is generally used for backbone service and is often referred to as multipair cable.

Copper cable pairs are usually attached to assigned positions on a terminal block. Methods of terminating copper cable include the following:

- Insulation displacement connection (IDC)
- Connectors
- Wirewrap

IDC hardware is the recommended termination method for UTP cable.

Category 6 and Category 7 cables have not been approved by a standards body. They have been characterized by TIA working groups to have the following criteria:

- Category 6 will support bandwidths of at least 200 MHz.
- Category 7 will support bandwidths of at least 600 MHz.

Screened twisted-pair (ScTP) cable

Category 5 screened twisted-pair copper cable. Screened twisted-pair cable (ScTP) is widely used in Europe, while in the United States its market is slowly growing. Although more expensive than UTP, screened twisted-pair cable offers a higher immunity to interference and may be used in areas susceptible to noise and interference.

With the exception of the mylar®/aluminum foil shield and drain wire, the construction and performance of ScTP cable is the same as UTP. Mylar foil blocks the reception and transmission of high-frequency signals through the cable jacket. Drain wire runs between the mylar foil and the outer jacket. It is used to ground the mylar foil, since the foil is not strong enough to be bonded directly.

ScTP cable has:

- A characteristic impedance of 100 Ω.
- Four pairs of 22–24 AWG solid conductors (normally).
- Mylar/aluminum foil shield around all conductor pairs.
- A drain wire that must be grounded.
- Similar electrical performance characteristics to Category 3, 4, or 5 UTP.
- May be used where extra protection from electrical interference is desired.

ScTP patch cords:

- Require ScTP patch cable.
- Use stranded conductors for flexibility.
- Are allowed to exhibit up to 20 percent more attenuation than solid conductors.
- Should meet the same standards as the horizontal cabling in use.
- Usually have 8-position, 8-contact (8P8C) connectors with a metal shield around plug housing.
- Must be twisted-pair construction.

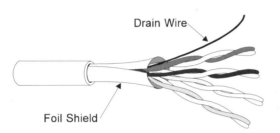

Figure 1.25 Screened twisted-pair (ScTP) cable.

Shielded twisted-pair (STP) copper cable

Shielded twisted-pair (STP) cable was developed at a time when unshielded twisted-pair cable was considered unsuitable for high-speed data transmission. By shielding the individual pairs of conductors, immunity to electromagnetic interference (EMI) was improved substantially, reducing the instances of interference-related network problems.

Transmission specifications for 150 Ω STP support transmission signals up to 20 MHz. With the demand for increased bandwidth capability, enhanced shielded twisted-pair cable and connectors (STP-A) that provide positive attenuation to crosstalk ratio (ACR) up to 300 MHz have emerged.

STP and STP-A cable is usually associated with token ring installations. STP-A has the highest bandwidth of any ANSI/TIA/EIA-568-A-approved copper medium. It can be used with most applications that do not require more than two pairs.

Shielded twisted-pair copper cable:

- Is composed of two pairs of wires.
- Contains individual mylar/aluminum foil shielded pairs to reduce high-frequency interference.

- Has an overall braid to reduce low-frequency interference.
- Conductor size is 22 AWG.
- Shield must be grounded.
- Has unique color code: green/red and black/orange pairs.

STP patch cords:

- Use patch cords with stranded 26 AWG conductors for flexibility.
- Are allowed to exhibit up to 50 percent additional attenuation than solid 22 AWG conductors.
- Use a hermaphroditic connector equipped with a pair of shorting bars inside each connector.

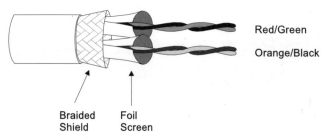

Figure 1.26 Shielded twisted-pair (STP) cable.

Coaxial cable

Coaxial cable provides a much higher bandwidth and degree of protection against EMI than twisted-pair wire. Coaxial cable consists of a centered inner conductor insulated from a surrounding outer conductor and an overall jacket. Coax is used for computer networks, CATV, and video systems.

Part numbers for coax cables are a series of characters that describe the cable's construction. RG stands for radio grade and the other characters represent variables such as:

- Center conductor diameter.
- Center conductor being solid or stranded.
- Dielectric composition.
- Outer braid percent of coverage (i.e., 80 percent or 95 percent coverage).

Ethernet networks require one end of the coaxial bus be bonded to ground.

Screw-on coaxial connectors have been added to the pending revision of the ANSI/EIA/TIA-570 standard for residential cabling. They are not recommended for commercial-building installations and should be avoided due to their susceptibility to failure. Though convenient, these connectors can cause intermittent problems that can be difficult to troubleshoot.

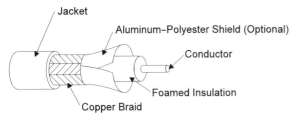

Figure 1.27 RG-6 coaxial cable.

RG-6 coaxial cable is used for video, CATV, and security cameras. It is similar to the popular RG-59 coaxial cable that, for years, has been used extensively in residential cabling. RG-59 has a distance limitation of approximately 61 m (200 ft), while RG-6 has a distance limitation of 152 m (500 ft) without the need for amplification.

Historically, contractors have used RG-6 for longer runs and the less expensive RG-59 for the shorter branch circuits. Today, the price difference is negligible, and most contractors use RG-6 exclusively. This cuts down on the number of tools, connectors, and cable that must be stocked.

RG-6 has:

- A characteristic impedance of 75 Ω.
- A mylar/aluminum foil shield over the dielectric to shield against high frequencies.
- A braided shield over the mylar foil to shield against low frequencies.
- A solid-center conductor.

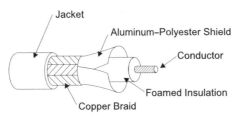

Figure 1.28 RG-11U coaxial cable.

RG-11U is used in video backbone distribution. It has less signal attenuation than RG-6, making it the preferred choice for longer runs. RG-11U has:

- A characteristic impedance of 75 Ω.
- A requirement for N-series connectors.
- A mylar/aluminum foil shield over the dielectric to shield against high frequencies.
- A braided bare copper shield over the mylar foil to shield against low frequencies.
- An 18 AWG stranded center conductor.

RG-8, IEEE 802.3 coaxial cable (thicknet)

RG-8 coax is used in 10BASE-5, thicknet, bus-type networks. The ANSI/TIA/EIA-568-A standard does not recommend that new installations use thicknet cable.

When adding new users, thicknet cable does not have to be cut. The cable has black stripes on its jacket every 2.5 m (8 ft) to indicate tap locations. A special tap (vampire tap) is used to bore through the outer jacket and shields to gain access to the center conductor. A transceiver must be attached to the cable only at these indicated tap locations.

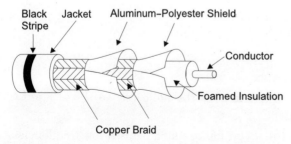

Figure 1.29 RG-8 Thicknet coaxial cable.

Depending on the manufacturer, a cable labeled RG-8 may not guarantee thicknet compatibility. Most manufacturers do not print RG-8 but color the cable's jacket yellow (PVC) or orange (plenum). A better identifier is IEEE 802.3 thicknet, the Ethernet standard. RG-8 IEEE 802.3 thicknet cable:

- Has characteristic impedance of 50 Ω.
- Has a solid-center conductor.

- Has two mylar/aluminum foil shields.
- Has two braided shields.
- Has black stripes on the jacket every 2.5 m (8 ft).
- Uses N-type connectors to join cable segments.
- Uses vampire taps.
- Has a maximum segment length of 500 m (1640 ft).

RG-58A/U, IEEE 802.3 coaxial cable (thinnet)

RG-58A/U IEEE 802.3 coaxial cable is used in 10BASE-2, thinnet bus-type networks. The ANSI/TIA/EIA-568-A standard does not recommend the use of thinnet cable in new installations.

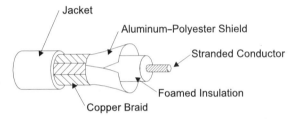

Figure 1.30 RG-58A/U Thinnet coaxial cable.

Depending on the manufacturer, RG-58 coaxial cable may not guarantee thinnet compatibility. The A/U in RG-58A/U describes a 20 AWG stranded center conductor with a 95 percent braided, tinned copper shield. This should indicate thinnet cable; however, the cabling installer should always verify that either thinnet or IEEE 802.3, the Ethernet standard, is printed on the cable.

RG-58A/U IEEE 802.3 cable has:

- Characteristic impedance of 50 Ω.
- A stranded-center conductor.
- A mylar/aluminum foil shield.
- A braided shield.
- BNC-type connectors with a center pin.
- A maximum segment length of 185 m (606 ft).

Optical fiber cable

Optical fibers allow the transmission of light impulses instead of electrical signals. The two key elements of an optical fiber are its core and cladding. The core is the innermost part of the fiber, through which light pulses are guided. The cladding surrounds the core to keep the light in the center of the fiber.

The core's refractive index is higher than that of the cladding. Light traveling down the core that strikes the cladding at a glancing angle is confined in the core by total internal reflection. In other words, the cladding acts like a mirror to reflect light that normally escapes through the sides of the optical fiber strand back into the center of the fiber. To reinforce and strengthen the strand, an acrylate coating is applied to the cladding. This increases the original 125 μm diameter of the cladding to 250 μm.

Optical fiber cable may be:

- Singlemode.
- Multimode.
- Composite.

All of the above types of cable are available in the following versions:

- Single fiber
- Multifiber

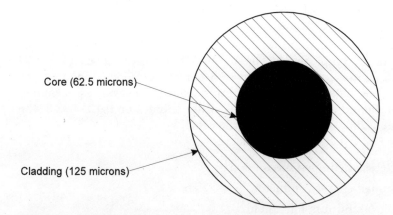

Figure 1.31 Multimode fiber.

Multimode optical fiber cable:

- Is the most common for backbones within buildings and in campus environments.
- Has a 50 μm or 62.5-μm core and 125-μm cladding diameter.
- Has a distance limitation of 2000 m (6560 ft) for structured cabling systems.

- Generally utilizes a light-emitting diode (LED) light source.
- LED light sources have wavelengths of 850 nm and 1300 nm.

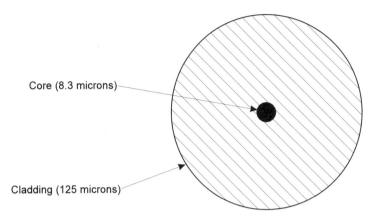

Figure 1.32 Singlemode fiber.

Singlemode optical fiber cable:

- Has an 8 to 9 μm core.
- May be used for distances up to 3000 m (9840 ft) for structured cabling systems.
- Normally uses a laser light source.
- Laser light sources have wavelengths of 1310 nm and 1550 nm.

Optical fiber cable may be:

- Tight buffered.
- Loose tube.

Tight-buffered optical fiber cable:

- Is mostly used inside buildings.
- Meets building codes.
- Protects the fiber by supporting each strand of glass in a tight-buffered coating, increasing the strand's diameter to 900 μm.
- Is easily connectorized.
- May be singlemode or multimode.

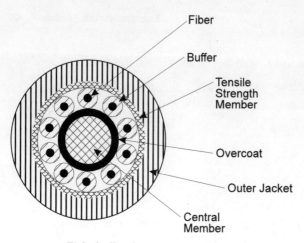

Figure 1.33 Tight-buffered optical fiber cable.

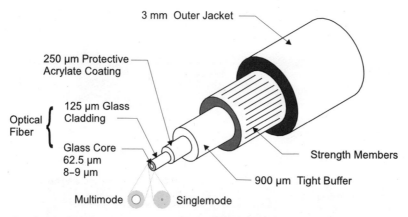

Figure 1.34 Single stranded tight-buffered fiber cable.

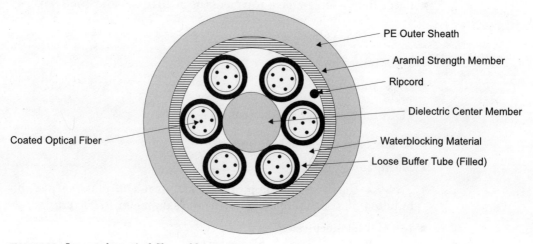

Figure 1.35 Loose-tube optical fiber cable.

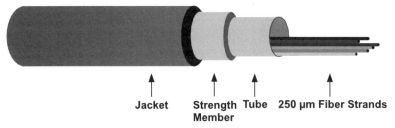

Figure 1.36 Side view of loose-tube optical fiber cable.

Loose-tube optical fiber cable:

- Is used primarily outdoors.
- Allows fiber to expand and contract with changes in temperature.
- May be singlemode or multimode.
- May require a furcating harness (breakout kit) to allow connectorization.

900 μm-OD Furcating Tubes

Figure 1.37 Loose-tube furcating harness.

ANSI/TIA/EIA-598-A, *Optical Fiber Cable Color Coding,* covers the color coding of optical fiber cables. The standard states:

- Strands 1–12 shall be uniquely color coded as shown in the following table.
- Strands 13–24 shall repeat the same color code as 1–12 with the addition of a black tracer.
- The black tracer may be a dashed line or solid line.
- The black strand has a yellow tracer.

Horizontal optical fiber cable shall be multimode.

Backbone optical fiber cable may be:

- Multimode.
- Singlemode.

TABLE 1.8 Optical Fiber Cable Color Code Chart

Fiber number	Color	Fiber number	Color
1	Blue	13	Blue/Black Tracer
2	Orange	14	Orange/Yellow Tracer
3	Green	15	Green/Black Tracer
4	Brown	16	Brown/Black Tracer
5	Slate	17	Slate/Black Tracer
6	White	18	White/Black Tracer
7	Red	19	Red/Black Tracer
8	Black	20	Black/Yellow Tracer
9	Yellow	21	Yellow/Black Tracer
10	Violet	22	Violet/Black Tracer
11	Rose	23	Rose/Black Tracer
12	Aqua	24	Aqua/Black Tracer

Multimode and singlemode optical fiber cables should perform as shown in the following tables.

TABLE 1.9 Horizontal and Backbone 62.5/125 µm Optical Fiber Cable Transmission Performance Parameters

Wavelength (nm)	Maximum Attenuation (dB/km)	Information Transmission Capacity (MHz–km)
850	3.5	200
1300	1.0	500

Bandwidth shown for 850 nm exceeds ANSI/TIA/EIA-568-A requirements and meets the requirements for the IEEE 802.3z standard for Gigabit Ethernet.

Note. Attenuation requirements are based upon ISO/IEC 11801.

TABLE 1.10 Backbone Singlemode Optical Fiber Cable Transmission Performance Parameters

Wavelength (nm)	Maximum Attenuation Optical Fiber Cable (dB/km)	Information Transmission Capacity (MHz–km)
1310	1.0	unlimited for premise application
1550	1.0	unlimited for premise application

Since the early 1970s, plastic fiber or polymeric optical fiber (POF) has been used in illumination, robotics, industrial, and automotive applications. Some common examples of crude implementations are warning indicators on automobile dashboards and novelty rainbow lamps.

Connectorization

Overview

A connector is a mechanical device that may be used to interface a cable to a piece of equipment or one cable to another. In the case of fiber, it allows light impulses to be transferred from one connector to another connector. In the case of copper, it allows electrical signals to be transferred from one connector to another connecter.

When two connectors are mated, there are minimum parameters that must be met to establish a good connection. A good connection requires:

- Physical alignment of connectors.
- Physical retention of components to prevent the connectors from unintentional separation.
- Light or electricity to be efficiently transferred from one connector to the other.

Durability is demonstrated by the connectors' ability to withstand hundreds of insertion and withdrawal cycles without failing. This is calculated as mean time between failures (MTBF).

Cable connectors are as essential to the integrity of a telecommunications network, as is the cable itself. The connector aligns, attaches, and decouples the medium to:

- A transmitter.
- A receiver.
- Another medium of same or similar type.
- An active telecommunications device.
- A passive telecommunications device.

Along the cable path, connections can contribute to the loss of signal from transmitter to receiver. The role of the connector is to provide a coupling mechanism that keeps loss to a minimum. Some of the key factors in minimizing loss of signal include:

- Fit.
- Alignment.
- Connector type.

Types of outlet interfaces

Suitable connecting hardware must be used to ensure proper signal transmission. The cabling installer must ensure the connectors being installed are designed specifically for the medium being connected. Currently, there are various types of outlet interfaces in use within the industry, including:

- Unshielded twisted-pair (UTP)
  - 8-pin modular connector
- Screened twisted-pair (ScTP)
  - 8-pin modular shielded connector
- Shielded twisted-pair (STP)
  - Hermaphroditic connector (type-A and type-B)
- Shielded twisted-pair-A (STP-A)
  - Enhanced hermaphroditic connector
- Coaxial
  - BNC series (Thinnet)
  - F-series (Video)
  - N-series (Thicknet)
- Optical Fiber
  - ST
  - SC

Additional connectors used to interconnect electronic equipment are:

- Telco 50-pin connectors.
- Subminiature-D connectors.

Unshielded twisted-pair (UTP) connectors. Twisted-pair cable requires insulation displacement connections (IDC). An IDC connection permits termination of an insulated conductor without stripping insulation from the conductor. As the insulated conductor is inserted between two or more sharp edges of the contact, the insulation is displaced, allowing contact to be made between the conductor and the connector terminal.

Connectors have IDC contacts designed for stranded, solid conductors, or connectors having universal contacts that accept either stranded or solid conductors.

Modular plug contacts for stranded conductors are designed as a single blade, intended to penetrate the conductors' insulation and slide between the individual conductor strands. When these contacts are used with a solid conductor, there are no strands to slide between, and the solid conductor is usually nicked or broken. Nicked conductors soon break from fatigue and become intermittent opens while providing electrical connection.

Note. It is rare to achieve a reliable connection using a connector designed for standard cable on a solid conductor.

Modular plug contacts for solid conductors are designed with three fingers that penetrate the insulation on both sides of the conductor. They provide an electrical connection by trapping the conductor between the fingers.

Some manufacturers market solid-conductor, modular-plug contacts as multipurpose connectors capable of terminating both stranded and solid conductors. These connectors work well for solid conductors. However, they do not always provide a connection as reliable as the single-bladed contact when used with a stranded conductor.

ANSI/TIA/EIA-568-A requires the use of stranded patch cords that are best served by connectors specifically designed for stranded cable. Any plug being installed on a solid conductor is outside the scope of the standard and will not be addressed in this manual.

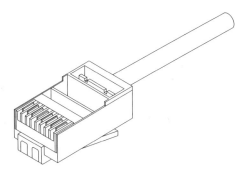

Figure 1.38 8P8C modular plugs nonkeyed.

ANSI/TIA/EIA-568-A requires that an 8-position, 8-contact (8P8C) modular connector be utilized for 4-pair UTP cable. Modular plugs and connectors are available in various sizes and shapes (keyed/nonkeyed). The number of positions (8P) indicates the connector's width, while the number of contacts (8C) installed into the available positions indicates the maximum number of conductors the connector can terminate.

A connector may be sized for eight positions but only have four contacts installed. This saves on manufacturing costs. For example, connectors are available as 8P2C, 8P4C, and 8P6C. They are all the same physical size but have different numbers of contacts to terminate conductors.

To help standardize all the various connectors and wiring configurations that were available, the Universal Service Order Code (USOC) was developed. The USOC codes were used for service ordering and customer billing.

Some typical designations for these connectors are "RJ-21X, RJ-45M, and RJ-11C" connectors. These terms indicate a specific type of jack or plug and its wiring configuration. "RJ" stands for registered jack, while the 21X indicates a demarcation or demarc configured with a 25-pair Telco connector.

Three common modular connectors used in telecommunications are:

- 4P4C—4-position and 4-contact connectors used primarily for telephone handset cords.
- 6P6C—6-position and 6-contact connectors used primarily for telephones and modem line cords (often referred to as an RJ-11 connector).
- 8P8C—8-position and 8-contact connector used for data communication line and patch cords (often referred to as an RJ-45 connector).

Since divestiture, USOC codes became irrelevant as each Bell operating company and the numerous independent providers began using their own billing codes and terms. Today's use of the terms RJ-45 and RJ-11 have come to represent their physical appearance more than any specific wiring configurations. The industry has adopted the term RJ-45 to describe all the 8-position connectors and RJ-11 to describe all the 6-position connectors. However, in the United States, FCC Part 68 has adopted some of these codes. Refer to FCC Part 68 for additional information.

To save costs, some line cords supplied with equipment may not have all the conductors expected. Telephone and modem manufacturers often supply line cords composed of 6P2C plugs with a two-conductor cable. This can be a problem if the cabling installer tries to use these cords on another type of equipment that requires more than one pair of conductors.

Outlets or jacks are female connectors, while the plugs are male connectors. Connector gender is determined by inserting the male plug into the female jack. Plugs and jacks are designed to withstand a minimum of 750 insertion and withdrawal cycles without failing. The connector, when prepared and terminated properly, ensures a good, reliable connection.

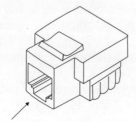

Locking-tab indentation

Figure 1.39 8P8C modular jack.

Modular jacks are designed to interface UTP cables with equipment lines or patch cords. UTP cables are terminated on the rear of the connector, while the connector's front provides access to the modular 8P8C jack.

The jack should be installed with the contact pins up and the locking-tab indentation down. When viewed from the front, the pin numbers run from left to right, consecutively 1 through 8. Unshielded twisted-pair outlet connectors are utilized in office environments. The illustration below shows UTP work area horizontal cables mated to UTP outlet connectors on a faceplate.

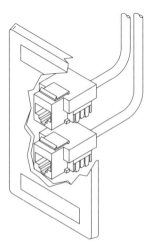

Figure 1.40 Telecommunications outlet and connectors.

Modular jacks are capable of accepting pin assignments compatible with all known data applications operating over 100 Ω twisted-pair cable. ANSI/TIA/EIA-568-A-approved jack and pin/pair assignments for these modular jacks are shown below.

Certain 8-pin cabling systems require the designation of T568B for optional pin/pair assignments. The optional 8-position assignments are shown below.

> **Note.** U.S. Government requirements often require T568A wiring as published in FIPS PUB 174 (known as the Federal Building Telecommunications Wiring Standard of 1992).

Another common pin/pair assignment for telephony is designated as USOC. There are many existing cabling systems that are still using the USOC wiring scheme.

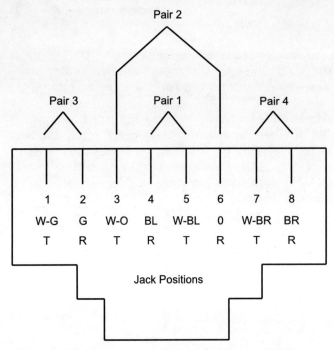

Figure 1.41 Eight-position jack pin/pair assignments T568A.

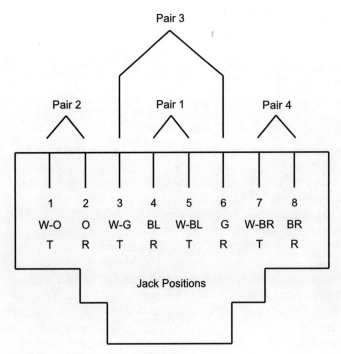

Figure 1.42 Optional pinout T568B.

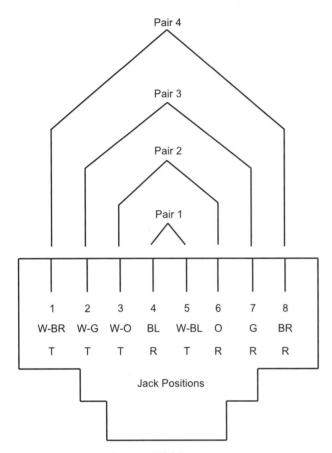

Figure 1.43 Optional pinout USOC.

UTP connectors may be 8P8C modular jacks or plugs and are designed to:

- Terminate 100 Ω, 4-pair, 22–24 AWG cable.
- Support up to 100 MHz, Category 5 standards.
- Use IDC terminations.

Many key and PBX telephone systems use 25-pair cables to interface the entrance facility and cross-connect fields. Manufacturers often specify the demarc to be an RJ-21X and the connectors used to interface their equipment to be 50-pin Telco connectors.

These connectors can be ordered as complete cable assemblies but are usually connectorized by the cabling installer. Field installation is less expensive and allows for custom lengths that can enhance the installation's appearance.

Note: It is common to buy double-ended cables twice as long as needed and cut them in half. This provides a factory connector and minimizes time to terminate.

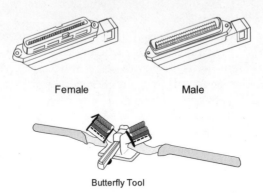

Figure 1.44 Telco 50-pin connector.

Telco connectors are available in both male and female versions and require a special tool, called a butterfly tool, for termination. During assembly, the tool's two actuating arms and conductor fanning strips extend outward and are then raised to terminate and cut the conductors.

Connections between data equipment often utilize "D-subminiature" connectors, so called because of the "D-shaped" metal skirt surrounding the connector's pins. These connectors are referred to as DB-## connectors. The "##" indicates the number of pins in the connector. There are both male and female connectors, but the name does not indicate gender and must be specified when ordering. The connectors are used for:

- DB-9—Standardized 9-pin serial port used on laptop computers.
- DB-15—Standardized 15-pin connector used for Ethernet transceivers, VGA monitors, and joy sticks (EIA-422-A and EIA-485).
- DB-25—Standardized 25-pin connector used for EIA-232-D and EIA-232-E serial data communications.

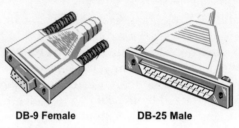

Figure 1.45 DB connector.

Screened twisted-pair (ScTP) connectors. Screened cable installations require a shield around every component of the cabling system. Shield continuity is maintained from the first connector in the closet to the telecommunications outlet through the user's equipment's line cord and to the equipment's chassis ground.

A metal shield surrounds each modular jack. A drain wire in contact with the cable screen must be attached to the connector's shield.

Shielded patch cords and line cords must be used to extend the shield from the screened modular jacks. Each cord is comprised of a stranded ScTP cable with a screened modular plug on each end. Each plug has a metal shield surrounding it. A drain wire in contact with the cable screen must be attached to each plug's shield.

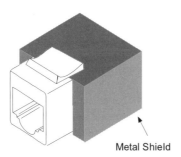

Metal Shield

Figure 1.46 ScTP modular jack.

An 8-position, 8-contact (8P8C) shielded modular connector is utilized for 4-pair ScTP cable. When prepared and terminated properly, the connector ensures a good connection and will serve to reduce electrical interference from external sources.

Outlets or jacks are female connectors, while plugs are male connectors. ScTP connectors are shielded 8P8C modular jacks or plugs that are:

- Designed to:
  - Terminate 100 Ω, 4-pair, 22–24 AWG cable.
  - Support up to 100-MHz, Category 5 standards.
  - Use IDC terminations.
- Covered by a metallic shield.
- Connected to the ScTP shield, using the drain wire.

Shielded twisted-pair (STP) cable data connectors. The data connector is a 4-contact hermaphroditic connector for shielded cable.

These are neither male nor female connectors; any STP can be mated to another STP connector.

Designed for token ring applications, these connectors worked by transferring the signal to the equipment through one pair of connectors. The signal then exited through the other pair of connectors. All equipment was connected in a large series loop. An opening anywhere in the circuit would cause the entire ring to fail.

A pair of shorting bars in each connector prevent an open in the ring when a connector is disconnected. When the connectors are separated, one of the bars shorts the two tip conductors together, while the other bar shorts the two ring conductors together. This loops the incoming pair to the outgoing pair and keeps the loop intact.

Capable of handling a bandwidth of 20 MHz, these connectors are designed to:

- Terminate 150 Ω, 2-pair, 22 AWG cables.
- Have a bandwidth of 20 MHz.
- Use a red/green and black/orange color code.

Enhanced shielded twisted-pair (STP-A) cable data connectors. STP-A cable and connectors are similar to the STP versions, but enhanced. They require an enhanced-data, 4-contact, hermaphroditic connector. These connectors are designed to handle a bandwidth of 300 MHz. The bandwidth improvement these connectors offer was accomplished through better shielding and the addition of a metal divider separating the cable pairs within the connector.

These connectors have the highest twisted-pair bandwidth recognized by ANSI/TIA/EIA-568-A and are designed to:

- Terminate enhanced 150 Ω, 2-pair, 22 AWG cables.
- Have a bandwidth of 300 MHz.
- Use a red/green and orange/black color code.

Coaxial connectors. Either male or female connectors can be attached to coaxial cable, but most installations utilize male connectors on cable ends. Coaxial patch cables should be purchased with connectors already attached, if possible.

"F" series coaxial connectors. F-type connectors do not have center pins. Instead, they use the solid-center conductor of the cable as a center pin. For this reason, F-type connectors cannot be used with cables that have a stranded center conductor.

"F" series connectors are:

- Economical.
- Used to connect RG-59 and RG-6 coaxial cable to video, CATV, and security cameras.

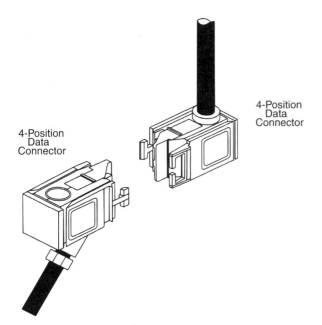

Figure 1.47 Two 150 Ω STP-A outlet connectors.

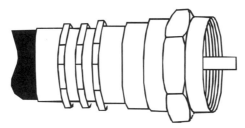

Figure 1.48 F-type connector.

- The most widely used connector for residential cable service.
- Available as one- or two-piece connectors. One-piece connectors have built-in crimp sleeves, while the two-piece connectors have separate crimp sleeves that must be crimped over the outer braid.
- Available as a screw-on connector. These types are not recommended for commercial use.

"N" series coaxial connectors. N-type connectors are used in data and video applications such as RG-8, Thicknet coaxial cable, and RG-11U coaxial cable, which are used for video backbone distribution.

N-type connectors have a center pin that must be installed over the cable's center conductor. The N-type male connector uses a threaded outer collar to mate with the female connector.

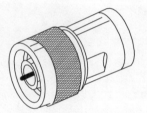

Figure 1.49 N-type connector.

BNC coaxial connectors. Named for its designers, the Bayonet Neill-Concelman (BNC) connector has been in use since World War II. BNC connectors have a center pin that must be installed over the cable's stranded-center conductor and are used with RG-6 (75 Ω), RG-58A/U thinnet coaxial (50 Ω), RG-59 (75 Ω), and RG-62 (93 Ω) coaxial cable.

Figure 1.50 BNC connector.

Three types of BNC connectors are common:

- Crimp
- Three-piece
- Screw-on (not recommended for commercial applications)

Crimping requires a special tool but no solder. The three-piece connector may require crimping or solder to hold the conductor to the center pin. The proper tool is especially important for obtaining an effective crimp on the sleeve of the connector.

Optical fiber connectors. Optical fiber connectors perform the same function as copper connectors but use a different medium (light impulses through optical fiber as opposed to electrical signals through a wire). Unlike electrons, which are forgiving when it comes to connector alignment as long as a good electrical contact is made, optical fiber connectors must be precisely aligned to allow maximum signal transfer between the connectors.

Standard optical fiber connectors do not have the male and female connectors common in electrical systems. Essentially, all optical fiber connectors are male and require an adapter to precisely align the two connectors tip to tip. The adapters are referred to as couplers or sleeves.

Connectors and adapters provide a low-loss coupling by ensuring alignment of the two elements being joined. Any misalignment of optical connectors increases the loss.

ANSI/TIA/EIA-568-A recognizes two types of optical fiber: multimode and singlemode fibers. Both fibers have the same cladding diameter.

ANSI/TIA/EIA-568-A recognizes two types of connectors to be used in either backbone or horizontal fiber installations. The ST-type connector is recognized but not recommended. The standard connector is the 568SC.

The ST-type connectors feature a bayonet-type coupling that employs a quick-release mechanism. The ST connector is keyed to prevent the ferrule from twisting during insertion, while the main connector housing requires only a quarter turn to engage or to disengage. In multimode and singlemode operation, the ST-type connector is used for:

- Premises cabling.
- Networking.
- Test equipment.

ST

Figure 1.51 Connector ST compatible.

The 568SC connector is the only connector approved for new installations under the ANSI/TIA/EIA-568-A standard.

The 568SC connector is the connector recommended by ANSI/TIA/EIA-568-A for new installations. However, facilities having an installed base of ST style connectors and adapters may remain with the ST style connector and adapters for both existing and future additions to their optical fiber network. The plugs mate in a connector housing which aligns the ferrules. An advantage is that two plugs can be joined together to form a 568SC duplex connector. Multifiber capabilities allow for building duplex (two-position) connectors. One fiber carries incoming information and the other fiber carries information in the other direction. Keyed connectors require a specific orientation of plug and jack and thus prevent mismating. The 568SC is preferred for:

- Network applications.
- Premises cabling applications.
- Space saving.
- Ease of use.

The 568SC connector housings shall be color coded as follows:

- Beige—multimode.
- Blue—singlemode.

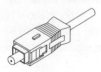
568SC

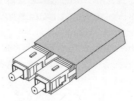
568SC Duplex

Figure 1.52 Connector 568SC compatible.

Matching connectors to media type

The following table shows the industry-standard connectors specified for different types of cable.

TABLE 1.11 Cable Connectors

Standard	Topology Logical	Topology Physical	Cable Type	Connector Type
IEEE 802.3:				
10BASE-5	Bus	Bus	RG-8 (50 Ω)	N-series
10BASE-2	Bus	Bus/Star	RG-58 IEEE 802.3 (50 Ω)	BNC-series
10BASE-T	Bus	Star	UTP (100 Ω)	8-Position Modular
10BROAD-36	Bus	Bus	Hardline RG-6, RG-11 (75 Ω)	F-series
10BASE-F	Bus	Star	Optical fiber Multimode	ST, SC
IEEE 802.4:				
Broadband	Bus	Bus	Hardline 75 Ω Coax	F, CATV
Carrierband	Bus	Bus	Hardline 75 Ω Coax	F, CATV
Fiber	Star	Star	Optical fiber Singlemode Multimode	ST, SC
IEEE 802.5:				
Token ring	Ring	Star	UTP (100 Ω)	8-position modular
Token ring	Ring	Star	STP, STP-A (150 Ω)	Hermaphroditic data connector
FDDI	Ring	Star/Ring	UTP, ScTP Optical fiber Multimode/ Singlemode	8-position modular MIC/FDDI SC, ST

Under IEEE 802.3 (CSMA/CD standard), the descriptor 10BASE-5 provides information about the circuit. The first set of numbers describes the amount of information or megabits per second the circuit can support.

The base describes the number of signals on the medium at the same time. Base refers to baseband, which has only one signal and no multiplexing.

The term *broad* refers to broadband signaling, which has multiple signals or multiplexing on the medium at the same time. Each transmission channel occupies a different frequency band on the cable. Each channel has a bandwidth of 6 kHz. An example of a broadband medium is CATV. Numerous channels are available at the same time, while the television's tuner selects the channel to view.

The last set of numbers describes the medium's distance limitations when the limitations are already part of the standard.

To determine the distance limitation, multiply the number by 100 m (328 ft). For example, the "5" in 10BASE-5 indicates a distance limitation of 500 m (1640 ft). In 10BASE-2, the distance limitation is 200 meters (actually 185 m [606 ft]). The 185 is rounded up to simplify the descriptor (i.e., 10BASE1.85 is more cumbersome).

If the last space is a letter and not a number (as in 10BASE-F or 10BASE-T), the letter represents the medium. For example, in the two previous examples, the "F" represents fiber and the "T" represents twisted-pair. Both types of media are in the standards with strict distance limitations (e.g., UTP cannot exceed 90 m (295 ft) when used for data transmission).

10BROAD-36 refers to 10 Mb/s over broadband coaxial cable to a maximum distance of 3600 m (11,808 ft).

Transmission

Overview

Transmission, as discussed in this chapter, relates to the movement of information as electrical or optical signals from one point to another via a medium. The medium can be air, water, copper, optical fiber, or whatever else might be used to carry the signal.

This section will deal with the use of copper conductors and glass strands for the passing of a signal in either analog or digital form.

The choice of a specific medium is influenced by economics and technical considerations—such as the physical diameter of the cable, the type of services to be provided (i.e., voice, high-speed data, video, etc.), the size of the network, and the transmission path distance. For copper systems, in which there is concern that electromagnetic interference (EMI) conditions exist which cannot be solved readily by means of an increased physical separation, optical fiber may be used, because

EMI and other electrical interference does not affect it, with the exception of electromagnetic pulse (EMP).

All media have certain characteristics that will limit its performance. Medium selection should be based on the specific network requirements. Once the proper selection has been made, the only common limiting factor is usually faulty or improper system design or installation techniques.

In order to assure compliance with industry cabling standards, manufacturers of copper and fiber cable have instituted design changes in the construction of their products as the standards continue to include higher transmission performance requirements. On the copper side, the assignment of a category performance-level rating (Category 3, Category 4, and Category 5) provides a simple means to select a cable construction suitable for the modest requirements of a voice circuit or one that can meet the strict demands of high-speed data transmission.

This section discusses the major terms associated with transmission. The objective of the discussion is to show the impact that improper installation can have on medium transmission characteristics.

Power, current, and voltage

Power (P) is a term which applies to the energy required, for example, to operate an electrical device such as a motor, an amplifier, a telephone transmitter, etc. Its main components are voltage (V) and current (I)—there is no power unless both are present. In an analogy to water, voltage may be thought of as the water pressure, while current is the quantity of water delivered. Millions of volts with no current will not provide any power and, thus, will not energize a device to perform the desired function.

Power is related to current and voltage in that power is equal to the voltage multiplied by the current, or $P = VI$.

In transmitting information from one point to another, the installer is primarily interested in transporting power over the transmission line. It is usually a very small amount compared to that of a commercial power line; nevertheless, it is power that is being transported.

ac and dc

The standard commercial power frequency in the United States is 60 Hz (cycles)—thus, this alternating current (ac) voltage completes 60 sine waves per second.

For each sine wave, the voltage begins at zero level and increases to its maximum positive value before dropping back to cross the zero level, where the voltage becomes negative or changes direction. The voltage continues in the negative direction to reach a maximum negative value; it then becomes less negative until it again reaches the zero level and completes the cycle (see Figure 1.53). Although the power has gone to zero twice each cycle, our eyes are not sensitive enough to perceive the rapid increase and decrease of the power and the accompanying flicker of our reading lamps.

Direct current (dc) refers to a steady value which does not change direction of voltage or current. It rises to its maximum value when switched on, and remains there until the circuit is interrupted. A battery is an example of a dc source.

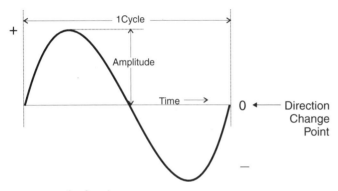

Figure 1.53 Analog sine wave.

Frequency

Frequency is defined as the number of cycles or "sine waves" (see previous figure) occurring in a given time. If the unit of time is equal to one second, the frequency is stated in hertz. One hertz (Hz) is equal to one cycle per second.

- 1 cycle in 1 second = 1 Hz (hertz)
- 60 cycles in 1 second = 60 Hz
- 1000 cycles in 1 second = 1 kHz (kilohertz)
- 1 million cycles in 1 second = 1 MHz (megahertz)
- 1 billion cycles in 1 second = 1 GHz (gigahertz)

Analog

Analog by itself means nothing. It is derived from the term *analogous,* which means "similar to." Analog transmission utilizes continuously varying electrical signals, which directly follow, or are similar to, the changes in loudness (amplitude) and frequency (tone) of the input signal (e.g., the human voice). For instance, the transmitter of a telephone converts sound waves to an analog electrical signal that varies in amplitude (signal strength) directly with the loudness of the talker and varies in frequency directly with the tone of the talker.

Digital

A digital signal is a discontinuous signal that changes from one state to another in a limited number of discrete (single) steps. In the simplest form, there are two states or signal levels—on and off, where the "on" translates to a digit "1" and "off" is seen as a zero level corresponding to the digit "0." Other means can also be used to represent the two states (i.e., a positive and a negative voltage).

The digital message is made up of a sequence of these digital pulses (bits) transmitted at regular and defined time intervals if the signal is synchronous. These bits are usually square in shape. Thus, digitally transmitted information is an encoded representation of input, unlike analog transmissions, which reproduce an analog input in both frequency and amplitude.

If the input is analog, the transmission of a digital signal generally involves a sampling process in which the analog voltage is sampled often enough that, when replicated, there is an acceptable facsimile of the original signal.

Megahertz and Megabits

It is important to highlight the relationship between hertz and bits, or Megahertz and Megabits.

Megahertz, as in 100 Megahertz Category 5 cable, is the bandwidth of the cable. This relates to the information-handling capability of the medium (the size of the basket).

Megabits, as in bits of information, refers to the number of bits of information that can be placed into the basket.

The number of bits that will fit in the basket is determined by the type of encoding scheme selected, such as Manchester, Return to Zero (RZ), or Non Return to Zero (NRZ).

Decibel

A measure of analog signal strength is the "bel," named in honor of telephone pioneer Alexander Graham Bell. The term is inappropriately large for telecommunications work, so decibel is used. It is not an absolute value but is the ratio of two signals and may be used to compare power, current, or voltage.

The decibel is a logarithmic function that allows large variations to be shown in very small increments. With respect to installation efforts, it is most important to understand that a 1 dB change, under ideal conditions, can be heard.

Increases or decreases of 3 dB will result in a doubling or halving of the power. Increases or decreases of 6 dB will cause a doubling or halving of the voltage.

Increases or decreases of 10 dB for power and 20 dB for voltage or current will cause a change of (×10). This moves the decimal point one place.

- Crosstalk = The larger the dB number the better.
- Attenuation = The smaller the dB number the better.
- Connector Insertion Loss = The lower the dB the better.
- Structural Return Loss = The higher the dB the better.

Copper Cable Media

Overview

With any copper cable media, the proper level signal must be coupled through the media from transmit to receive end and still drive the receiver.

The most effective and efficient transfer of energy will occur in systems in which all individual parts have the same characteristic impedance.

This section will deal with the electrical properties that make up characteristic impedance and how characteristic impedance can be impacted by design and installation methods or procedures.

Resistance

For metallic cables, the electrical energy of the signals is transmitted over copper conductors. Copper is preferred since it is a better conductor of electricity than most other metals and is relatively economical. Further, copper is readily converted into a round wire through a drawing process whereby it is drawn, or pulled, through a series of successively smaller round holes (dies) until the desired diameter is reached.

In the United States, the copper conductors are categorized through an accepted numbering system called the American Wire Gauge (AWG). The AWG number assigned to a particular wire size approximates the number of steps required to draw it to that diameter. For example, a 14 AWG wire is drawn through 14 different-sized die openings during the drawing process. Thus, the smaller the wire diameter, the larger the gauge number. Conductors larger than an AWG 1 are 0, 00, 000, and 0000 and are called one aught, two aught, three aught, and four aught, respectively. Conductors larger than 0000 are measured by their diameter referenced in circular mil.

Copper is highly conductive and offers little resistance to the flow of electrical energy when compared to metals of less conductivity. A larger diameter wire will present a greater cross-sectional area of copper to the electrical energy. This wider path (like a wider highway) permits an increased energy (traffic) flow at lower frequencies—its ability to conduct has increased; therefore, the resistance is lowered. Larger diameter wires provide a lower resistance than small ones. Remember though, larger AWG numbers are associated with smaller wire diameters.

The resistance to the flow of electrical energy is measured in ohms. By definition, one ohm is the resistance that will allow one ampere (electric current unit) to flow when one volt is applied. This is known as ohm's law and can be stated mathematically as,

$$V = IR$$

where: V = Voltage in volts
       I = Current in amps
       R = Resistance in ohms

This formula can be restated to calculate any one value when the other two are known,

$$I = \frac{V}{R} \quad \text{or} \quad R = \frac{V}{I}$$

Figure 1.54 Ohm's law.

- Ohm's law is usually shown as in the figure above. Values are easily solved for reference. Place your finger over the **I** and you see **V** is over **R**. Place your finger over the **R** and you see **V** is over **I**. Place your finger over the **V** and you see that **I** is next to **R**.
- Higher temperatures increase the conductor resistance by approximately two percent for each 5.5 °C (10 °F) rise.
- The resistance is increased in proportion to any increase in length (e.g., doubling the length of the cable doubles its resistance).
- Although unaffected at frequencies in the audio range, considerably higher frequencies cause the resistance to increase (e.g., increasing the frequency by four times results in a doubling of the resistance).

Inductance

The diameter of the conductors of a pair is maintained at a close tolerance in order to keep any resistance unbalance to a minimum. Such unbalance contributes to undesirable effects such as crosstalk and noise, which will be discussed later.

Inductance is the property of an electrical force field built around a conductor when current flows through it. When direct current (dc) flows through a conductor, the field is steady. An alternating current (ac) causes the lines of force to constantly build and collapse.

Simply put, inductance is a resistance to change in current. Inductive coupling is the transfer of energy from one circuit to another. For example, power lines on a utility pole can inductively couple a power surge onto telephone cables.

Capacitance

Capacitance is the tendency of an electronic component to store electrical energy. Capacitance is the resistance to the change in voltage. Pairs of wire in a cable tend to act as a capacitor, which has two conductors, or plates, that are separated by a dielectric.

Cable normally exhibits some level of capacitance. Typically, unshielded twisted-pair will have a value of 15 to 20 pf/ft (picofarads per foot).

Capacitance in cable is considered undesirable and must be minimized.

If cable is improperly installed, capacitance will change. Since it is a very important element of the cable's characteristics impedance, it will degrade the quality of transmission through the cable.

ANSI/TIA/E IA-568-A
Mutual Capacitance Specifications

Category 3 = 20 pf/ft Max
Category 4 = 17 pf/ft Max
Category 5 = 17 pf/ft Max

Characteristic impedance

Characteristic impedance is defined as the input impedance of a uniform analog transmission line of infinite length. Characteristic impedance is made up of three things:

- Capacitance
- Inductance
- Resistance

Every cable or transmission line has characteristic impedance. The value of the characteristic impedance is determined by the diameter of the conductors and the insulating value (dielectric constant) of the materials separating them.

Cables are designed to achieve constant characteristic impedance independent of cable length. For example, if a 30 m (98 ft) length of cable has characteristic impedance of 100 Ω, it will still be 100 Ω if the length is doubled or tripled for a given frequency.

In general, the most important thing to remember about characteristic impedance or impedance is that all components of a system must have the same value of impedance if maximum signal (energy) transfer is to occur.

Impedance matching becomes even more critical at higher frequencies.

Unshielded twisted-pair and screened twisted-pair cables utilized for building cabling exhibit a characteristic impedance of 100 Ω ±15% at 1 MHz (megahertz) and above. The characteristic impedance for shielded twisted-pair defines STP and STP-A at 150 Ω ±10% from 3 to 300 MHz. Shielded cable normally has more capacitance (15–20 pf/ft).

Attenuation

The attenuation of a line is the difference between the input power and the output power as a ratio. This ratio is expressed in terms of decibels (dB). However, attenuation is really a special case and only applies when the load and source impedances match (equal) the characteristic impedance of the wire pair. For all other terminations, measurements of loss are properly referred to as insertion loss and will be higher than the attenuation value for a given transmission line.

Since attenuation is actually a loss of usable signal or power at the load or receiver, the lower the dB number the better. Higher attenuation means less available signal.

There are times when attenuation is added so that the signal is not too high at the receiver end. This is called padding down the circuit.

Return loss

When the termination (load) impedance does not match (equal) the value of the characteristic impedance of the transmission line, some of the signal energy is reflected back toward the source and is not delivered to the load. This signal loss contributes to the insertion loss of the transmission path.

The amount of reflected energy is affected by the degree of mismatch between the load and the line. The better the match, the less energy is reflected. The degree of reflection is termed the "reflection coefficient."

The higher loss in the reflected signal allows the maximum power or signal level to be transferred to the load.

For cabling systems, reflections also result when impedance discontinuities exist along the line, as well as at the end termination. Such discontinuities result from connectors as well as variations within and between cables. Such variations are expressed in terms of return loss.

To summarize, closely matching the source and load impedances to that of the characteristic impedance and minimizing impedance variations along the transmission line reduces the power reflected back into the line from the load and minimizes reflections as well as circuit power losses.

Insulation

The copper wires must be physically separated from each other. In the case of a single pair of wires, contact of the two wires (short circuit) will prevent the signal from traveling down the transmission line. The choice of material to cover the copper is one that involves economics as well as a trade off in the characteristics desired for the application and installation environment. An electrically efficient insulation is nearly always desired, but a trade-off may be required (e.g., to obtain insulation capable of meeting plenum-cable requirements). In like fashion, less than the most efficient insulation may have to be utilized to secure more physically robust characteristics.

> **Note.** Efficient insulation is defined as a material in which any losses of the transmitted signal due to losses associated with the insulation itself are minimal.

Insulation common in building wiring are such materials as polyethylene (PE), polyvinyl chloride (PVC), and fluorinated ethylene propylene (FEP). Combinations of these materials are also sometimes used.

The type selected can affect not only the physical size of the completed cable but also determines two of the four primary electrical cable characteristics:

- Mutual capacitance (C) of a pair
- Conductance (G)

The mutual capacitance is dependent not only on the insulating material employed, but also the diameter (thickness) to which it coats the conductor.

The conductance value of a cable is an ac phenomena and only becomes a concern at higher frequencies. However, at such frequencies the conductance may be the major contributor to the cable losses.

The insulation is directly responsible for the speed at which the signal travels over the transmission line, known as the nominal velocity of propagation (NVP): The time it takes a signal to travel from one point to another on a circuit is called propagation delay. This becomes important to the installation when used for high-bandwidth or high-bit-rate applications. This difference in propagation delay between pairs in a cable is known as delay skew.

Pair twist

In order to comply with the transmission quality level demanded of telecommunications cables installed in buildings today, the two wires of an insulated copper pair are uniformly twisted together. This twist (or lay) usually varies from approximately 13 mm (0.5 in) up to 180 mm (7 in) long. The telecommunications cabling designer selects these lengths such that no two pairs in up to a 25-pair group will be the same lay lengths. The twisting is closely controlled to ensure one wire of a pair is not longer than the other.

Both the effects of capacitance unbalance and electromagnetic induction are improved by twisting of the pairs.

By proper design and twisting of the pairs, detrimental crosstalk and noise (see later paragraphs) are minimized. It is necessary that the cabling installer recognize the importance of the pair twist—for example, for Category 5 installations, the pair twist must be maintained up to within 13 mm (0.5 in) of the termination point.

Shielding

Much of the copper cable installed today in the horizontal section does not involve a metal covering (shield) over the pairs. However, certain systems do make use of shielded pairs in this area. Backbone cables of 25 pairs and larger may also utilize an overall shield.

The shield is employed to:

- Reduce the level of radiated signal from the cable.
- Minimize the effect of external electromagnetic interference (EMI) on the cable pairs.
- Provide electrical hazard protection when properly grounded.
- Provide physical protection.

The specific shielding material used, its thickness, and relative coverage (number of holes) influences the degree of effectiveness in meeting these goals. Methods of application include copper braids, metal foils, and solid tubing.

Coaxial cables derive their name from the fact that there is an insulated conductor (centrally located) surrounded by an overall metallic covering. The geometry of such a construction inherently provides reduced external interference and radiation protection; however, the metallic covering is not a shield—it is a conductor in the circuit.

Shields are effective only when continuity is assured and when they are properly grounded.

Digital transmission speed

The basic unit of digital information is the bit (a contraction of the term binary digit). It is utilized to indicate the existence of two (binary) states or conditions (e.g., current flow or no current, on or off—a logical one and a logical zero). Several such bits grouped in a specific coded way make up a character, which provides the building block for an information system much as does our alphabet.

The correct term for the rate of flow (speed) of binary digit information over a transmission line is bits per second (bps). Unfortunately, some incorrectly use the term *baud* in describing this information flow rate. The baud technically is the signal-change rate on the line and is the rate at which the signal changes from one state to the other. While in a simple binary digital system the baud and the bit rates are equal, the terms are not interchangeable. For example, a system that uses one timing bit for every four bits of data may be set to operate at 125 megabaud while only transmitting 100 Mb/s of information.

The transmission of digital data may be accomplished on a direct digital basis or with modulators/demodulators (modems). These modems first convert the digital 1's and 0's into two separate frequencies representing a 1 and a 0 for serial transmission over telephone lines, and then reverse the process at the destination. Totally digital lines, on the other hand, do not need any conversion and transport the information bits in the native digital (1 and 0) mode.

Crosstalk

The unwanted transfer of intelligence from one or more circuits to other circuits is called crosstalk. This transfer may be between pairs in close proximity, or between adjacent cables. Crosstalk is decreased by twists, cable lay, shielding, and physical separation done during the cable manufacturing process.

The crosstalk effect increases with an increase of frequency and length of the circuit in the case of far-end crosstalk (FEXT), although the increase is not directly proportional.

There are two types of crosstalk of interest for copper cable systems. Near-end crosstalk (NEXT) is associated with the signal transfer between circuits at the same (near) end of the cable, while far-end crosstalk (FEXT) is the signal level at the near end of the disturbing pair compared to that transferred to the disturbed pair and measured at the opposite (far) end of the line.

Crosstalk is expressed in dB. For NEXT, it is the ratio of the transferred signal level measured on the disturbed (receiving) pair at the near end to the level of signal, also at the near end, of the disturbing (transmitting) pair. The ratio will be less than one and the dB value will be negative, indicating a loss. The lower the amount of signal transferred between circuits, the smaller the received-to-transmitted signal ratio and the larger the dB crosstalk loss number.

As an example of crosstalk limits, the TIA/EIA TSB-67 lists NEXT losses for a worst-case link for unshielded twisted-pair (UTP), four-pair cables at frequencies in the range of 1 MHz to 100 MHz.

The figure below illustrates both NEXT and FEXT crosstalk paths for two pairs.

When a cable contains more than four pairs, the calculation for crosstalk performance must take into account that these additional pairs are also carrying electrical signals and could provide multiple disturbing paths.

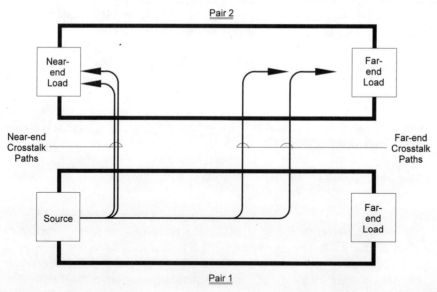

Figure 1.55 Crosstalk paths.

Hardware

Although much of the previous information has been concerned with the characteristics of copper cable transmission medium, attention should be paid to the connecting and terminating hardware used to establish the total circuit. Crosstalk, signal loss, and other detriments to the performance of the network can be attributed to these devices.

Optical Fiber Cable Media
Overview

Today, optical fiber cables are found in intrabuilding backbone and horizontal applications. They are economical and offer characteristics that can make them the medium of choice; a requirement for high data rates or if an all-dielectric construction is desirable are two such instances.

Transmission of information through optical fiber cables is free from some of the detrimental effects inherent with copper cables, such as crosstalk, noise, lightning, and EMI problems. However, attenuation (loss) of the transmitted signal along the cable route and environmental considerations are of concern for both copper and fiber systems.

The primary difference between copper and fiber as a transmission medium is that, with fiber, pulses of light containing photons are injected into the fiber instead of electrical energy in a copper-based cable. On and off light pulses are generated by either lasers or a light-emitting diode (LED) in accord with the digital information to be transmitted.

The inner portion of the glass fiber utilized to carry the light pulse is called the core, and the surrounding glass layer is called the cladding. The purpose of the cladding is to confine the light within the core by creating a reflective zone.

Fibers are classified as either singlemode or multimode. Singlemode fibers have a relatively small diameter featuring a core of 8–9 µm (micrometers or microns) and a cladding diameter equal to 125 µm. Lasers have a narrow light beam and can focus 100 percent of the light beam down the core of the fiber. Multimode has a larger core diameter, for example, 62.5 microns with the cladding at 125 microns—usually written 62.5/125 µm. The light is restricted to a single path, or mode, in singlemode fibers, while the larger diameter multimode permits many paths, or modes, to exist in the fiber. Multimode construction is addressed here, since it is the type normally selected for the short distances applicable within buildings (< 2 km).

There are two characteristics of particular importance in the transport of information over fiber media—attenuation and bandwidth. Bandwidth, attenuation, and a few other aspects that relate to transmission of light pulses over fiber cables are reviewed in the following paragraphs.

Bandwidth

The bandwidth of an optical fiber provides a measure of the amount of information a fiber is capable of transporting within a given error-free bit rate. It is described in terms of megahertz·kilometer (MHz·km). Thus, increasing either the length of the cable or the modulation frequency of the light source decreases the bandwidth and, therefore, the information-carrying capacity of the transmission path.

Dispersion is the widening or spreading out of the modes in a light pulse as it progresses along the fiber. Should the pulse widen too much, it can overlap at the receiver and make it impossible to distinguish one pulse from another. As errors occur in reading the pulses (bits), there is an increase in the bit error rate. Bandwidth is, therefore, limited by total dispersion, which is the sum of modal dispersion and chromatic dispersion.

Modal dispersion in multimode fibers is the major factor in the total dispersion. It is the result of different lengths of the light paths taken by the many modes (multimode) as they travel down the fiber from source to receiver. Since portions of the signal arrive ahead of or are delayed from other portions of the signal, an individual light pulse may be spread out, making the identification of the light pulse at the far end questionable.

Chromatic dispersion begins at the light source. The sources utilized to create the light pulses are either a laser or an LED and do not furnish a perfectly monochromatic (single wavelength) light; thus, the light injected into the fiber medium contains a number of slightly differing wavelengths. Inasmuch as the index of refraction of the glass fiber is not the same for different wavelengths, each travels through the fiber at a slightly different speed. Modal dispersion tends to further broaden these pulses, and together, these dispersion factors increase the bit-error rate and lower the effect of bandwidth.

Attenuation

Light pulses in optical fiber cables are subjected to a loss in power as they travel along the fiber. This attenuation occurs:

- As a result of absorption of power by impurities within the glass itself, and
- Scattering of the light.

Losses due to scattering are caused by such factors as:

- The glass material.
- Core size variations.
- Variations in the interface between core and cladding.
- Macrobends and microbends in the strands.

Fiber attenuation is referenced in decibel per kilometer (dB/km). It is proportional to length and is affected by the wavelength of the light contained in the pulse.

Grounding and Bonding

Overview

The electrical protection of telecommunications systems is an essential part of every installation.

Ground is defined as a conducting connection, whether intentional or accidental, between electrical circuits or equipment and the earth, or to some conducting body that serves in place of the earth.

Bonding is the permanent joining of the conducting parts of equipment and conductor enclosures to assure an electrically conductive path between them. Effective equipment bonding helps equalize potentials caused by lightning and electrical system faults that would otherwise be damaging to equipment and harmful to personnel.

Grounding system components

The two areas of grounding that apply to the telecommunications equipment are the:

- Grounding electrode system (also known as the earthing system).
- Equipment grounding system (also known as the safety ground).

The grounding electrode system consists of a:

- Grounding field (earth).
- Grounding electrode conductor.
- Grounding electrode.

The equipment grounding system consists of:

- An equipment grounding conductor.
- A main bonding jumper.

The telecommunications bonding and grounding system consists of:

- A telecommunications main grounding busbar (TMGB).
- A telecommunications grounding busbar (TGB).
- A telecommunications bonding backbone (TBB).
- Lightning protector grounding system connections.
- Grounding electrodes.
- A grounding electrode conductor.
- A telecommunications bonding backbone interconnecting bonding conductors (TBBIBC).

There are several types of electrical protection systems within every building. Although the telecommunications installer is not usually responsible for the other systems, it is not safe to work in a building without recognizing and understanding the purpose for each system. A clear understanding of these systems also provides a basis for further training and working with other trades when necessary.

This chapter will look at the various electrical protection systems found in most of today's commercial buildings, such as:

- Lightning protection systems.
- Grounding electrode systems.
- Electrical bonding and grounding systems.
- Electrical power protection systems.
- Surge protection devices.
- Telecommunications bonding and grounding systems.
- Telecommunications circuit protectors.

Safety

A primary responsibility of the cabling installer is safeguarding personnel, property, and equipment from "foreign" electrical voltages and currents. *Foreign* refers to electrical voltages or currents that are not normally carried by, or expected in, the telecommunications distribution systems.

The results of such disturbances could be:

- Death or injury.
- Destruction of electronic equipment.
- Downtime.
- Work or process degradation.

Avoiding electric shock

The most common electric shock occurs from inadvertent, accidental contact with energized devices or circuits. Other common conditions that create potential shock hazards are:

- Touching a faulty or improperly grounded electrical component.
- Standing on a damp floor.
- Poor clearance or lighting.
- Using or being near conducting material during a lightning storm.

Preventing electric shock

Until verified, always assume equipment is energized. Cabling installers must be especially observant for irregular or abnormal conditions during the construction phase of a project. Use prudent electrical safety measures, such as insulated rubber gloves for personal protection, and use appropriate test equipment to verify the presence or absence of dangerous voltages on all exposed:

- Cables.
- Strands.
- Wires.
- Metal.

Properly installed telecommunications cabling is almost never dangerous. During installation, verify that exposed conductors, cable shields, and metal equipment are grounded or free of fault potentials (and otherwise generally safe).

Overall protection must consider:

- Direct lightning strikes.
- Ground potential rise.
- Contact with power circuits.
- Induction.

Examples of some conditions causing these disturbances are shown in the following four illustrations.

References

National Electrical Code® (NEC)	A useful reference is the expanded version of the code—*National Electrical Code-1996 Handbook* (it is the same code but has additional explanatory comments).
NEC Chapter 8	"Communication Systems" covers general requirements for grounding, bonding, and protecting low-voltage communications equipment.
NEC Article 250	"Grounding" covers electrical power circuits and low-voltage control and signaling systems. Even though NEC Chapter 8 is a standalone chapter, it refers the reader to Article 250 for some specific grounding concerns. Much of this chapter

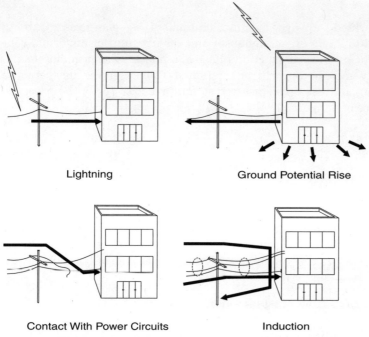

Figure 1.56 Electrical hazards.

	is based on the 1996 edition of the NEC, although many local jurisdictions still require adherence to earlier editions.
Canadian Electrical Code (CEC)	The Canadian Standards Association (CSA) publishes the *Canadian Electrical Code* (CEC), a *CE Code Handbook,* and a number of coordinated product test standards. These are comparable to the NEC (based on the same system grounding and protection methods) and have many similarities, but the two codes are not identical or interchangeable.
ANSI/TIA/EIA-607	This document covers grounding and standard bonding requirements for telecommunications applications within commercial buildings. It is available along with the ANSI/TIA/EIA-568-A and 569-A documents as a set of *Telecommunications Building Wiring Standards.* The ANSI/TIA/EIA-607 standard does not replace NEC requirements but pro-

vides additional standards for grounding and bonding. Note that it covers only grounding and bonding but relies on other standards and codes for many of the other important protective measures.

Electrical exposure (telecommunications)

NEC Section 800-30(a) covers safety code requirements for protectors. Additionally, it defines "exposed" as when a "circuit is in such a position that, in case of failure of supports or insulation, contact with another conductor may result."

The NEC requires a listed primary protector (at both ends) whenever outside plant cable may be exposed to lightning or accidental contact with power conductors operating at over 300 volts (see Figure 1.59).

Exposure refers to an outdoor telecommunications cable's susceptibility to electrical power system faults, lightning, or other transient voltages. A cable is also considered exposed if any of its branches or individual circuits are exposed.

Lightning exposure

Lightning strikes can cause severe damage to telecommunications systems that have not been properly installed. Even with a properly installed grounding infrastructure, there are no guarantees that a direct lightning strike will not damage a system. However, it will have a better chance of not sustaining interruptions or damages than a poorly installed system.

A Lightning Exposure Guideline is included as a Fine Print Note (FPN) in the NEC Section 800-30(a). It states, "Interbuilding circuits are considered to have a lightning exposure unless one or more of the following conditions exist:

1. Circuits in large metropolitan areas where buildings are close together and sufficiently high to intercept lightning.

2. Interbuilding cable runs of 140 ft (42.7 m) or less, directly buried or in underground conduit, where a continuous metallic cable shield or a continuous metallic conduit containing the cable is bonded to each building grounding electrode system.

3. Areas having an average of five or fewer thunderstorm days per year and earth resistivity of less than 100 ohm-meters. Such areas are found along the Pacific coast."

If cable exposure is in question, consider it to be exposed and protect it accordingly.

Two additional exposure factors are:

- Aerial cable usually has power cables routed above it that will intercept and divert direct lightning strikes. This can help but does not negate the need for protectors.
- Buried cable collects ground strikes within a distance determined by soil resistance (typically 2–6 m [6–20 ft]). High soil resistance intensifies this problem. Without the proper protection, a system could be receiving repeated ground strike surges without the evidence of damaged cable associated with an aerial strike.

Lightning is so powerful and unpredictable that the best insurance against damage is a properly grounded and bonded system. Lightning may strike at any time with the potential of:

- A direct current charge, pulsating between 100 kHz and 2 MHz.
- Greater than 10 million volts.
- An average 40,000 amps and can peak as great as 270,000 amps.
- Temperatures in excess of 27,760 °C (50,000 °F).

Earthing systems must be designed and installed that can safely carry these unwanted voltages to the earth.

Electrical exposure (building structure)

American National Standards Institute/National Fire Protection Association (ANSI/NFPA) 780 covers lightning protection systems for buildings and defines "exposed" as anything above ground and outside a "zone of protection" (an area under or nearly under a lightning protection system). This does not include communications cable.

These lightning protection systems intercept and ground lightning strikes to the building. They are generally recognized by the metal spikes on top of the building.

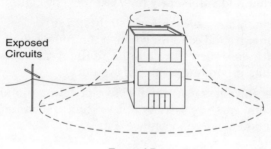

Figure 1.57 Zone of protection.

ANSI/NFPA 780 also describes a simplified zone of protection that can be considered for small buildings (commonly called cone of protection).

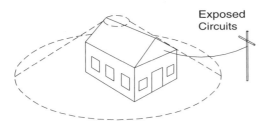

Figure 1.58 Cone of protection.

Ground

NEC Article 100 defines "ground" as "a conducting connection, whether intentional or accidental, between an electrical circuit or equipment and the earth, or to some conducting body that serves in place of the earth."

Bonding

NEC Article 100 and Section 250-70 define bonding as "the permanent joining of metallic parts to form an electrically conductive path that will ensure electrical continuity and the capacity to conduct safely any current likely to be imposed."

Bonding conductors are not intended to carry electrical load currents under normal conditions, but must carry fault currents so that electrical protection (circuit breakers) will properly operate.

Another important safety application for bonding is to limit hazardous potential differences between different systems during lightning or power faults as described earlier. This protects against arcing between different system (or equipment) grounds and protects personnel who may come in contact with both systems at the same time.

American Wire Gauge

The American Wire Gauge (AWG) has become a standard for specifying and measuring conductor diameter. Smaller gauge numbers represent larger wire diameters. A gauge change of three, doubles or halves the conductor's resistance. The gauge change of six, doubles or halves the conductor's diameter.

TABLE 1.12 American Wire Gauge (AWG).

AWG	Diameter		Weight		D.C. Resistance	
	mm	in	kg/km	lbs/1000 ft	ohms/km	ohms/1000 ft
27	0.361	0.0142	0.91	0.610	169	51.4
26	0.404	0.0159	1.14	0.765	135	41.0
25	0.455	0.0179	1.44	0.97	106	32.4
24	0.511	0.0201	1.82	1.22	84.2	25.7
23	0.574	0.0226	2.31	1.55	66.6	20.3
22	0.643	0.0253	2.89	1.94	53.2	16.2
21	0.724	0.0285	3.66	2.46	41.9	12.8
20	0.813	0.032	4.61	3.10	33.2	10.1
19	0.912	0.0359	5.80	3.90	26.4	8.1
18	1.02	0.0403	7.32	4.92	21.0	6.4
17	1.144	0.045	9.24	6.20	16.3	5.0
16	1.296	0.051	11.65	7.82	13.0	4.0
15	1.449	0.057	14.69	9.86	10.4	3.2
14	1.627	0.064	18.09	12.4	8.1	2.5
13	1.83	0.072	23.39	15.7	6.5	2.0
12	2.059	0.081	29.50	19.8	5.2	1.6
11	2.313	0.091	37.10	24.9	4.2	1.3
10	2.593	0.102	46.79	31.4	3.3	1.0
9	2.898	0.114	59.00	39.6	2.6	0.8
8	3.254	0.128	74.50	50.0	2.0	0.6
7	3.66	0.144	93.87	63.0	1.6	0.5
6	4.118	0.162	118.46	79.5	1.3	0.4
5	4.626	0.182	149.00	100.0	1.0	0.30
4	5.186	0.2045	187.74	126.0	0.8	0.25
3	5.821	0.229	236.91	159.0	0.7	0.20
2	6.558	0.258	299.29	201.0	0.5	0.16
1	7.346	0.289	376.97	253.0	0.4	0.13
0	8.261	0.325	475.31	319.0	0.3	0.10
0	9.278	0.365	600.47	403.0	0.26	0.08
000	10.422	0.41	756.92	508.0	0.20	0.06
0000	11.693	0.46	955.09	641.0	0.16	0.05

Protection Systems

Lightning protection system

Described in ANSI/NFPA 780, *Standard for Installation of Lightning Protection Systems,* a lightning protection system consists of:

- Multiple rooftop air (lightning) terminals.
- Down conductors.
- Equalizing conductors.
- Ground terminals that surround a building for the exclusive purpose of intercepting, diverting, and dissipating direct lightning strikes.

Some systems are designed as part of the building structure so that the structural steel performs the down-conductor and equalizing functions.

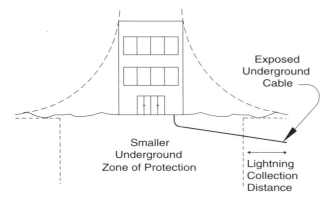

Figure 1.59 Lightning protection system.

Lightning is substantially prevented from directly striking within and under the tips of the lightning terminals, thus forming a "zone of protection." These systems are usually out of the scope of a telecommunications cabling installer's responsibility, but the cabling installer needs to observe the following:

- ANSI/NFPA 780 specifies the requirements for spacing and interconnection of the lightning system and the communications system. Communications ground must be bonded to the lightning protection system grounding within 3.7 m (12 ft) of the base of the building.
- If the communications ground relies on the electrical service grounding electrode system, then the more common grounding practices will apply. Communication grounding practices will be discussed later in the chapter.
- NEC Section 800-13 requires that "where practicable, a separation of at least 6 ft (1.83 m) shall be maintained between open conductors of telecommunications systems on buildings and lightning conductors." If structural steel is used as the lightning down conductors, then such separation is impractical and usually not necessary.

Electrical power systems

An electrical power system provides the electrical infrastructure necessary to distribute electricity throughout the building. All the appliances, lighting circuits, and equipment are fed through the network of branch circuits. The heart of the electrical power system is its being referenced or grounded to earth. In most telecommunications grounding systems, the ground reference is established by bonding the telecommunications system grounding conductor to the electrical service ground at the electrical service equipment.

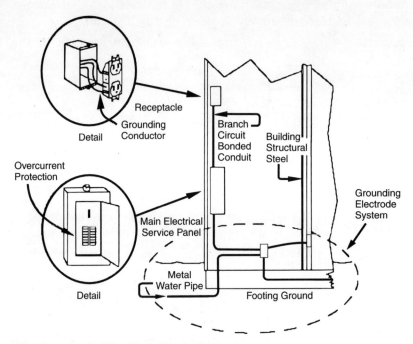

Figure 1.60 Typical (small) electrical power system.

Grounding electrode system

Described in NEC Article 250 Part H, a grounding electrode is a metallic conductor (e.g., rod, pipe, plate, ring, or other metallic object) in contact with the earth, used to establish a low-resistance-current path to earth.

A grounding electrode system is a network of electrically connected ground electrodes used to achieve an improved low resistance to earth, and in many cases, to aid equalization of potentials around a building.

Basically, all ground electrodes can be divided into two groups:

- Underground piping systems, metal building framework, well casings, steel pilings, and other underground metal structures installed for purposes other than grounding.

- Electrodes specifically designed for grounding purposes (i.e., driven and buried ground rods, buried ground rings, metal plates).

A properly functioning ground is essential to electrical protection because it:

- Conducts any excess electrical energy to the earth without causing hazardous arcing, heating, or explosion during lightning.
- Establishes the voltage reference for electrical systems in the building.

NEC Section 250-54 requires a common grounding electrode system for different electrical systems within a building. Where two or more electrodes are effectively bonded together, they shall be considered as a single grounding electrode system.

NEC Section 250-71(b) requires an intersystem bonding connection accessible at the electrical service equipment. This connection is a prime choice for establishing a telecommunications ground. (See "Using the Electrical Service Ground" in this chapter.)

Electrical bonding and grounding

Throughout NEC Article 250, electrical bonding and grounding is described as metallic panels and raceways that are bonded to an equipment grounding conductor, which are then all bonded to the grounded electrical service neutral at the service equipment. An equipment grounding conductor shall be routed with the power and neutral conductors.

Electrical systems and metallic apparatus are bonded and grounded to limit hazardous voltages due to:

- Electrical power faults.
- Lightning.
- Other electrical transients.

This facilitates overcurrent protection operation in the case of electrical power faults that would otherwise place hazardous voltages at dangerous points. Inadvertent shorting of the power conductor to the equipment ground, or to other bonded metal or conductors, will cause a circuit breaker (overcurrent protection) to operate and disconnect power.

These systems are not usually the responsibility of the telecommunications installer, but they should be recognized and understood, since most sites have power protection designed specifically for telecommunications equipment.

Electrical power protection

Electrical power protection is covered throughout NEC Chapter 2, which includes electrical bonding and grounding requirements.

The following are required for complete electrical protection:

- Surge arresters (divert surge current coming in on utility power conductors)
- Service disconnecting means (main service breaker: provides a method for overall power shutdown based on emergency or maintenance)
- Overcurrent protection (circuit breakers: open individual power circuits if the current reaches a predetermined, unsafe level)

> **Note.** Other components that improve the overall quality of electrical power include power line conditioners, uninterruptible power supplies (UPS), and power backup systems.

These systems are not usually the responsibility of the telecommunications cabling installer, but they should be recognized and understood, since most sites have power protection designed specifically for telecommunications equipment.

Telecommunications bonding and grounding

Telecommunications bonding and grounding is additional bonding and grounding installed specifically for telecommunications systems. From a safety code standpoint, the NEC and ANSI/NFPA 780 already cover such bonding and grounding.

There are many situations in which these minimal safety codes can be interpreted or implemented in different ways. Some methods may not be as suitable as others for sensitive equipment reliability and performance. Manufacturers and designers use a wide range of differing solutions to adapt to:

- Site variations.
- Prewired buildings.
- Manufacturer interpretation.
- Multivendor applications.
- Equipment design.
- Sensitive high-speed electronics.

Most situations cannot rely solely on safety-grounding methods. Instead, telecommunications systems require direct and dedicated grounding and bonding.

Telecommunications bonding and grounding is used to:

- Minimize electrical surge effects and hazards.
- Augment electrical bonding.
- Lower the system ground reference impedance.

This does not replace the requirements for electrical power grounding but supplements them with the additional bonding that generally follows communication pathways between telecommunications entrance facilities, equipment rooms, and telecommunications closets.

Telecommunications bonding principles

Most buildings have low overall impedance (many are designed as part of a lightning protection system to safely conduct lightning strikes to earth). However, significant differences in ground potential can exist throughout a building during electrical transients. Additional bonding conductors are an effective way to improve marginal situations, especially in buildings that lack an overall bonded structure.

If continuous structural steel exists along the same path, there may be little actual improvement. But even here, a certain assurance is gained by having explicit bonding conductors that can be verified and inspected.

Three specific principles behind telecommunications bonding conductors are:

- *Equalization.* Potentials between different ground points are very dependent on the impedance between them. Ground equalization is improved because the additional bonding lowers the impedance between different ground points. The shortest and most direct path using large conductors provides a low impedance (both resistive and inductive). Multiple conductors or wide straps will provide a lower impedance.

- *Diversion.* Because the bonding conductor follows the telecommunications cable and is directly connected to system grounds at each end, electrical transients that are forced down the cable path may be diverted (carried) by the bonding conductor and are less likely to influence the telecommunications conductors.

- *Coupling.* The closer the bonding conductor is to the telecommunications cable, the greater the mutual electromagnetic coupling. During electrical transients, this coupling tends to partially cancel the transient when it reaches the telecommunications equipment at the end. A tightly coupled bonding conductor or a backbone cable shield is often called a coupled bonding conductor (CBC).

Each of these three effects is achieved in varying degrees, depending on a wide range of factors. It is often difficult to predict or measure specific results, but any combination of the three is usually beneficial to telecommunications equipment.

Telecommunications Grounding Practices

Overview

Establishing a suitable telecommunications ground is critical in properly grounding telecommunications equipment. A telecommunications ground is always required, and is typically found in the following:

- Telecommunications entrance facility, for sites with exposed cable
- Equipment rooms
- Telecommunications closets

Grounding choices

Direct attachment to the closest point in the building's electrical service grounding electrode system is preferred because telecommunications cabling and power cabling must be effectively equalized.

Select the nearest accessible location on one of the following:

- The building ground electrode system
- An accessible electrical service ground (more information in this section)

If no electrical service exists, use one of the following:

- Driven ground rod (described in "Installing a Grounding Electrode" in this section)
- Another grounding electrode system installed for the purpose

Using the electrical service ground

A direct electrical service ground is one of the best points for grounding telecommunications systems. In new construction, an electrical contractor must provide accessible means. NEC Section 250-71(b) requires an intersystem bonding connection accessible at the electrical service equipment, such as:

- Exposed metallic service raceway (using an approved bonding connector).
- Grounding electrode conductor.

    Warning. This conductor is critical to the safety of the electrical power system. Do not move, modify, or disconnect without the direct participation of personnel responsible for that system.

- Approved external connection on the power service panel. The NEC allows direct connection to a minimum size of 6 AWG copper conductor provided by a licensed electrical contractor.

Note. Ensure that the grounding electrode system is properly installed. Verify with the electrical contractor or have the system tested by a licensed electrician.

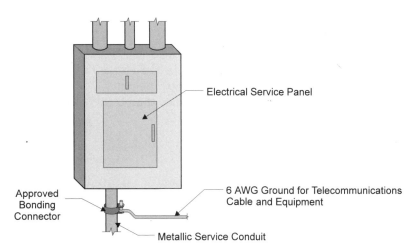

Figure 1.61 Typical telecommunications ground.

Installing a grounding electrode

The resistance of the grounding electrode system should be as low as possible (25 Ω or less) and measured annually with an earth megger. Many equipment manufacturers require less than a 10-Ω resistance between the grounding electrode system and the earth. The resistance of the grounding electrode system is not as important as the bonding between the different grounding electrode systems that may exist on site.

The NEC Handbook, Section 250-83, provides a few examples for the installation of a grounding electrode or electrode system. The installation of a telecommunications grounding electrode is allowed if:

- There is no electrical service ground.

  or

- Additional grounding is needed (NEC Section 250-91[c]). If so, the installed electrode must be bonded to the existing ground electrode system.

As a last resort, NEC Section 800-40(b) (3) specifies a minimum 13 mm (0.5 in) diameter, 1.5 m (5 ft) ground rod driven completely into the ground.

BICSI recommends:

- A minimum 2.5 m (8 ft) copper-clad ground rod.
- A 6 AWG solid grounding conductor.
- If the connection to the grounding conductor is made below grade, use an exothermic weld.

Another alternative is a ground ring electrode system. These are noninsulated conductors that are buried in the shape of a ring. The conductors are buried at a minimum depth of 760 mm (30 in) for a minimum of 6 m (20 ft) and should not be smaller than 2 AWG. The following conditions must be strictly observed:

- Any installed grounding electrode must be at least 1.83 m (6 ft) away from other existing electrodes.
- Electrodes or down conductors that are part of a lightning protection system are not allowed for use as an electrode for this purpose.
- Regardless of what alternative is selected for installing a ground rod, all other electrical system grounds, structural building steel, and metallic piping systems must be bonded together. This is required of the electrical service and is usually already accomplished, but it should be verified.
- Gas pipes, steam pipes, or hot water pipes are not allowed as a grounding electrode.

Three components determine the resistance of an electrode to earth:

- Resistance of the earth itself
- Contact resistance of the electrode to the surrounding earth
- Resistance of the electrode itself and its connections

Physical protection

If a chance of damage to the grounding conductor exists, then some form of physical protection is required. If this is metallic conduit or raceway, it must be bonded to the grounding conductor at both ends [NEC Section 800-40(a)(5)].

CATV

Bond the established ground, the intended ground termination, or the outer conductive shield of a CATV coaxial cable in the same manner as other telecommunications cables to help limit potential differences between these systems and other metallic systems (NEC Section 820-33).

Water pipes

Historically, the first choice for a grounding electrode was a metallic water pipe connected to a utility water distribution system. This is no longer true. There is an increased use of nonmetallic pipe, and electrical systems should not rely on plumbing systems. Water pipes must be bonded to another electrode type.

For a similar reason, caution must be exercised when water pipe is used as an intersystem bonding conductor. NEC Sections 250-80 and 250-81 cover such usage; however, avoid this practice and use a minimum 6 AWG copper bonding conductor.

Telecommunications Bonding Practices

Bonding connections

Bonding connections should be made directly to the points being bonded. Avoid unnecessary connections or splices in bonding conductors, but when necessary, use an approved connection and position it in an accessible location.

Many aluminum and stainless-steel connectors are approved (NEC Section 800-4-[a][2]) and available for use, but it is recommended that connectors and splices should be one of the following:

- Tin-plated copper
- Copper
- Copper alloy

Copper and copper-alloy connections should be cleaned and coated with an antioxidant prior to establishing the connection.

Typical connections are made by using:

- Bolt or crimp/compression connectors, splices, clamps, or lugs. Bolt-type connectors have a tendency to loosen over time from vibration and repeated temperature fluctuations. Listed mechanical crimp/compression connectors are virtually maintenance free and are not susceptible to vibration and temperature.

  Note. Use listed hardware that has been laboratory tested to eliminate most field problems.

- Exothermic welding (see NEC Section 250-81).

The exothermic weld process uses a special heat-resistant mold to bond metallic conductors together.

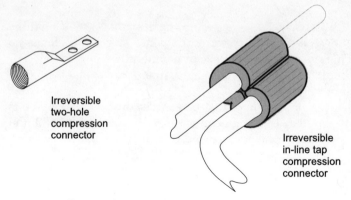

Figure 1.62 Compression connectors.

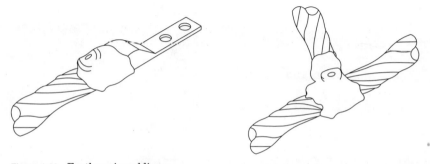

Figure 1.63 Exothermic welding.

There are a variety of molds available to meet the requirements of:

- Conductor size and shape.
- Ground-rod material: diameter, treated, or plain.
- Application: pipe, rebar, building steel.
- Configuration: tee, tap, inline splice.

The exothermic weld process is commonly applied:

- Within the ground electrode system.
- To parts of a grounding system that are subject to corrosion or that must reliably carry high currents.
- To locations requiring minimal maintenance.

To complete an exothermic weld:

- Determine what type of connection is needed.
- Obtain the proper mold for conductors to be bonded.

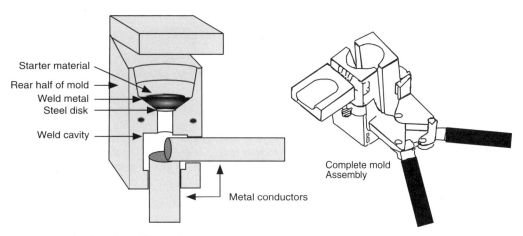

Figure 1.64 Exothermic welding mold.

- Follow safety procedures and wear leather gloves and safety glasses.
- Open the mold and place the metal objects into the mold, following the manufacturer's instructions.
- Close the mold around the conductors to be bonded.
- Place the steel disk in the base of the crucible.

  Note. The disk keeps the weld metal in the crucible until the disk melts and the weld metal flows down into the weld cavity.

- Fill the crucible with the weld metal.

  Note. Each canister of weld metal is sized for a specific mold.

- Sprinkle the starter material over the weld metal.

  Note. The starter material has a very low flash point to start the exothermic process.

- Close the cover and spark the starter material to ignite the weld.
- Allow the mold to cool, then open and clean it for the next weld.

  Note. Always adhere to the manufacturer's instructions when using exothermic welds. The most common exothermic welds are for outside use only. The process described above is for outside or well-ventilated areas. Inside welds are similar but use a nontoxic welding material, a battery, and small spark igniter with a special filtering material to cut down on smoke.

All bonding connections should be made using the proper tools and following manufacturer's guidelines.

Bonding conductors

Bonding conductors must be copper. To avoid unintentional ground connections, bonding conductors within buildings should be insulated. Bonding conductors must be routed with minimum bends or changes in direction.

NEC Section 800-40(a) requires at least a 14 AWG (stranded or solid) insulated conductor for connecting the telecommunications protectors and associated metallic cable sheaths to the selected ground. Other NEC ground requirements generally indicate a 6 AWG minimum (for ground-rod and for intersystem bonding), but a 6 AWG stranded conductor should be used, since this accommodates different code requirements and allows for future changes.

In most applications, the bond type selected depends on the application and the fault current-carrying capacity needed, but a minimum 6 AWG copper conductor is generally used throughout typical commercial buildings. According to ANSI/TIA/EIA-607, consideration should be given to sizing conductors as large as 3/0 AWG.

An exception to the minimum 6 AWG bonding conductor is in smaller entrance facilities. UL 497 requires the protection unit to determine the minimum-sized conductor based on anticipated current flow from the number of conductor pairs. Units that are designed for:

- One and two pairs of protection require a minimum 12 AWG.
- Three to six pairs of protection require a minimum 10 AWG.
- Greater than six pairs of protection require a minimum 6 AWG.

Some bonding conductors must be guarded against physical damage. The preferred physical protection is nonmetallic. If metallic conduit is used, the bonding conductor must be bonded to the conduit at both ends.

Inspection

Ground systems require scheduled maintenance. They should be checked for tight connections annually. Critical telecommunications systems should be tested annually to ensure low resistant connections throughout the system and to the earth.

Visual inspection can usually reveal problems, such as:

- Loose connections.
- Corrosion.
- Physical damage.
- System modifications.

During any service work, the installer should visually inspect bonding connections. Once a ground system has been installed properly, 90 percent of all grounding troubles are the result of loose connections.

Exposed cables

Bond and ground exposed cable shields and metallic sheath members according to manufacturer's installation instructions. They should be grounded as close as practicable to the point of entrance, and grounding conductors should be routed directly to the closest approved ground. This may also apply to optical fiber cables with metallic members.

System Practices

There is a wide range of differing (but current) practices concerning system grounding. Each manufacturer, service provider, inspector, or customer may have unique requirements. The system practices outlined here are commonly applied to commercial buildings and were derived from PBX installation practices.

Small systems

In small equipment rooms and entrance facilities, the ground connection terminals on the installed equipment (protector panels, PBX cabinets, etc.) are usually directly connected to the closest approved grounds (typically made available by electrical service personnel).

All exposed cable shall be directly terminated at the associated protectors. Any exposed cable shields are bonded directly to the closest:

- Protector.
- Protector's ground terminal.
- Approved protector ground.

Beyond this, equipment manufacturers' guidelines must be followed. Typically, a bonding conductor is installed directly between the protector's ground terminal and the PBX (or other equipment) ground terminal.

If the protector and PBX are colocated and rely on the same approved ground, then the bonding conductor may not be needed.

Large systems

A grounding busbar or multiple busbars are used for sites with extensive equipment rooms and separate entrance facilities in which there are usually many connections. Requirements and guidelines for this method are detailed in the ANSI/TIA/EIA-607 standard.

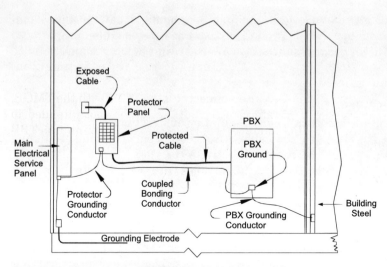

Figure 1.65 Typical small-system arrangement.

Busbars should be:

- Positioned so that telecommunications cabling will generally follow the related bonding conductors.
- Positioned near associated equipment.
- Insulated from their support.

The busbar designated for protectors, the Telecommunications Main Grounding Busbar (TMGB), must safely carry lightning and power-fault currents. The TMGB is directly bonded to the electrical service ground. It should be positioned adjacent to the protectors and directly between the protectors and the approved building ground for protector operation. The minimum dimensions of the TMGB are 6 mm (0.25 in) thick, 100 mm (4 in) wide, and variable in length.

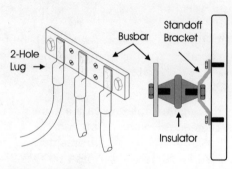

Figure 1.66 Busbar.

A Telecommunications Bonding Backbone (TBB) conductor is connected from the TMGB to the Telecommunications Grounding Busbar (TGB) in telecommunications closets within the building. The minimum dimensions of the TGB are 6 mm (0.25 in) thick, 50 mm (2 in) wide, and variable in length.

A TBB is a conductor that interconnects all the TGBs with the TMGB. The TBB is designed to interconnect busbars and is not intended to have equipment bonding conductors spliced on to it. The minimum TBB size shall be a 6 AWG and could be as large as a 3/0 AWG.

Each TBB should be a continuous conductor from the TMGB out to the farthest TGB. Intermediate TGBs should be spliced on to the TBB with a short bonding conductor.

The TBB shall be connected to the busbars with a two-bolt attachment. This provides a secure connection as per ANSI/TIA/EIA-607.

Larger closets may require additional TGBs. Multiple TGBs within a closet are allowed, provided they are all bonded together.

In larger, multistory buildings that have multiple telecommunication closets on each floor, the potentials between closets need to be equalized. When applicable, a Telecommunications Bonding Backbone Interconnecting Bonding Conductor (TBBIBC) is installed between TGBs to interconnect all the closets on the same floor. The TBBIBC is not required for every floor, but is installed on every third floor and the top floor.

To facilitate identification, busbar connections should be grouped with protector, busbar bonding, and approved building grounding conductors toward one end and equipment grounding conductors should be grouped toward the other end.

Telecommunications closets

In a telecommunications closet, suitable ground options include:

- Building structural steel.
- An electrical receptacle box or approved conduit connection.
- A combination of the above that are accessible.
- An already established telecommunications ground.

For telecommunications closets requiring a large number of connections or requiring flexibility for equipment changes, a grounding busbar should be installed.

Backbone cable protection

Cables inside a building are generally considered nonexposed, but there are situations that call for protective measures.

120   Chapter One

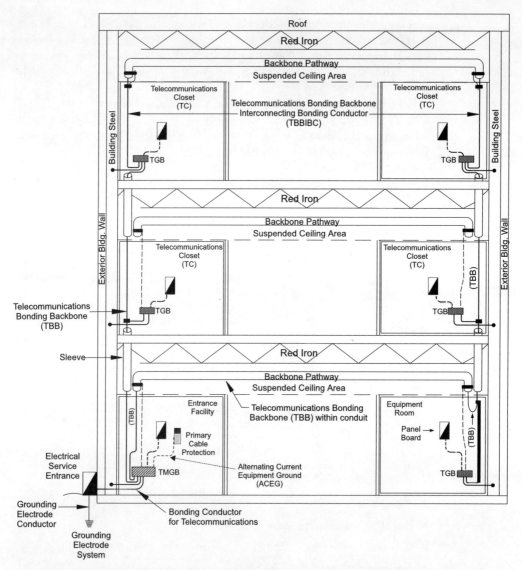

Figure 1.67 Typical building-grounding infrastructure.

The following guidelines should be applied:

- Electrical power cabling should not be routed directly alongside telecommunications cable. Electrical cabling is usually in conduit, providing additional shielding (see NEC Section 800-52 for separation and exceptions).

- The telecommunications cable should be routed near the middle of the building. Surrounding the cable with structural building steel or a lightning protection system provides shielding and usually diverts transient currents.

- A lightning protection system should be installed.
- The other exposed cables that enter the building must be protected and grounded.
- Some form of bonding conductor should be installed along each backbone cable pathway (see "Large Systems" in this section and ANSI/TIA/EIA-607).

In high-rise buildings and low-wide buildings (particularly if located in an area with high lightning activity or high soil resistivity), protective measures are vital. The same may be true for buildings that are close to an electrical substation or heavy industrial facilities.

Telecommunications bonding backbone (TBB)

A TBB is a 6 AWG or larger bonding conductor that provides direct bonding between different locations in a building, typically between equipment rooms and telecommunications closets. A TBB is usually considered part of a grounding and bonding infrastructure (part of the telecommunications pathways and spaces in the building structure), but is independent of equipment or cable.

ANSI/TIA/EIA-607 provides standard requirements for TBBs and associated hardware.

Coupled bonding conductor (CBC)

A coupled bonding conductor (CBC) is a bonding conductor that provides equalization like a TBB and also provides a different form of protection through electromagnetic coupling (close proximity) with the telecommunications cable (see figure 1.65). The CBC is generally considered part of an installed telecommunications cable and not part of a grounding and bonding infrastructure. Some PBX equipment manufacturers specify a CBC between their equipment and exposed circuit protectors.

There are two basic forms of CBC:

- Cable shield
- Separate copper conductor tie wrapped at regular intervals to an unshielded cable

Typically, the CBC is specified as 10 AWG. To be consistent with ANSI/TIA/EIA-607, it is recommended that a minimum 6 AWG conductor be used. This also provides the potential for the TBB to be used as a CBC. To work properly, the CBC must be connected directly to both the protector ground and to the PBX ground.

Unshielded backbone cable

For unshielded backbone cable, a coupled bonding conductor (CBC) should be installed as follows:

Step	Installing coupled bonding conductor (CBC)
1	Route a 6 AWG copper conductor along each backbone cable route. Ensuring a minimal separation between the conductor and the cables along the entire distance may satisfy equipment manufacturer's requirement for a CBC.
2	Bond each end at the nearest approved ground in the area where the associated cables terminate or are spliced/cross-connected onto other cables. Such bonding should be done with a grounding busbar.

Note. A TBB may also be required for installations of shielded backbone cable. This depends on customer requirements (refer to ANSI/TIA/EIA-607).

Backbone cable shields

Many indoor and outdoor backbone cables have an overall cable shield. These shields serve as:

- Physical protection.
- A coupled bonding conductor.

They should be directly bonded following the manufacturer's guidelines, typically to the nearest approved ground at each end. Cable shields do not satisfy the requirements for a TBB.

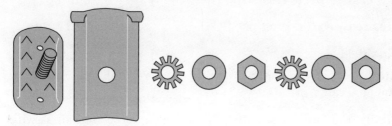

Figure 1.68 Bullet bond-clamp kit.

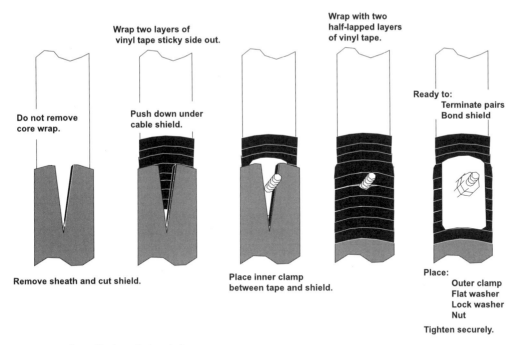

Figure 1.69   Installation of a bond clamp.

Shielded cabling systems

Some indoor cabling systems rely on shielding as an integral factor in their signal transmission performance, most notably those with coax, twinax, shielded twisted-pair wire, or screened twisted-pair. Through standard cable connectors, the cable shields are typically grounded to a connector/administration panel at each end so that even after administration changes, the cable shields are grounded at both ends.

The administration panels should be bonded to the nearest approved ground with a direct minimum length grounding conductor. At the user terminal end, these cable shields are commonly terminated by the user terminal, which relies on the nearest power plug's green wire (safety ground) instead of direct bonding. The bond is completed through the equipment's chassis ground connection.

Use manufacturers' instructions and apparatus for terminating and grounding these cable types.

Exposed cable sheath terminations

Outside plant cables usually have metallic sheath members or a metallic shield that are bonded together at the point of entrance and grounded at the nearest approved ground.

Under high lightning incidence, an isolation gap is sometimes designed. The cable sheath routed within the building is isolated from the cable sheath that enters the building. Both sides of the isolation gap are grounded with individual grounding conductors.

Under conditions causing dc ground currents, an insulating joint is sometimes designed. An insulating gap is made in the cable sheath similar to the isolation gap, but the field (outside) side is not grounded directly. Instead, it is grounded through a capacitor that blocks dc current.

Equipment Grounding

Single-point equipment grounding

Some PBX manufacturers and other large equipment suppliers require single-point equipment grounding. When equipment is grounded through only one point, surges that are conducted through the building ground will not pass through the equipment. Manufacturer's documentation should provide details for acceptable implementations.

Receptacle outlet grounding

Discussed in NEC Sections 250-45(d)(2) and 250-59(b), some equipment relies solely on a receptacle (safety) grounding conductor, power cord, and plug. This is typical of small equipment and may provide for suitable performance and reliability if the receptacle ground contact is adequately bonded and free of electrical disturbances.

Additionally, the receptacle outlet ground should be bonded to other metallic systems if the receptacle is not located close to the grounded electrical service equipment. Equipment grounding through a receptacle outlet is not suitable for large equipment or equipment manufactured with bonding terminals.

Telecommunications Circuit Protectors

Overview

NEC Article 800 Part C covers telecommunications circuit protection, a primary responsibility of the telecommunications distribution designer. The basic functions of protectors are:

- Arresting surges or overvoltages (diverting them to ground) that come from exposed circuit pairs.
- Protecting against sustained hazardous currents that may be imposed.

Based on Underwriters Laboratories (UL) standards, there are three types of telecommunications circuit protectors:

- *Primary protectors as qualified by UL 497.* These are intended for application on exposed circuits according to NEC requirements. These must be installed as near as possible to the point at which the exposed cables enter the building, and the grounding conductor must safely carry lightning and power-fault currents.
- *Secondary protectors as qualified by UL 497A.* These are not required by the NEC, but are typically used for additional protection behind primary protectors. In addition to voltage protection, this type must also protect against sneak current. Sneak current can be caused by:
  - Power faults that are too low in voltage to operate primary protectors.
  - Station equipment that will draw excess current and overheat station wire.
  - Induction from power lines.
- *Data and fire alarm protectors as qualified by UL 497B.* These modules are more sensitive with a lower fault threshold because they do not have to allow for the 90 volts associated with a telephone's ring voltage. Though not required by the NEC, these modules must perform primary protection against lightning transients. They do not have the ability to protect against power faults. These should be used according to manufacturer's guidelines.

There can be some overlap in function with available products. The manufacturers should be consulted for specific applications; however, the following rules generally apply:

- A primary protector is required where a circuit is exposed to electrical power faults and lightning. Other protector types are not qualified to protect under these conditions.
- Where a circuit is exposed to lightning surges, a primary protector or a data/alarm protector is required, as dictated by equipment manufacturers.
- Where sneak currents are hazardous, a secondary protector or primary protector with secondary protection is required (as directed by equipment manufacturers). The basic secondary protector function (sneak current protection) can be included by manufacturers in some primary protectors.
- Manufacturers may include additional protection functions, sometimes called enhanced protection, for specific applications.

Protector Technology

Primary protectors

There are many different components used by manufacturers to implement protection functions. The following components are typical of primary protectors:

Carbon blocks. These are the original protectors. An air gap between carbon elements is set to arc at about 300 to 1000 volts and conducts surge current to a grounding conductor. When the surge current finally drops low enough, the arc stops and the protector resumes its normal isolation of ground.

A fail-safe function causes carbon blocks to short permanently to ground when an extended hot surge or permanent fault current overheats them. They are typically installed as pairs and are the lowest cost option; however, carbon blocks tend to wear out quickly under extreme conditions and can cause leakage and noise in voice circuits.

Gas tubes. These are improved arresters that basically operate the same as carbon blocks, arcing over a gap to a grounding conductor. They have a wider gap because of special gas, and therefore, a higher reliability. They also have tighter tolerances on arc breakdown voltage and are typically set to arc at a lower voltage, providing better protection than carbon blocks.

Another type, the dual gap, provides a common arc chamber that grounds both wires of a pair together and minimizes metallic surges that would otherwise occur from individual arrester operation.

Solid state. This is the newest type of arrester. It relies on high-power semiconductor technology. Though more expensive than either carbon block or gas tube, the cost is recovered over the extended life of the protector. They are fast acting, well balanced, and do not deteriorate with age below a rated maximum surge current.

Fuses and fuse links

The NEC identifies the two types of primary protectors as fused and fuseless. In case of extended overcurrent situations, the exposed side must fuse open without damaging the ground conductor or indoor circuit. The fused type accomplishes this with an integral line fuse. The fuseless type must be installed with fine-gauge fuse wire (a fuse link) on the exposed line side (provided by the manufacturer). Both types, when installed, will operate in the same way.

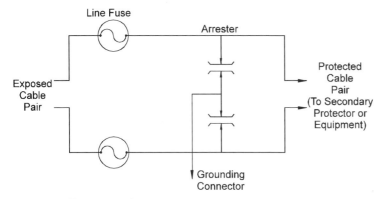

Figure 1.70 Fuse protection.

Secondary protectors

Secondary protectors are required to coordinate with lightning transient and power-fault requirements of primary protectors. In some cases, secondary protectors may include one of the previously described arrester components. For this reason, cost-effective secondary protection is typically available as an option on primary protectors (qualified to both UL 497 and UL 497A). Secondary protectors must handle sneak currents. The components for sneak current are different than those used for primary protection.

Heat coil. The coils detect sustained low-level current by heat. The heat melts a spring-loaded shorting contact that permanently shorts the line to ground, requiring manual inspection and replacement.

Sneak current fuse. This is a fuse that opens the station circuit wiring under sustained low-level current, requiring manual inspection and replacement. This fusing is on the station side of the arresters and should not be confused with the primary protector line fusing.

PTC resistors. These positive temperature coefficient (PTC) resistors are used in place of a sneak current fuse and will limit sustained current as they heat. They do not require replacement after the sneak current fault is cleared.

Enhanced protection

Enhanced protection uses additional components to provide protection that is typically suited for specific low-voltage data circuits. In most cases, equipment manufacturers have designed such protection into their equipment, but many prefer to use available external protectors.

In some cases, these protectors are UL 497-qualified and can substitute in a primary protector panel for special applications of individually exposed circuits.

In other cases, protectors may be qualified according to UL 497A (or even UL 1863 as telecommunications circuit accessories) for use in office locations where the protected station or terminal is powered, or behind primary protectors.

Primary Protector Installation Practices

Overview

Suggested installation practices:

- Primary protectors must be installed immediately adjacent to the exposed cable's point of entry. The associated grounding conductor must be routed as straight as possible directly to the closest approved ground.
- A noncorrosive atmosphere is required for long-term reliability.
- Adequate lighting is important for personnel safety at protector locations.
- When a protector is installed in a metal box, bond the box with an approved grounding conductor directly to the protector ground.
- Make certain there are no obstructions around or in front of protectors and that protector locations will not be used for temporary storage.

Warning. Do not locate primary protectors near any hazardous or easily ignitable material.

Common Safety Practices

Overview

It is every organization's responsibility to provide safety training for its employees and ensure that safety procedures are strictly followed. Safety training should be given on a regularly scheduled basis, not as a result of an injury.

Due to the daily hazards to which cabling installers are exposed, it is vital that all cabling installers have a complete understanding of rescue and first aid procedures.

Cabling installers must know their company's safety policies and practices—and follow them while working. Be aware of any site-specific safety issues that affect a task. Pay close attention and ask questions during every company or job-site safety meeting.

A safe work environment is much more likely when each worker makes ensuring workplace safety a part of the job. Do not depend only on the efforts of others to ensure job safety.

When working, try to consider the possible effects of every action. This is especially important for actions that could have consequences in remote locations (for example, turning power on or off or activating distant machinery).

OSHA

Passed by Congress in 1970, the Occupational Safety and Health Act attempts to ensure a safe and healthful environment for every working person in the United States. Under this statute, the Occupational Safety and Health Administration (OSHA) was created within the U.S. Department of Labor. The provision and requirements of OSHA are set forth in the Code of Federal Regulations (CFR).

While OSHA is responsible for the administrative work relating to the statute, most field work has been passed down to each state's Department of Labor. As a result, each state is responsible for field inspections and enforcement.

OSHA is responsible for job-site inspections and has the authority to shut down a job site and levy fines against individuals and companies for noncompliance to OSHA regulations. Additionally, OSHA is responsible for the development, publication, and enforcement of thousands of safety standards. The two main standards sections with which the telecommunications industry is concerned, but to which they are not limited, are:

- OSHA Regulations (Standards-29 CFR)—1910, Occupational Safety and Health Standards.
- OSHA Regulations (Standards-29 CFR)—1926, Safety and Health Regulations for Construction.

First aid

All cabling installers should take courses in and be capable of providing:

- Basic first aid.
- Cardiopulmonary resuscitation (CPR).

The American Red Cross offers local courses in standard first aid and community CPR around the country.

First aid is the emergency aid or treatment given before medical services can be obtained. Training in first aid prepares individuals to act properly and help save lives in the event of an emergency.

CPR is the emergency procedure used on a person who is not breathing and whose heart has stopped beating (cardiac arrest).

First aid, CPR, and the law. Legally, a victim must give consent before a person trained in first aid can provide assistance. Therefore, it is important to ask a conscious victim's permission. However, the law assumes that an unconscious person would give consent. Each state has "Good Samaritan" laws that provide legal protection to rescuers who act in good faith and are not guilty of gross negligence or willful misconduct. Laws may vary from state to state and should be verified as part of first aid training.

First aid and CPR certifications should be kept current. Certification cards have expiration dates that require refresher courses to renew.

First aid kits. First aid kits and portable eye wash stations must be a part of the equipment for every job. Fresh water to rinse out debris or toxins may not be available during construction periods.

Figure 1.71 First aid kit.

Check to see that first aid kits are restocked after each use and ensure eye wash stations have not passed their expiration dates. Promptly report any use of supplies from the first aid kit to the proper supervisor.

> **Note.** Many companies keep additional first aid kits. At the end of each month, kits that have been used on the job are swapped with fully stocked kits. The used kits are then restocked and prepared for reuse.

First aid kits and eye wash stations should be accessible to all personnel on the job site. Prevent eye wash stations from freezing, since a frozen station is of little use in an emergency.

Written copies of the first aid procedures for exposure to a hazardous substance, such as Material Safety Data Sheets (MSDS), should be brought to any job where cabling installers might be exposed to that substance. Review these procedures before work begins.

Emergency rescue

When there is an emergency, there is no time to ask questions or learn from mistakes. There is only one chance to make the correct choice to save a life. An untrained rescuer will often become a victim of the situation that caused the emergency. For example, if a rescuer attempts to assist an unconscious victim who is lying across an energized electrical circuit, the rescuer can become part of the circuit and a victim, too.

Training in emergency rescue and first aid are often combined into one comprehensive course.

There are six basic steps to safely assist others without endangering yourself:

- Survey the scene—Check for fire, toxic fumes, heavy vehicle traffic, live electrical wires, a ladder, or swift-moving water. If the victim is conscious, ask questions to get information.
- Notify someone—It is imperative to let someone know that you need help and where you are. If you attempt a rescue alone and become overwhelmed from smoke, electrocution, or unseen gases, you will need help yourself.
- Secure the area—Make the area safe for you and the victim. Locate and secure the power to the energized circuits and turn off gas or water mains. Move the victim to a safe area only if it will not further complicate their medical condition.

   Note. Do not move someone with a neck or back injury unless they are in a life-threatening situation. Remember that you cannot help anyone if you become part of the problem.

- Primary Survey of Victim—Check the victim's ABCs. "A" is for opening the victim's "airway." This is the most important action for a successful resuscitation. "B" is to check for "breathing," and "C" is to check "circulation" or pulse.
- Phone Emergency Medical Services (EMS) 911—Direct someone to call EMS and relay all the information you have collected in your initial surveys.
- Secondary Survey of Victim—Perform CPR as needed and check for secondary minor injuries that may have been overlooked previously.

   Caution. The above rescue techniques are a basic outline that should only be used after receiving the proper training. Some rescues require specialized training and equipment.

Communication

Communication is an important part of any safety program. Attend and pay close attention to all safety meetings and safety-equipment training. Ask questions.

On the job, cabling installers must communicate freely and clearly with everyone affected by their work and those whose work may affect them. These people include:

- Coworkers.
- Supervisors and building management.
- Building occupants (if any).
- Other workers (construction, electric utility, etc.) on site.

When work is being performed in two locations (for example, an electrical circuit is being switched off from one location to allow a cabling installer to work safely in another location), the worker(s) in each location should repeat each message and get confirmation that it was heard correctly before acting on the message. Never assume that related tasks have been performed; always get confirmation.

Note. Portable radios use a limited number of frequencies; therefore, it is very likely that different crews will be using the same frequencies. When using radios to communicate between two locations, workers should always confirm that they are talking to the correct person.

Figure 1.72 Communications headset.

Be alert and read any warning signs or markings. Bring them to the attention of coworkers who may have missed them. Encourage communication by accepting repeated information politely; it is better to be notified about the same hazard several times than it is not to be notified at all.

If a cabling installer discovers any defective or damaged equipment or facilities, the cabling installer should report them promptly to the supervisor or directly to persons qualified to handle the problem. For example, report damaged electrical power lines to electrical workers on site or to a building or construction supervisor who will contact electrical workers. If the defect or damage poses an immediate hazard, the cabling installer should do whatever is safely possible to ensure that others are not harmed by the hazard before qualified personnel arrive to fix the problem. This may involve notifying other workers in the area, putting up signs and barriers, or standing guard until qualified personnel arrive.

Cabling installers should promptly report all accidents or injuries to their employer.

Designating work areas

Always use safety cones to designate work areas and to restrict access. Yellow caution tape and folding A-frame signs may also be used. Be sure to leave enough room inside the cone perimeter to do the required work.

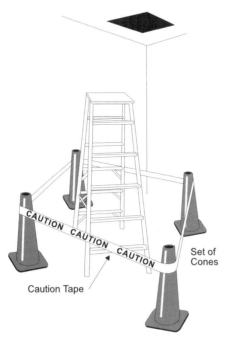

Figure 1.73 Designated work area.

Consider the needs of the building occupants whenever possible. Try not to block a doorway or hallway for which there is no alternate route any longer than necessary. When working near doors or hallway corners, try to ensure that oncoming pedestrians can tell there is a work area ahead.

Do not leave open floor systems unattended. Do not leave open ceiling systems with dangling access panels or equipment unattended. Do not leave work areas cautioned off longer than required. The customer and coworkers will soon realize the work is complete and start disregarding your warnings. This could cause an accident in the future when a real danger exists.

Tools and equipment

Use only those tools with which you are trained or certified. Manufacturers require users to be certified to use certain tools, such as power-actuated nailers.

Never be afraid to ask questions or say that you are uncomfortable using a tool.

Use tools only for the purpose for which they are intended. For example, use a tool designed for stripping to strip cable insulation—not a pocketknife. A screwdriver is not a scraper, chisel, pry bar, sheet-rock saw, hole punch, or a drill for wood.

Examine both hand tools and power tools regularly to ensure that they are in safe working condition. Broken tools must be immediately tagged as "BROKEN" and removed from the job site. A detailed description of the problem should accompany the tag. This will prevent an injury to someone using the tool prior to its repair or replacement.

Figure 1.74 Tag broken tools.

The wooden or plastic handles of hand tools must be kept free of splinters, sharp-edged cuts, or other surface damage that could injure a worker's hand. Do not use a hand tool if its handle is loose. Loose-handled tools can give way suddenly, causing injuries to both people and equipment.

When using power nailers, always verify that the area behind the work area is clear.

Do not attempt to drive nails or other fasteners into:

- Very brittle or very hard materials (glazed tile, glass block, face brick, etc.). The shattering of the material (or the fastener) can scatter dangerous shards across a wide area.
- Very soft or easily penetrated materials. The nails or fastener can pass through the material and create a hazard for people on the other side.

> **Note.** These precautions apply to both manual and powered nail driving.

Power tools should be inspected regularly to ensure that automatic cut-offs, guards, and other safety devices work properly. Follow the manufacturer's recommended maintenance schedule to ensure reliable operation.

Before each use, examine power tools to ensure that all guards are in place and securely attached.

The NEC requires a Ground Fault Interrupter (GFI) when temporary wiring is used. Temporary wiring is an extension cord or even an entire building's internal wiring prior to final inspection and acceptance.

Power tools that require a three-conductor power cable must be grounded. Never use a power tool if the ground prong of the plug has been cut off. Never use a power tool with an extension cord or adapter that eliminates the ground prong before the cord reaches the outlet.

Never use a tool's power cord to lift or lower the tool.

Carefully follow all manufacturer's instructions when mounting, securing, and using potentially dangerous mechanical equipment, such as tuggers (for cable pulling), tension arms, cable wheels, cable brakes, power-actuated guns, etc. Do not set up or operate this equipment without first receiving adequate training and having access to the manufacturer's instructions.

> **Note.** Keep original instructions for tools in a file at the office. Provide a photocopy of the instructions with each tool. This ensures that there is always a set of clean instructions at the office that can be used to make additional copies for the job site.

Ensure that adequate lighting is available to safely and efficiently perform work. Proper lighting will help prevent accidents and rework. Use portable lighting and keep it away from combustible materials.

Ladder safety

Cabling installers must know how to choose a ladder, place it securely, and climb and work on it safely. The location of telecommunications cabling and equipment requires that ladders often be used for both installation and repair work.

Note. Laws require that manufacturers print ladder-use guidelines on their ladders. Read and follow these guidelines.

OSHA does not require workers to wear fall-arresting safety equipment while working on portable ladders. However, it is a good safety practice to belt into a secure anchorage when working aloft with heavy equipment or over a prolonged period.

Figure 1.75 Stepladder.

Use the correct type of ladder. Never use a metal ladder where there is a chance that the cabling installer or the ladder will touch energized electrical cables or equipment. Use ladders made of wood or nonconductive synthetics in these situations. Most construction sites will not allow metal ladders on site for safety and insurance reasons.

Ladders should be examined before each use. Check to ensure that:

- Joints between the steps and side rails are tight.
- Anti-skid feet are secure and operating properly.
- Any moving parts operate freely.
- Rungs are free of dirt, liquids, or other substances that could cause slipping.
- Side rails are not excessively bent or dented.

Choose a secure location to set up the ladder, such as flooring or ground that is solid, level, and offers adequate traction for the ladder's feet. If adequate traction is not available, the ladder must be lashed in place or held in position by another worker or workers. Never set a ladder on top of a box, furniture, or any other unstable surface.

Place an extension ladder so that both side rails are supported at the top, unless the ladder has a single support attachment at the top. For stepladders, verify that the supports that link the ladder rails to the back rails are fully extended and locked into place.

Confirm/verify that the extension ladder is set at the proper pitch (angle). The distance from the base of the ladder to the supporting wall should be one-quarter (25 percent) of the length of the ladder. For example, a ladder extended 6 m (20 ft) up a wall would have its base 1.5 m (5 ft) from the wall.

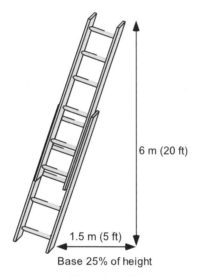

Figure 1.76 Extension ladder.

Extension ladders should always overlap between sections by at least three rungs. The top of the ladder should extend up to the work area and 0.91 m (3 ft) above catwalks or lofts. This allows the cabling installer to easily find the steps when getting back onto the ladder from the catwalk.

Never paint a ladder. Doing so will hide any stress cracks or damage.

Try to place the ladder where it will be out of traffic. Use safety cones to designate a restricted area around the base of the ladder. Never set a ladder in front of a door that opens toward the ladder, unless the door is locked or can be blocked or guarded from the other side.

When using a ladder:

- Never exceed the ladder's weight rating. Most ladders are designed for one person only.
- Always face the ladder when climbing up or down.

- Never stand on the top two rungs of a ladder.
- Never leave any object (tools, gloves, etc.) on any rung of a ladder.
- Never straddle a ladder or stand on the rear rungs. The rear rungs are narrower than the front steps and are not designed to support weight.
- Never intentionally drop or throw down anything (tools, excess wire, scraps, etc.) when on a ladder. Use a hand line and a "grunt sack" to raise and lower items.

When the job requires a ladder, use a ladder. Do not stand on furniture, boxes, or any other makeshift ladder substitute.

Personnel lifts

A personnel lift is required when a ladder cannot be used safely because of the required working height or weight of personnel and equipment. There are two types of lifts:

- A bucket lift—This is a fiberglass bucket mounted on the end of an extendable arm in which the user stands. The articulating arm allows the user to approach the work area from several angles and to avoid obstacles or possible safety hazards. These units are usually large and can be used in limited areas.
- A scissor lift—This is a working platform mounted on a large scissor jack. The scissor lift is very stable but not very flexible in its use. As the scissors are extended, the platform moves straight upward. If there are any obstacles above it, it is unable to maneuver around them.

Factors that determine if a lift will work include:

- The maximum working height of the lift.
- The size of the work area.
- Obstacles that may obstruct extending the lift.

Lifts must be secured by setting brakes and using stabilizing legs or outriggers, if equipped.

Personnel must be surrounded by side rails or the bucket's outer walls and should wear a full body harness with two lanyards, one of which must be attached to the lift at all times.

As of January 1998, OSHA prohibited the use of safety belts and lanyards equipped with the "non-locking type of self-closing keepers" (metal clips).

Personal Protective Equipment

Overview

Personal protective equipment is safety equipment worn by the telecommunications cabling installer. When used correctly, personal protective equipment greatly decreases the cabling installer's risk of injury. When it is used incorrectly—or not used—it can leave the cabling installer exposed to a wide variety of dangers.

The personal protective equipment that a cabling installer is required to wear when performing a task depends on:

- The hazards of the task.
- The hazards at the work site.
- Local, state, and national safety requirements.

Personal protective equipment must fit well and be as comfortable as possible. Equipment that fits properly and comfortably ensures that the cabling installer and the protective equipment can work at the same time.

Pay careful attention to the training for each item of personal protective equipment. Be sure to learn:

- When the equipment must be used.
- How to put on, adjust, and take off the equipment.
- What the equipment can and cannot protect against.
- Care and maintenance of the equipment.

It is important to inspect personal protective equipment each time it is used. Look for wear, cracks, tears, punctures, weak joints, or other signs that the equipment may not be able to provide protection. Report any problems to the proper supervisor. Never use defective protective equipment.

Remember that no amount of protective equipment can provide complete protection. Often the best personal protection comes from using caution, proper procedures, and common sense when working.

Headgear

Cabling installers must wear protective headgear (hardhats) when working in any area where there is danger of:

- Falling or flying objects.
- Electric shock.
- Striking their heads.

Figure 1.77 Hardhat.

Generally, the hardhats provided for telecommunications cabling installers afford both physical and electrical protection. Cabling installers should ensure that their hardhats provide electrical protection before working around power lines or equipment.

The hardhat must fit securely enough to ensure that it will not slip and block the cable installer's vision or fall onto the equipment the installer is working on. Cabling installers may choose to use a chin strap to secure the hardhat only if the chin strap is thin enough to give way easily if the hardhat catches on something during a fall.

Before putting headgear on, inspect it for cracks, weakness of the internal support structure, or other defects.

Eye protection

Cabling installers must wear eye protection (full-face shield, goggles, or glasses) whenever there is a potential hazard to the eyes. If the cabling installer must wear prescription glasses to correct vision, prescription safety glasses with side shields may be required, or goggles that fit over personal prescription glasses can be used. All eye protection must meet or exceed OSHA requirements. A wide variety of work situations require eye protection, including:

- Working with batteries.
- Using power-fastening tools.

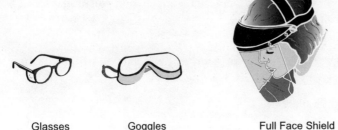

Glasses   Goggles   Full Face Shield

Figure 1.78 Eye protection.

- Working with optical fibers.
- Any situation in which the cabling installer is working above eye level and looking up at the work.

Wearing a full-face shield is required when there is danger of splashing chemicals, such as when working with batteries.

Wear protective goggles or glasses that provide side protection as well as front protection when the hazards involve flying objects. Cabling installers who wear prescription lenses must have eye protection that either fits over the prescription lenses or includes the prescription in the protective lenses.

Breathing protection

Cabling installers must wear a respirator or filter mask whenever harmful dust, gas, smoke, chemical vapor, or other pollutant is present at the work site.

Never work without the proper breathing protection. The effects of breathing some harmful substances may not show up until hours, weeks, or years after exposure.

Filter masks are used in cases where the atmosphere is moderately hazardous. Very hazardous atmospheres require the use of gas masks or even compressed-air respirators.

To provide the proper protection, each mask must seal itself to the user's face. This may require the removal of facial hair.

> **Note.** Several chemical manufacturing plants will not allow personnel with any facial hair to work on site before shaving.

Cabling installers should not try to work while wearing a breathing protection device unless:

- They have been fully trained in how to use the device.
- The device has been carefully fitted.
- They have been found physically fit to work while wearing the device.

Although employers are required to inspect and maintain breathing protection devices, the cabling installer should also inspect the device every time it is used. Report any problems to the supervisor in charge of breathing protection devices.

Lifting belt

A lifting belt does not give the user any added strength. The belt is designed to be worn around the wearer's abdomen and help support the stomach muscles while encouraging proper posture.

142 Chapter One

Use correct lifting techniques when lifting any object on the job (i.e., lifting with the legs and not with the back, turn with your feet and not at the waist). Whenever possible, wear a lifting belt when lifting or moving heavy objects or equipment.

Always know your path and do not block your vision when carrying items.

Most government contracts require that workers wear lifting belts.

Protective footwear

Wear protective footwear on work sites where feet could be injured by falling objects, rolling carts, or by stepping on sharp objects. A good pair of shoes will protect feet from injury and fatigue. Steel toes will keep feet from getting smashed, while steel or fiberglass shanks will offer protection from stepping on sharp objects. Steel or fiberglass shanks also help distribute weight across the base of the shoe. This reduces foot fatigue while standing on the thin rungs of an extension ladder. Leather-soled shoes are not advisable, since leather conducts electricity when wet. Tennis shoes are not acceptable footwear.

Gloves

Wear physically protective gloves when performing any work that has the potential for hand or forearm injuries.

- Leather gloves provide protection from cuts, abrasions, and extreme temperatures.
- Rubber gloves provide protection from harmful chemicals.

   Note. Rubber and leather gloves are not for high-voltage use. All high-voltage situations should be referred to qualified persons.

Leather   Rubber

Figure 1.79 Gloves.

Detection badges/exposure monitors

Some work sites may require that cabling installers wear a detection badge or use a monitor to ensure that exposure to a hazardous substance does not exceed safe levels. Understand how the badge or monitor works before entering the hazardous area.

Some monitors or badges will provide real-time feedback. That is, they show actual exposure levels as they happen. They may change colors or have a dark stripe that gets longer with the amount of exposure. Other types will only absorb the toxins at the same level the cabling installer is being exposed to them. These need to be inserted into a scanner to determine the amount of exposure. It is extremely important to check these types of monitors frequently to limit the cabling installer's exposure levels.

Always observe any time limits specified for working in a hazardous environment.

Hearing protection

Hearing loss is one of the most frequent injuries encountered in the construction trades. The victim does not feel any pain, but after years of exposure to high levels of construction noise, varying frequencies of their hearing may be lost.

Wear hearing protection while working in the vicinity of loud noises. Even the sound of a hammer striking a metal clamp onto red iron requires hearing protection. If the cabling installer experience a ringing in the ears, adequate hearing protection is not provided.

There are three major types of hearing protection:

- Disposable, foam plugs—These plugs can be rolled between the fingers and slipped into the ear canal.
- Reusable rubber earplugs—These may be on a breakaway cord or individually housed in a pocket-sized plastic container. These are convenient because they can be attached to a hardhat or around the cabling installer's neck and tucked inside the installer's shirt. It is vital that they be on a breakaway cord to prevent strangling.
- Aural—These types resemble ear muffs. They are available in passive or active models. When wearing the active models, normal con-

Aural

Rubber Reusable

Foam Disposable

Figure 1.80 Hearing protection.

versation can be heard; however, when a loud noise occurs, the protection automatically dampens the louder sound.

When working in a noisy work site (with or without earplugs), be careful not to rely on hearing to detect the location of machinery, coworkers, or other hazards.

Safety harness

To prevent falls, cabling installers must wear a full-body safety harness and two lanyards any time they use an elevating device (scissor lift, bucket lift, etc.) to reach their work. Using two lanyards allows the wearer to always have one lanyard attached to a safe support. If the user has to move along a catwalk, but runs into an obstacle, he or she simply attaches one lanyard beyond the obstacle prior to disconnecting the second one, allowing the user to be safely attached to the structure at all times.

Safety lanyards are available as simple nylon ropes with self-closing and locking keepers (metal safety hooks) on each end, or they can incorporate a shock absorber into the line. If a fall were to happen, the shock absorber would reduce the force of the sudden stop to the victim.

As of January 1998, OSHA prohibited the use of safety belts and also lanyards equipped with the "non-locking type of self-closing keepers" (metal clips).

Many work sites do not allow the use of safety belts. They do not provide the same level of protection as the full body harness.

Before going up:

- Inspect the harness and its hardware carefully for signs of wear or damage.
- Ensure that the harness is properly secured to the elevating device's anchoring point, never the guard rail or platform that is supporting you.

Whenever securing the lanyard, always check the connection. The cabling installer must ensure that the metal clip has captured the lanyard's rope or the equipment's safety ring and is securely fastened.

Clothing

Work clothing should be reasonably snug but must allow the cabling installer to move freely. Do not wear dangling or floppy clothing, which may get caught on tools or surroundings. Keep shirttails tucked in, cuffs (if any) buttoned or neatly rolled up, etc. This is especially important when the cabling installer is working in a confined space, on an elevating device, or near operating machinery.

Do not wear metal jewelry or metal watchbands when working on telecommunications circuits or equipment. Field-damaged clothing shall be repaired immediately or the individual must leave the job site. Duct tape may be used to temporarily mend torn clothing.

Grooming

Long hair can be extremely dangerous when working around operating machinery and while working aloft. Hair can easily be pulled into machinery or become caught on ceiling grids. Pulling the hair back in a ponytail usually provides adequate protection while allowing the worker to wear safety equipment.

> **Note.** The safest way to ensure that your hair does not catch on something is to completely tuck it under a hat.

Hazardous Environments—Indoor

Overview

Although there are hazards involved in any telecommunications installation work, some indoor situations are especially hazardous. These situations require extra safety precautions and often require extra protective equipment.

A complete study of hazardous indoor environments is beyond the scope of this manual. The following sections give a brief overview of several hazardous environments and some of the extra precautions they require.

Always carefully follow all safety procedures when working in an indoor hazardous environment.

Electrical hazards

The presence of electrical power cabling and electrical equipment is probably the most common environmental hazard faced by cabling installers. Like telecommunications cables, electrical power cables run in walls, under floors, and over ceilings. Power is also required for telecommunications equipment in closets and equipment rooms.

All electrical systems are potential killers; therefore, all personnel should be aware of the dangers and have electrical safety training. Use power tools and equipment only for the purposes for which they were made. Use tools only according to the manufacturer's instructions.

Wear rubber-soled shoes and remove all metallic jewelry. Most jewelry is made of gold or silver, which are two of the best electricity conductors.

Never intentionally expose yourself to an electrical shock. That is, do not run your finger down a termination block to check for ringing current.

Physical effects of current are as follows:

- 2–3 mA: produces a tingling of the skin.
- 10 mA: produces a painful shock and the muscles cannot release the contact.
- 50–100 mA: breathing becomes difficult.
- 100 mA: ventricular fibrillation occurs, causing the heart to repeatedly relax and violently clamp shut. This action destroys the heart and usually results in death.
- 200 mA and above: the heart clamps shut, severe burns occur, and a sickening smell is produced as the skin and hair burn away. At this level, the damage to the heart may actually be less than at 100 mA, allowing the victim to survive if medical treatment is given in time.

Treat all electrical circuits as if they were live (energized). Even after the circuits have been turned off and tested to ensure that they are off, treat electrical circuits as if they were likely to become live again at any moment. Always lock out all electrical circuits that have been turned off. Continue to maintain the proper clearances, wear the proper personal protective equipment, and take all the proper precautions.

Cabling installers must be especially careful in situations where electrical circuits or equipment may be contacted blindly (drilling into walls, fishing conduits, etc.).

Warning. Do not use a metal fishtape in a conduit if the exit point is unknown.

Avoid working in standing water. If cabling installers must work in standing water (e.g., in a basement tunnel), take extra care to ensure that there are no electrical power circuits near the water or the work area.

Never cut the ground prong off of a power tool plug. Removing the ground prong creates a serious possibility of severe electrical shock for the worker using the tool.

Avoid working on energized equipment. If the cabling installer must work on an energized circuit, such as performing an alignment in a microwave radio or troubleshooting a telephone system, have a qualified safety person standing by. A qualified safety person will know:

- Where and how to secure electrical circuits.
- First aid and CPR.
- Where and how to get help.

Many equipment rooms are outfitted with an emergency electrical safety board.
This board is mounted to the wall and may have a:

- First aid kit.
- Static grounding wrist strap. This is used to take the built-up static charge from the cabling installer to the ground and avoids damaging sensitive circuits.
- Safety grounding wand. This is an insulated handle with a metal tip that is connected to an insulated 1.8 m (6 ft) cable with a large metal clip on the opposite end. The clip is connected first to a ground source. The metal tip is used to short any transient voltages left on a de-energized circuit.
- High-voltage rubber gloves and protective leather outer shells. The rubber gloves should be inspected regularly for holes and cracks, while the leather outer shell must show no signs of wear. Only cabling installers trained in high-voltage rescues should use these gloves.
- Wooden cane. The cane should be lacquer free so as to be a nonconductive rescue device. It may be used to pull live wires off a victim or to pull the victim to safety.
- Class C fire extinguisher. The Class C extinguisher does not have any chemicals with conductive properties and is used for electrical fires.

Figure 1.81 Fire extinguisher.

In the case of an electrical fire, it is most important to protect people. Protection involves four steps easily remembered by the word RACE. This acronym stands for:

- Rescue—Get people out of danger.
- Alarm—Sound the alarm, call for help.
- Confine—De-energize all electrical circuits involved with the fire. Close windows and doors and secure the HVAC.
- Extinguish—Control the fire with the correct type of firefighting equipment. All telecommunications closets and equipment rooms should have access to a fire extinguisher designed to fight an electrical fire. The extinguisher must be Class C.

When working in telecommunications closets, make it a habit to check for an approved fire extinguisher. If the letter "C" appears somewhere on the label, it is approved. For example, "ABC," "AC," "BC," or "C" are approved. "A" is for combustibles: paper, wood, or anything that leaves ashes. "B" is for liquids, gas, oil, or alcohol. "C" is for electrical fires. These letter designations are important.

Lightning hazards

Although lightning is generally thought of as an outside plant hazard, it can also endanger indoor workers. This is especially true during construction or renovation, when protective systems may be incomplete or disconnected. Exercise caution when working on premises cables that are electrically connected to outside plant cables during an electrical storm.

Access floors

Raised computer flooring is becoming very popular in equipment rooms. There are potential dangers associated with access flooring:

- These floors are usually part of the HVAC distribution system. When a floor tile is removed, pressure is released while dirt, dust, and debris flies at the installer. Safety glasses are a must, while a face filter is suggested.
- Floor systems are usually constructed with metal pedestals interconnected with stabilizing stringers joining them. The floor tiles are then placed on top so that each pedestal will support the intersection of four separate tile corners. These systems are very stable, even with several tiles removed. Another construction method is seen without the stringers. The floor maintains its stability when all the tiles are in place. Never remove all the tiles in a single row if stringers are not used for stability. There is danger of the entire floor collapsing if it shifts towards the open row from which the tiles were removed.
- When tiles are replaced, it is often like a jigsaw puzzle. If the tiles are not placed exactly in the same spot, the floor will often have loose tiles that wobble when walked on.

    Note. Prior to removing tiles, put a piece of tape in the corner of every tile to be removed. Place the tape in the same corner of each tile (i.e., northeast corner) and put a unique number on the tape. This ensures that the tiles will go back exactly as they came out.

Catwalk hazards

In large buildings or industrial facilities, a catwalk may be provided to help workers reach utilities. When working from a catwalk, stay on the catwalk. Do not climb or walk onto beams or other support structures.

Never intentionally drop or throw anything (tools, excess wire, scraps, etc.) from a catwalk or other elevated work location. Raise and lower equipment with a hand line and grunt sack.

Crawl-space hazards

Telecommunications cabling often runs over suspended ceilings, below raised floors, and in other spaces where cabling installers cannot stand upright. These areas are called crawl spaces.

It is a good idea to wear protective headgear (hardhat) when working in crawl spaces, especially when electrical wiring is present. The hardhat will also protect the cabling installer's head from the hard surfaces and sharp edges that may be found on the supporting hardware for the floor or ceiling system.

Ensure that lighting is adequate to see the work clearly. If it is not, use a flashlight or other work light for extra light.

Before beginning work in any crawl space, take the time to locate and identify any other facilities that are routed through the crawl space (electrical power wiring, pipes, HVAC ducts, etc.). Identifying surrounding hazards can keep the cabling installer from accidentally damaging another system or endangering himself/herself.

A filter mask or other breathing protection may be required if dust, fibrous insulation, or other breathing hazards are present in the crawl space. Check with employers and the building management to determine the nature of the hazard and the protection required.

When moving through a crawl space, walk or crawl only on surfaces designated to support walking or crawling. The cabling installer should never put weight on ceiling support hardware that is not designed to support crawling or walking. Before putting full weight on a walk or crawl surface, the cabling installer should ensure that the surface is strong enough to bear weight. Never put weight on cable support devices (cable trays, etc.).

Never intentionally drop or throw anything (tools, excess wire, scraps, etc.) from a crawl space above a suspended ceiling. Do not drop, place, or throw anything on top of the ceiling tiles.

Crawl spaces may be considered as confined spaces that require additional precautions. See "Confined Spaces" for more information.

Confined spaces

According to OSHA 1910.146-Permit-Required Confined Spaces, "a confined space:

1. Is large enough and so configured that an employee can bodily enter and perform assigned work; and

2. Has limited or restrictive means for entry or exit (for example, tanks, vessels, silos, storage bins, hoppers, vaults, and pits are spaces that may have limited means of entry); and
3. Is not designed for continuous employee occupancy."

Maintenance holes, splice pits, crawl spaces, and attics can fall under the OSHA definition of a confined space.

Confined spaces may require testing for a hazardous atmosphere:

- Safe oxygen levels of 19.5 to 23.5 percent.
- Flammable gas, vapor, or mist.
- Combustible dust.
- Any toxic substance in a concentration greater than deemed safe by OSHA standards. Nonlethal or incapacitating toxins are not covered by this provision.
- Any other atmospheric condition that is immediately dangerous to life or health.

When toxins, gases, or combustibles are detected, the cabling installer shall:

- Provide continuous forced-air ventilation to purge the contaminants from the space.
- Periodically monitor the air quality within the space.
- Evacuate the space immediately if contaminates return.
- Determine why the contaminates returned and take corrective action prior to re-entering the space.

Certain confined spaces may require:

- Breathing apparatus.
- Protective clothing.

Figure 1.82 Breathing apparatus.

- A trained safety person stationed outside the space.
- A lifeline attached to the worker inside the space.

Hazards from toxic or flammable gas are rare when working inside a building. Telecommunications closets and equipment rooms should be completely free of gas hazards. However, there may be some situations in which a sealed vault-type structure (for example, an entry facility or splice pit) may accumulate gases.

Always use caution when entering any room or work area that is marked with warning signs that prohibit open flames or indicate other potential gas hazards.

When opening a vault-type structure, treat the vault like a maintenance hole. Before entering the vault, use a gas detector to determine whether any dangerous gases are present. If dangerous gases are present, the gases must be cleared from the vault before any worker enters it. Testing a maintenance hole for gases and exhausting gases out of a maintenance hole are usually outside plant procedures and are outside the scope of this manual. Detailed descriptions of these procedures are available in OSHA 1910.146, and 1910.146, appendixes A–D.

Optical fiber hazards

Optical fiber systems involve some hazards that copper cabling systems do not. Most of these different hazards involve the optical fiber or the transmission light source.

Optical fibers are very thin but surprisingly strong. Small scraps can easily penetrate skin, causing irritation or infection. Ensure that cleanup is thorough after any optical fiber splicing or termination. Many workers use a loop of sticky tape or a container to collect fiber scraps after each cut. This ensures that all scraps can be disposed of properly at the end of the job. Never eat or drink while terminating optical fiber. Small pieces of glass may fall into the food or drink.

Never throw bare fiber scraps into community trash containers. Always seal fiber scraps in a container, tape it closed, and mark it as optical fiber glass scraps. Take the container directly to the dumpster to avoid accidents to the unsuspecting customer.

Warning. Never look into the end of an optical fiber cable. Most optical fiber transmission light is invisible but can burn the retina of the eye before the cabling installer realizes that the light is on. Light sources for test equipment may be just as hazardous as the regular system light source. Ultraviolet (UV) light used to cure UV adhesives may be harmful to the skin and cornea of the eye. Avoid direct exposure to skin and eyes.

Always wear eye protection when handling exposed fibers. Small fragments of optical fiber can easily fly into the eyes during cleaving. Exposed fiber ends can injure the eyes when cables twist, flip, or fall.

Battery hazards

Working with or around flooded (wet) cell batteries requires:

- Training in handling electrolytes.
- Full eye protection (front and side).
- Acid-resistant gloves and apron.
- Training in emergency procedures for spills.

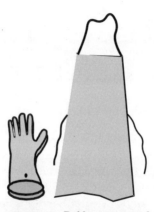

Figure 1.83 Rubber apron and gloves.

Always use care when working around batteries. Batteries are always live. Batteries release hydrogen gas as they are charged. Hydrogen is extremely combustible and must be vented outdoors. Some batteries are vented into the room and an exhaust fan pulls the hydrogen and oxygen outdoors. If the fan fails to operate, the gases will build up and create a potential hazard.

Neutralize all acid spills with baking soda and clean up with damp rags.

Flush electrolyte burns to the skin with large quantities of fresh water. Apply a salve such as petroleum, boric acid, or zinc ointment to the skin and seek medical attention.

Asbestos hazards

Asbestos is a fibrous mineral substance that was used in many buildings as an insulation material between the mid-1940s and 1978.

Asbestos was widely used in acoustical ceilings, wall and ceiling insulation, fireproofing for structural steel, and pipe and boiler wrapping. Asbestos often looks like plaster and cloth tape wrapped around pipes or an expanding insulation that is sprayed on boilers and structural steel. The difficult part of identifying asbestos is that it resembles other forms of insulation and must be checked in a laboratory.

The use of asbestos was banned in 1978, after it was learned that breathing asbestos fibers could cause cancer of the lungs, stomach, colon, esophagus, and other organs.

Schools and other public buildings were required to locate asbestos-containing materials and determine the threat they posed. Then the asbestos-containing materials were either:

- Removed.
- Cleaned and sealed to prevent fiber releases.
- Labeled as asbestos and left intact.

The mere presence of an asbestos-containing substance is not hazardous, as long as it is not releasing fibers into the air. However, disturbing a substance that contains asbestos (by sawing, drilling, breaking, rubbing, etc.) may cause it to release fibers and create a serious breathing hazard.

Working in an asbestos atmosphere requires full breathing protection (respirators), protective suits, specialized training, and many other special precautions.

Laboratory tests are required to determine whether a substance contains asbestos. If cabling installers are working in a building that was built or renovated between 1945 and 1978 and encounter a substance which may contain asbestos, they should check with building management and maintenance personnel before disturbing the substance. Building managers/owners are required to maintain records of any known or suspected asbestos-containing substances in their building.

If cabling installers encounter a labeled asbestos-containing substance and cannot perform the job without disturbing it, they should stop work immediately and consult their employer about alternative plans.

Chemical hazards

Many products used in the telecommunications industry contain chemicals that can be hazardous to people and the environment. To help the cabling installer work safely with commercial products, manufacturers are required to provide Material Safety Data Sheets (MSDS).

MSDS are provided for all products used in a commercial environment that can be absorbed through the skin, inhaled, ingested, or require special handling for disposal.

MSDS must be readily available at the job site where the products are being used. The sheets should be indexed and kept in a three-ring binder in alphabetical order by product name.

Chemical vapors can be very dangerous. Without proper ventilation:

- Toxic vapors can overwhelm the user, causing immediate and long-term effects. Fumes may cause nausea, headache, or vomiting. Prolonged exposure can cause disease to internal organs.
- Vapors may be flammable and create a fire when exposed to a spark.
- Vapors may be explosive when concentrated in a confined space.

When working with products that produce toxic or flammable vapors, it is always best to use them outdoors. When this is not practical:

- Notify other workers and have a safety person keep checking in.
- Open windows.
- Restrict air flow to other areas where people are working.
- Blow toxic air directly outdoors paying attention not to send the vapors into an unsuspecting office next door.
- Blow fresh air into work area.
- Take frequent breaks to keep the levels of toxin low.

Hazardous Environments—Outdoors

Outside plant

Hazardous outdoor working environments (maintenance holes, tunnels, ditches, aerial facilities, etc.) are outside the scope of this manual. Special outside plant training is required for working in these environments.

Professionalism

Overview

The telecommunications cabling installer's role has grown and changed as rapidly as the industry. Requirements of the industry demand that the cabling installer be a highly skilled professional, a team member, and an ongoing learner. The cabling installer's role is affected by:

- Current and future technology.
- Complex equipment.

- Advanced tools.
- Constant refining of standards and codes.
- Intricate designs.
- Customer requirements and expectations.

The ability to plan, organize, work on a team, and communicate effectively becomes just as important as the actual installation performance.

Customer relations

Customers are not dependent on cabling installers; however, cabling installation companies are dependent on customers. Each year a part of a company's customer base is lost because:

- They discontinue business, move, or merge with another company.
- Some do not stay with any vendor for an extended period of time.
- Some are intentionally abandoned by the cabling installation company.

However, the majority of the losses occur because of lack of attention to the customer's needs.

As the on-site representative of an employer, each employee carries certain responsibilities with respect to the relationship between the employer and the customer. Because of this, it is important to:

- Make a powerful and positive first impression.
- Maintain a professional image.

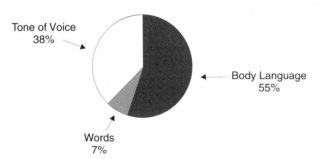

Figure 1.84 Communications perceptions.

Either positive or negative, an impression is formed within the first 10 to 15 seconds of a first meeting. Therefore, it is in the employee's best interest to do everything possible to ensure the first impression is positive. The way in which others perceive an employee can be broken down into three distinct areas.

Because of the impact that first impressions have, it is important to maintain a professional image at all times.

Figure 1.84 shows how communications perceptions are broken down in a "face-to-face" meeting with the customer. When talking to the customer on the phone, the tone of voice and the choice of words become much more critical. When communicating with the customer in writing, the only perception that the customer gets is through the choice of words and how these words are presented. Spelling, handwriting, and grammar become important to a good perception.

Project team

A telecommunications cabling installer is a part of a team. The full team consists of the installer's company and organization as well as the people and organizations for whom and with whom they work. A cabling installer may be required to talk and work with many different people:

- Supervisors
- Coworkers
- Customers
- Vendors
- Administrative and clerical staff
- Regulatory and government staff
- Other trades

Team membership has its own set of requirements. The cabling installer must take responsibility to know, understand, and communicate all aspects of the job. This involves not only getting and receiving information but asking the right questions, such as:

- What is expected?
- What is the project environment?
- What is the expected schedule?
- What are the critical time frames and why?
- What other projects are being completed simultaneously or during the project that may impact the project?
- Why was a particular cabling installer chosen for this job? Is there a particular skill, knowledge, or task that the cabling installer is to contribute or learn?
- What new skills might the cabling installer acquire?
- Is there training that the cabling installer is expected to provide?

Supervisors may provide information about job location and work area, customer requirements, project schedule, materials, safety issues, and code requirements. However, the cabling installer is responsible for knowing:

- Objectives of the project.
- Background of the project and project relationships.
- Project milestones.
- Specific role within the project.
- Specific tasks expected to be performed.
- Challenges and issues.
- Methods and techniques planned to be used.
- Applicable standards: local, state, national, company, and customer.
- Security requirements.
- Safety requirements.
- Evaluation of work process.
- Recordkeeping process and requirements.
- Status and monitoring process, both company and customer.
- Required meetings.

Feedback and follow-through are the keys to effective customer and coworker relationships. The customer must be kept informed. Any potential problems or concerns must be reported to the cabling installer's supervisor immediately. When changes are made, they must be recorded and acknowledged with the customer and the cabling installer's organization. The cabling installer's ability to follow through is the bond which promotes trust, protects reputation, and helps to ensure that time and resource commitments are met.

Courteous communication skills

The cabling installer's role is impacted by the effectiveness of the communications received and the communications delivered, whether verbal or written. Each person involved in the project expects all communications to be timely, clear, specific, unbiased, truthful, and free from unacceptable language. This is especially true when it comes to project meetings. Take time to prepare. Distribute copies of the presentation to all attendees.

The following table relates the person's position and communications that may be required.

As a courteous communicator, the cabling installer should:

- Listen attentively and take notes as appropriate.
- Provide complete information.

TABLE 1.13 Project Communications

Position	Required communications
Coworker	Safety issues, pull methods and techniques, cable tests, clarify work responsibilities.
Customers	Project schedules, test results, safety and security issues, progress reports, and potential problems or delays. Also, may include cutover issues, access requirements, and clarification of requirements.
Contractor	Coordination of schedules, blueprints, building access, storage, and required meetings.
Vendors	Project supplies, delivery schedules, verification of specifications, tools, materials, equipment requirements.
Administrative clerical staff	Job location, ordering and delivery of materials, and work schedule.
Other trades	Work-areas arrangements, schedules, and safety issues. Also, work coordination such as grounding, space in closets, and shared pathway space.
Local/state/federal agencies	Permits and inspections, regulations and codes clarification.

- Respect the values, traditions, and beliefs of others.
- Always find at least one positive thing to say.
- Praise in public, correct in private.
- Seek understanding of others.

Professional appearance

The cabling installer's appearance presents to the customer, coworkers, and supervisors an attitude about the quality of the work to be performed. Appearance must be appropriate for the task to be performed. Do not overdress, as this may be perceived as one who is trying to avoid certain tasks.

TABLE 1.14 Appearance

Appearance	Expectation set
Organized	The job is understood; lost time will not be spent looking for materials or tools.
Job-site cleanup	The client's site will be left neat, the cables will be treated with care, and the unseen work areas will be safe and correctly installed.
Free from interference	A respectful attitude toward coworkers. Appropriate clothing is: • Safe—nothing that will be tripped over or caught in equipment. • Nondiscriminating—such as inappropriate slogans on a tee shirt.
Safety	The cabling installer and the company represented considers safety for everyone. Hardhats, goggles, personal protective equipment, and gloves used appropriately protect the cabling installer, the cabling installer's organization, and the customer.

The cabling installer must recognize that appearance is a communication which can establish positive or negative expectations. Therefore, an organized, neat appearance, free from offensive or distracting items is a benefit to cabling installers and the organizations for which they work. Having and using appropriate safety equipment is part of that image. The safety section of this chapter reviews the available equipment and safety considerations.

Chapter 2

Planning

Designer's Drawings	163
Materials List	164
Labor List	179
Project Time Line	180
Project Schedule	181
Site Survey Checklist	182
Job Change Orders	183
Scope of Work (Example)	184

Overview

Planning for the proper installation of a cabling system involves a multitude of tasks. All of these tasks must be completed before any wire or cable is installed. When all of these planning tasks are completed, a project plan will be developed and the telecommunications installation team (see Chapter 1, Background Information) will be prepared to proceed with the project.

Safety concerns need to be addressed in the plan. Cabling system installations may be scheduled for existing structures (retrofit) or in buildings under construction. In the retrofit, it is recommended that all existing cable that is no longer used be removed to clear pathways and eliminate fire hazards.

When a cabling system installation is requested, the customer has already developed an idea of what they want. Through a request for proposal, formal contract, purchase order, or some other official document, these plans are transmitted to the cabling bidding company. These documents should contain the drawings and specifications that determine which materials are to be used, when they are to be installed, and under what standards they will be installed. These documents should be further enhanced by the development of a set of installation drawings, a list of materials, installation scope of work, and an overall project schedule. The larger the project, the more complex and important these documents become to a successful installation.

A member of the selected telecommunications installation team must conduct a site survey to identify all aspects of the cabling installation and how each must be addressed by the team. The results of the site survey will be used by the team to develop the schedule and the plan for meeting the customer's requirements.

Developing a telecommunications cabling installation project plan must involve all aspects of the installation. An understanding of telecommunications installation standards, national and local codes, BICSI manuals, manufacturer's specifications, and basic telephony principles will contribute to the success of the installation plan. The complexity of the plan is directly proportional to the size of the installation.

Various documents that already exist should be brought together with the new documents for the project. The telecommunications installation team must have available to them such documents as the customer's drawings, the designer's drawings, a list of materials, the customer's scope of work, and a copy of the customer's contract that includes any other documents which will dictate the progress or method of installation for the project. The installation team should also be provided with a copy of the final bid response, minus the pricing.

Customer's Drawings

Usually, these drawings are architectural blueprints prepared by a licensed architect. The following can be used to determine which blueprints are necessary:

- "A" Architectural drawings show a plan view of each floor in each building. They are referred to as "A" drawings. Architectural drawings cover all aspects of the physical construction of the building. They provide the detail for such items as walls, windows, ceilings, doors, flooring, cabinetry, and furnishings. If the architect employs an interior decorator, their work will also be reflected on these drawings.

- "M" and "P" Mechanical drawings include drawings for both heating, ventilating, air conditioning (HVAC), and plumbing. The mechanical drawings are identified as "M" drawings while the plumbing drawings are "P" drawings. The "M" drawings indicate the size and route of the various mechanical structures. In addition, the "M" drawings also give an indication of what obstructions are present to the installation of wires and cables inside the building. The "P" drawings indicate the locations and routing of the various plumbing systems for the building, including roof drain systems. Additionally, "P" drawings indicate whether or not the main water pipe serving the building is metallic.

- "E" Electrical drawings are very important to telecommunications installers. They indicate not only where the electrical services are installed inside the building, but also what pathways are to be installed by the electrical contractor for use by the telecommunications installation team. They also indicate the grounding system designed by the electrical engineer. The drawings indicate the location of supporting structures installed by the general contractor, such as plywood backboards. Some electrical engineers also create separate drawings for telecommunications. These drawings may be "T" sheets, "C" sheets, "V" sheets, or some other designation used by the architect or electrical engineer to indicate telecommunications. In large complex installations, such as school buildings and hospitals, there may be separate drawings for electrical, telephone, data, video, fire alarm, and intrusion detection systems.

- "S" Structural drawings are important in that they indicate the location of all of the various components of the building structure itself—the steel beams, the concrete floors, exterior walls, and various other components that make up the building's basic structure. These elements contain detailed sectionalized drawings that give both a plan (top) view and an elevation view of the structures. Many times, a sectionalized view of specific building components is contained on these drawings. These drawings indicate whether the building steel is welded, bolted, or riveted together. This is important in determining whether the entire building's steel skeleton can be used as part of the building grounding system.

- Site drawings will indicate the location of any exterior pathways that are being installed for use by the telecommunications contractor. They will indicate the size, quantity, and route of the various pathways and what service each will support. These drawings will usually indicate the service entrance pathways which will be used by the local service provider in bringing service to the building from the public network.

Designer's Drawings

A Registered Communications Distribution Designer (RCDD®) or RCDD/LAN Specialist may be employed by the customer, the architect, or the cabling installation company to develop the required telecommunications drawings and specifications. If not, a member of the telecommunications installation team may have to develop them. These drawings indicate the size, quantity, description, and route of the cables to be installed, as well as the type of hardware used to support the cables.

The drawings should indicate the type of cable to be installed, the splicing sequences (if any), and the cable pairs to be extended to all telecommunications closets (TCs). The type of connecting hardware by size, quantity, and configuration is also shown on the drawings. Most of the time, separate drawings will be prepared for each type of telecommunications system to be installed. That is, separate drawings are prepared for copper cables, optical fiber cables, coaxial cables, and low-voltage cables. However, there should be an overall drawing showing all the cables and sleeves so that the backbone routing can be determined. If the project is small enough, all of this information may be contained on a single drawing.

Elevation details of the closets and the arrangement of the various items of equipment should be part of these drawings. A detail of each wall and rack in each closet, as well as a plan view of the floor-mounted hardware, should be included.

The drawings must also indicate the supporting structure to be installed by or for the telecommunications installation company, how it is to be installed and identified, and how it is to be used during the installation of the cables.

An example of this drawing is in Attachment A of this chapter.

Materials List

A list of materials should have been prepared during the bidding phase. This list should contain all of the items to be installed by: (a) description, (b) catalog number, (c) quantity, (d) unit price, and (e) total price. Many contractors prepare a similar list for installation labor units, which indicate the amount of time allocated for each portion of the installation. In addition, the list should contain the differing labor hours allocated for the project by work process. Examples of these lists are shown in Attachments B and C. It should also have a column for indicating what materials were received and what materials were dispensed. Field personnel can use these two columns to manage inventory during the actual installation work.

Scope of Work

A scope of work for the project is a guideline to the telecommunications cabling installation team. The scope of work is the document that lists all of the elements of the installation. It can be generated by the customer, the designer, the installation company when bidding or responding to a customer's request for proposal (RFP), or by the telecommunications installation team. The scope of work should indicate:

- What work is to be performed.
- What materials are to be installed.
- What methodology is to be employed.
- The identification, labeling, and documentation system to be employed on the project.
- The testing methods to be employed.
- When and how the installation is to be turned over to the customer.
- Clarifications or understandings that elaborate on the various items involved in the installation. An example of a scope of work is shown in Attachment H. This scope of work is an example and not intended to be all-encompassing.

Contract

A contract is written to document the entire understanding between the customer and the contractor. Some customers generate a purchase order that refers to the other documents associated with the project. If a contract is available, ensure that any documents listed therein are available. Familiarize yourself with the contract form that your company uses. Contracts may also list penalties associated with not completing the work or delays in completion. Pay particular attention to liquidated damages. Performance bonds and insurance may be required as part of the contract.

Project Schedule

Companies use a variety of different project management styles and software. Two of the more prominent software packages in the marketplace are Microsoft® Project and MacProject Pro from Claris® Corporation. Various charts and graphs are employed that allow the telecommunications installation team to track materials receipt and disbursements, labor items completed, and the overall status of the project on a day-to-day basis. Manually generated project schedules can also be employed, especially when the project is small and not complex.

Examples of three project schedule documents are shown in Attachments D and E.

Project Log

The person in charge at the project site should also maintain a daily project log. This log should reflect any work done, whether it is complete, and the plans for the following day. Sometimes, copies of this log are required to be turned over to the customer or general contractor on a periodic basis. Accuracy is critical to the credibility of this log. This log should be used to record all activities associated with the project, especially such occurrences that may result in changes in financial impact to your company.

Site Survey

Once all of the initial project documents are obtained, a site survey is performed. A member (or members) of the telecommunications installation team will visit the place of installation. While there, they should observe all of the various locations where installation work will be performed. When making this site survey, carry the designer's documents with you. This will allow you to identify specific locations related to the project and the work to be performed there. The drawings may indicate hidden obstacles not visible from floor level.

A checklist should be employed on each project to ensure that all items of concern are addressed during the site survey. When problems are found, plans can be formulated to overcome them while still on site, rather than having to return to the site.

Examples of checklists for new construction and retrofit construction are shown in Attachment F. Employ a checklist, based on the criteria for each project, to ensure that everything is taken into consideration during the site survey and to double-check development of the project plan.

All information gathered during the site survey should be placed into the project file. This information will become invaluable later, especially if new team members are assigned to the installation after it starts.

The first stop at a job site should be at the general contractor's site office. While at the office:

- Introductions can be made.
- The work to be performed for the customer can be explained.
- Other contractors working for the general contractor can be identified and the impact of their work discussed.
- A copy of the general contractor's construction progress schedule can be obtained.

   Note.   This document can be used to determine how the installation schedule can be coordinated with the contractors working on the project.

When visiting a project site, determine who is responsible for the construction of the pathways and spaces. The customer's documents should state whether they are being provided by the general contractor or by others. Most of the time, the pathways and spaces of a new building are part of the responsibility of the general contractor or the electrical subcontractor.

If you (as contractor) are responsible for the pathways, determine how to install these pathways and what obstacles must be overcome in order to install them. Failure to make this determination might result in additional site surveys.

It is important to review your responsibilities with the general contractor in order that he or she understands the role your company will play in the completion of the overall project. Remember, the general contractor owns a new building until the owner accepts it. What the general contractor says becomes "de facto" law on such a site.

Be sure to bring the tools required to perform any work required for a thorough site survey. These tools might include personal protective equipment such as hardhat and safety glasses, leather gloves, leather boots, and hearing protection. Additional items that may prove useful are a ladder, flashlight, measuring wheel, a handheld tape recorder, a digital camera, or a video camera.

Determine the physical location of all closets, their size, type of construction, configuration of utilities within the confines of their walls, and responsibilities required to interface with other trades.

Locate the existing pathways which have been or will be constructed by the general contractor or the subcontractors. Determine their state of completion. Ask, as a minimum, the following questions:

- Have the subcontractors adhered to the architect's and designer's drawings and specifications?
- Are there any change orders that will affect the pathways and spaces? If so, how do they affect the project?
- How is the building grounding infrastructure installed?
- Does it comply with ANSI/TIA/EIA-607 and the National Electrical Code?
- When will the telecommunications rooms and closets be completed?
- When will the pathways be completed?
- When will the inspector or fire marshall be on site to perform the certificate of occupancy inspection?
- When will the project be turned over to the owner (tenant)?

The answers to these and other questions will determine how to plan to implement this project.

If the project is a retrofit, identify all of the existing pathways and spaces being utilized for telecommunications, their size, capacities, usability, congestion, and compliance with code. Some questions to be answered are:

- Are new telecommunications closets required?
- Are new pathways required?
- Are any existing pathways vacant?
- Does any pathway have usable space?
- Can existing facilities be utilized to assist in installing the new cable?
- How large are the existing closets?
- How much space is available in them?
- Will the new hardware fit within the confines of their spaces?
- Are new closets required?
- What, if any, are the requirements for maintaining dual service?
- Does any existing cable, hardware, or equipment need to be removed?
- Do any hazards exist (i.e., inadequate or nonexistent firestops, etc.)?

Most of the answers to these questions should come from the materials contained in the customer's documents and the designer's documents. Do not leave anything to chance. Always review all of the requirements of the project before concluding additional pathways and spaces are not required. If they are, and they have not been included in the original plans and specifications, job change orders may be required. At the very least, it may change the approach to the installation methodology.

Project plan

A good project plan is essential for the successful completion of the work. This plan should reflect all aspects of the work to be performed from a priority schedule. Project plans for new construction can be simple documents on smaller projects (such as a checklist) to complex documents on larger projects.

Every project plan should reflect the company's effort to ensure the work will be performed in a timely manner. It should include:

- Resources required for compliance with the schedules.
- Permit acquisition.
- Material staging.

- Coordination with other trades on the project.
- Scheduled meetings.
- Overall job schedule.
- Security and safety plan.
- Materials list.
- Tools list.
- Task list and description.
- Job requirements.
- Labor estimates.
- Acceptance plan.
- Inspection schedules.
- Special circumstances.
- Labeling system.

Routing for horizontal and backbone cables and their respective pathways should be identified. All labeling schemes should be discussed. Closet layouts should be reviewed to ensure that all items can be installed in their respective places, on the floor or on the wall spaces. Grounding and bonding methods should be discussed to ensure compliance with the scope of work, codes, and standards. Safety plans should be reviewed for applicability on the project. Appropriate personal protective equipment should be itemized.

The telecommunications installation team should review the project plan as part of the initial construction meeting. If changes are required, they should be made at this time.

Initial Construction Meeting

After completing the site survey and formulating the project plan, the telecommunications project manager should hold an initial meeting with the entire telecommunications installation team. At this meeting, the project manager and team leader should lay out the responsibilities of everyone involved. This ensures that everyone is "on board" and no one is kept in the dark regarding any aspect of the project. The project plan can be reviewed and updated as necessary. Questions can be asked and answered. Communications between all personnel involved in the project are critical to its success.

Minutes of all meetings should be maintained and a printed copy provided to each person attending and to additional persons having a position of responsibility on the project.

Materials Ordered

The person responsible for ordering the materials should place the order. Easily overlooked are items termed "exempt materials." These are materials small enough not to be detailed as a line item on the material list but are normally shown as a lot. Items could include tape, screws, tie wraps, etc.

The purchase order (PO) for the materials should be in writing. A copy of each material PO should be kept in the project file and used to develop the inventory for the project. The materials received can be checked against materials ordered.

Materials Received

Receiving materials is one of the most important tasks on a project. It can affect the cost of the project just as much as the labor employed. Determining where the materials are to be stored, how they are to be dispensed, how they are to be secured from theft or damage, and how unused materials are to be disposed of is critical to the project's success.

As the materials arrive, the responsible team member must receive the materials, inventory them, and stage them in preparation for transportation to the job site. If the materials are to be delivered to the site, an on-site representative will be responsible for receiving, documenting, and storing them in a secure location.

All items received should be inspected and inventoried upon receipt. Each package should be checked against the packing slip for quantity, identity, and condition. If packages are damaged in transit, the contents on the interior of the package may be damaged. This is especially true with large reels of copper cable and optical fiber cable. Verify this with the delivering agent prior to signing for them. If there is visual indication that reels of cable are damaged, consider refusing them or immediately notify your supervisor to determine the proper course of action. Even if the materials are kept, damaged or not, indicate the extent of damage on the packing slip and the shipper's manifest for future claims processing. When optical fiber cables are received, check the cable even when no damage is visible, using an optical fiber flashlight, power meter/light source, or some other type of testing device to determine the continuity of the glass strand end-to-end.

If other types of materials delivered to a job site are visibly damaged upon arrival, they should be refused and the delivery service instructed to return them to the distributor or manufacturer. If materials are accepted and then found to be defective, they should be stored separately from other materials and returned to the appropriate source via prearranged instructions from that distributor or manufacturer. If it is not possible to identify the defect and document it, enclose a copy of the documentation with the materials to aid the distributor or manufacturer in properly replacing them and correcting the problem that caused the defect.

Storage of Project Materials

There are three basic locations for staging materials for a project. Each has its own advantages and disadvantages. The three are the job site, your company location, and the distributor from whom the materials were purchased.

Job site

The job site offers immediate availability of materials. However, there are risks associated with storing materials at a job site. Is a secure space inside the building available? Is an exterior space (i.e., trailer or other building) available and secure? Most of the time, the security of materials is the first concern. Until the materials are installed and accepted by the customer, they are the property of and the responsibility of the contractor installing them.

While insurance is available to cover the loss, the delays in obtaining replacement materials may make it undesirable to store them on the job site unless significant security can be assured. Some customers may make space available for storing job material but few, if any, will agree to accept liability for loss or damage until they are installed and accepted.

When storing materials at the job site, other factors affect the use, distribution, and security of the materials. The general contractor may allow the materials to be stored inside of the building. Ensure that the space will be secure, dry, and lockable. Unauthorized persons should not have access to this space. Risk of loss insurance may be needed to protect against loss of materials and tools.

If space is not available for storage inside of the building, a job site trailer might be needed. In most locations, a permit is required prior to locating a job trailer on a construction site. If the trailer will be used only for material and tool storage, then temporary utilities will not normally be required. Office trailers requiring temporary electrical power, telephones, and sanitation utilities can require additional permits or coordination with the local utility companies providing these services. In most cases, local telephone and power companies will bill the contractor requesting the temporary service the actual cost of constructing the facilities up front. The contractor's company may also be required to sign a contract for a minimum period of time that the service will be used.

If large cable reels are to be used on the project, a security fence may be required. Fencing can be rented, and the rental company will usually install the fencing. The degree of security associated with job-site fencing will depend on what is to be stored inside the fence and the social environment associated with the job-site location. In most locations, chain-link fencing will be adequate. In other locations, concertina wire may be needed for maximum security. In very unsecure locations, private security might have to be hired during nonworking hours.

Company location

If the project is in the same city as the contractor's office, materials can be stored there and sent to the job on a daily basis.

Adequate space may be available at the contractor's location for storing materials utilized at the project. Security may be less of a concern at the contractor's home location. Break-ins, however, occur even at the most secure location. The telecommunications installation company should have insurance to protect against losses when materials are stolen or lost. The project may be of such size that materials might have to be stored at the job site.

Distributor

Most distributors are in the business of stocking materials. Also, most distributors are set up to deliver materials to a job site or to your company location on demand. Distributors use a process called "assemble and hold" to allow companies to order materials for a specific project and then stage them at their closest branch location until picked up by the contractor or delivered by the distributor.

Security is not an issue for the contractors since the materials are the property of the distributor until picked up or delivered. In addition, distributors have adequate space to hold materials for specific jobs and can deliver them by pallet or by partial order. Normally, your company is not billed for the materials, regardless of how long they stay in the staging area, until they are picked up or delivered.

Depending on the size of the project, it is necessary to consider all of these options. The best alternative may be a combination of all options. These options should be of prime concern when considering what to do to plan for material distribution and use on a job.

Regardless of which method is employed, eventually the materials will be received on the job. Upon receipt of the materials, they should be inventoried and identified against the packing slip. The packing slip should be retained and returned to the project manager to allow accounting of materials, quantities, defective items, etc.

Distribution of Materials on Site

Control of access to job materials will determine who is allowed to distribute them to the installer. When materials are distributed, some record of accountability should be made to ensure that the materials can be located. Excess materials should always be accounted for at the end of each work day and stored for use later on the job or returned to the company storage area for use on another project. They may eventually be returned to the distributor or manufacturer for credit after the project is completed and accepted.

Only designated persons should be allowed to distribute materials on the job site or receive them at the end of the work day. Allowing full access to the job materials by the entire work crew invites abuse and theft. In addition, the records of the distributed materials should be returned to the company project manager to ensure proper accounting.

Plan for a staging area on site, regardless of where the bulk of the materials and tools are being staged.

Development of a Project Schedule

Once all of the items associated with planning an installation have been identified, a project schedule needs to be developed. If the job is at a building under construction, the first schedule to obtain and reference is the general contractor's construction schedule. It includes all of the trades working on the project and indicates their specific time frames for accomplishing work on the project. Of particular concern are the schedules for completion of the supporting structure inside of the building. The backbone cables or the horizontal cables cannot be installed until the electrical contractor or the general contractor has completed installation of the pathways and spaces used to house these cables.

With new building construction, it is necessary to identify the finishing schedules for other trades. For example, faceplates cannot be installed until the wall covering is completed. Racks may not be installed until the floor covering is installed.

The project schedule should begin with the award of the contract and is complete upon acceptance by the customer. The detail required is directly proportional to its importance in completing items that precede or follow it. The project schedule should indicate the planned time required for each item, as well as provide space for inserting the actual time to perform the job task. The telecommunications project schedule should complement the general contractor's project schedule. Failure to coordinate the telecommunications project schedule with the general contractor's schedule will result in conflict between the two companies and could jeopardize timely completion of the project for both companies. Even after the original project schedule is created, it should be verified with the other construction schedules at least once a week for accuracy.

Project scheduling software is now available to simplify this task and provide a wealth of information relating to the status of a project. Most types of scheduling software track resources, materials, tools, and expenditures, as well as the actual schedule of progress of the work.

Copies of the completed schedule should be provided to all concerned parties. The schedule should be updated daily, indicating the progress of the day's work and whether the project is on schedule, ahead of schedule, or behind schedule. Any supporting documentation that will lend credibility to delays encountered in the project should also be referenced in the project schedule updates.

Project Log

In addition to a project schedule, a person on site should keep a project log during all work operations. All activities of relevance to the work should be logged. This document could prove valuable when others have caused delays. Contractual obligations that are affected by the work of others should be logged and accompanied by detailed notes. This log should begin on the first day of activity on the project site and continue until your company vacates the site completely.

Preinstallation Meeting

Once the project schedule is compiled, an internal meeting, called a preinstallation meeting, should be convened by the project manager, the telecommunications installation team, and appropriate contractors. All aspects relating to the project should be addressed, discussed, and adjustments made to the project schedule based on the results of the meeting. The project checklist should be reviewed, and everyone should be completely aware of their individual responsibilities. To ensure that the work is performed in a timely and professional manner, the project should be reviewed in detail so that each team member can work in concert.

Meetings

Periodic meetings should be held for every project. These meetings can occur as often as once a week or as necessary to ensure that everyone knows what is going on and what is expected of them. Work progress, as well as roadblocks and ways of overcoming them, should be discussed and agreed on by all concerned parties. Each team can contribute to the success of the project by thoughtful participation in these meeting.

On a project in which the building is under construction, the general contractor and subcontractors should be requested to attend and participate in these meetings. The project manager should also attend the general contractor's construction meetings to ensure that all concerns are addressed. This will ensure proper coordination between the telecommunications cabling installation team and other contractors working on the project.

Project Safety Plan

Safety is the first item of importance on a project. The safety of workers, customer personnel, and the subcontractors is of paramount importance. Workplace accidents can disrupt the best-planned job and cause costly delays. To lose a good employee during a project could adversely affect the planned schedule of the project.

Your company should have an approved safety plan. Before beginning any work operation, the contents of that safety plan should be reviewed with each employee working on the project. Each employee should fully understand how the rules of safety should be implemented as each installation task is performed. Take the time to ensure that each employee is equipped with the proper safety equipment and has the knowledge to use the equipment safely. As a contractor or subcontractor, attending periodic safety meetings may be contractually required.

It is better to halt a work operation if questions of safety come up, rather than risk an on-the-job accident.

All employees should attend an initial safety meeting prior to work commencing. All safety plans should be reviewed. Many general contractors and customers have safety orientation and drug screening programs that all employees must complete prior to beginning work on a project. Only through knowledge and understanding can each employee perform their work safely.

Job Change Orders

Few projects are completed without change to the original work plan. Even on small projects, changes occur. The changes may be insignificant, but they must be documented. When it comes to accounting for all materials and work operations at the completion of the project, unless the changes are documented and approved by duly authorized agents, compensation may not be received. A contractor should never perform additions, deletions, or material changes to the scope of work without proper written authorization.

An example of a job change order form is provided in Attachment G. Your company may already have forms prepared for this purpose. If not, and if the attached form is not used, obtain copies of an approved American Institute of Architects (AIA) change order form for use on the project. For many projects, this may be the only approved change order form.

The original change order form should be maintained by the contractor, with copies provided to the customer and to any other interested parties. Be aware of the consequences of change orders prior to implementing them, especially their impact on the project schedule.

Strategy

All of the items contained in the preceding paragraphs are an integral part of the success of the project. Ensure that everything is taken into consideration and roadblocks eliminated prior to occurring, if possible. When unforeseen circumstances occur, plan how to overcome them and implement a backup strategy in case your initial plan meets with obstacles. Proper project planning is essential for the successful completion of a project.

Planning

Attachment A Designer's Drawings

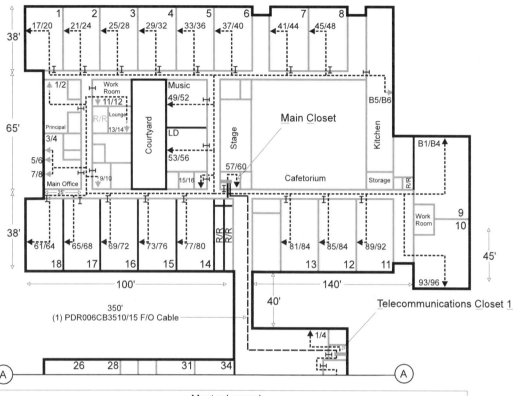

Attachment B   Materials List

Project Number: XXXXXXXXXXXXX
Project Name: Anywhere Elementary School
Address: Anywhere, USA

Item	Material description	Manufacturer	Catalog Number	Quantity Each	Unit Price	Total Price
1	Wire, 4 pair, 24 AWG, UTP, Cat 5, CMR	Mfg. A	530141-TP	54000	0	$0.00
2	Surface raceway, small	Mfg. B	MT1ABC5	350	0	$0.00
3	Surface mount box, single gang	Mfg. C	WT12DB	56	0	$0.00
4	Cat 5 relay rack E/W wire management	Mfg. D	55053-703	2	0	$0.00
5	Vertical wire management hardware	Mfg. E	11374-703	2	0	$0.00
6	Patch panel, 96 port, T568A	Mfg. F	49485-C96	4	0	$0.00
7	Patch panel, 24 port, T568A	Mfg. G	49485-C24	2	0	$0.00
8	Faceplate, single gang, dual port	Mfg. H	41080-2AP	9	0	$0.00
9	Faceplate, single gang, quad port	Mfg. I	41080-4AP	47	0	$0.00
10	Modular jack, Cat 5, orange, T568A	Mfg. J	41108-RO5	208	0	$0.00
11	Horizontal wire management panel	Mfg. K	49253-BCM	16	0	$0.00
12	Rear cable bars	Mfg. L	49258-TWB	36	0	$0.00
13	Firestop compound	Mfg. M	AA529	10	0	$0.00
14	Electrical Metallic Tubing, ¾″	Mfg. N	34EMT	150′	0	$0.00
15	Electrical Metallic Tubing, 2″	Mfg. O	2EMT	160′	0	$0.00
16	Cable, fiber optic, 6 str., OFNR	Mfg. P	PDRCB3510/15	300′	0	$0.00
17	Fiber distribution panel	Mfg. Q	4R130-OTA	2	0	$0.00
18	Connector panels	Mfg. R	4F100-6TM	2	0	$0.00
19	ST connectors	Mfg. S	95-100-01R	12	0	$0.00
20	10BASE-T 24-port hubs	Mfg. T	AT3624TR-15	1	0	$0.00
21	10BASE-T 24-port hubs	Mfg. U	AT3624TRS-15	3	0	$0.00
22	Chassis	Mfg. V	AT-36C3	2	0	$0.00
23	Fiber optic transceivers	Mfg. W	AT-MX26F-05	2	0	$0.00
24	Duplex, STST, 3 m F/O jumpers	Mfg. X	STST3M	2	0	$0.00
25	Data patch cords	Mfg. Y	42454-03O	104	0	$0.00
26	Data patch cords	Mfg. Z	42454-050	104	0	$0.00
27	Data line cords	Mfg. AA	42454-10O	208	0	$0.00
28						
29						
30						
31	Exempt materials	various	various	1 lot	NA	$0.00
32	Transportation					$0.00
Total Materials						$0.00

Attachment C Labor List

Project Number: XXXXXXXXXXXXX
Project Name: Anywhere Elementary School
Address: Anywhere, USA

Item	Material description	Units/Quantity	Unit rate	Total price
1	Installing horizontal wires (2 per run)	9	0	$0.00
2	Installing horizontal wires (4 per run)	47	0	$0.00
3	Installing faceplates and jacks	208	0	$0.00
4	Installing relay racks	2	0	$0.00
5	Installing patch panels	6	0	$0.00
6	Terminating wires at patch panels	208	0	$0.00
7	Certifying Cat 5 wires	208	0	$0.00
8	Installing surface raceway	56	0	$0.00
9	Installing surface mount boxes	56	0	$0.00
10	Installing backbone fiber optic cables	300	0	$0.00
11	Installing F/O connecting hardware	4	0	$0.00
12	Terminating F/O cables	12	0	$0.00
13	Testing F/O cables	6	0	$0.00
14	Install fire-/smoke-rated partition pentrations	59	0	$0.00
15	Mount hubs on relay racks	4	0	$0.00
16	Installing backbone conduit from main bldg to portables	120	0	$0.00
17	Installing horizontal conduit between backbone conduit and portables	48	0	$0.00
18				
19				
20				

Total labor				$0.00
Total cost				$0.00
Materials markup		50%		$0.00
Labor markup		50%		$0.00
State sales tax on materials		6%		$0.00
Total price to customer				$0.00

180   Chapter Two

Attachment D   Project Time Line

Name	Month Starting July 1, 1995				Month Starting August 1, 1995			
	7/3/95	7/10/95	7/17/95	7/24/95	7/31/95	8/7/95	8/14/95	8/21/95	8/28/95
Contract Award									
Site Survey									
Installation Team Meeting									
Initial Construction Meeting									
Project Schedule Compiled									
Materials Ordered									
Materials Shipped									
Materials Received									
Materials Stored/Staged									
Install Project Infrastructure									
Install Backbone Cables									
Install Horizontal Wires/Cables									
Installation Progress Meeting									
Install Backbone Termination Hardware									
Install Horizontal Wire/Cable Termination Hardware									
Installation Progress Meeting									
Terminate Backbone Cables									
Terminate Horizontal Wires/Cables									
Installation Progress Meeting									

Attachment E   Project Schedule

Name	Start Constraint	Finish Constraint	Actual Start	Actual Finish	Percent Done
Contract awarded	* 7/12/95	* 7/12/95	* 7/12/95	* 7/12/95	100%
Site survey	* 7/14/95	* 7/17/95	* 7/15/95	* 7/17/95	100%
Installation team meeting	* 7/17/95	* 7/17/95	* 7/17/95	* 7/17/95	100%
Initial construction meeting	* 7/20/95	* 7/20/95	7/20/95	7/20/95	100%
Project schedule compiled	* 7/17/95	* 7/17/95	7/17/95	7/17/95	100%
Materials ordered	* 7/17/95	* 7/17/95	7/17/95	7/17/95	100%
Materials shipped	* 7/18/95	* 8/4/95	* 7/18/95	* 8/4/95	100%
Materials received	* 7/20/95	* 8/19/95	* 7/20/95	* 9/25/96	100%
Materials stored/staged	* 7/20/95	* 10/20/95	7/20/95	10/20/95	100%
Install project infrastructure	* 7/21/95	* 8/10/95	* 7/21/95	* 8/10/95	100%
Install backbone cables	* 8/11/95	* 9/8/96	8/11/95	9/8/96	100%
Install horizontal wires/cables	* 9/11/95	* 12/22/96	9/11/95	12/22/96	100%
Installation progress meeting	* 9/11/95	* 9/11/95	9/11/95	9/11/95	100%
Install backbone connecting hardware	* 1/2/96	* 1/18/96	1/2/96	1/18/96	100%
Install horizontal wire/cable connecting hardware	* 1/19/96	* 2/26/96	1/19/96	2/26/96	100%
Installation progress meeting	* 2/26/96	* 2/26/96	2/26/96	2/26/96	100%
Terminate backbone cables	* 1/19/96	* 2/9/96	1/19/96	2/9/96	100%
Terminate horizontal wires/cables	* 2/27/96	* 3/29/96	2/27/96	3/29/96	100%
Installation progress meeting	* 3/29/96	* 3/29/96	3/29/96	3/29/96	100%
Label all facilities as per ANSI/TIA/EIA-606	* 3/22/96	* 4/5/96	3/22/96	4/5/96	100%
Test backbone cables	* 4/5/96	* 4/12/96	4/5/96	4/12/96	100%
Test horizontal wires/cables	* 4/15/96	* 5/3/96	4/15/96	5/3/96	100%
Compile all test results	* 5/6/96	* 5/8/96	5/6/96	5/8/96	100%
Installation progress meeting	* 5/3/96	* 5/3/96	5/3/96	5/3/96	100%
Punch list	* 5/9/96	* 5/10/96	5/9/96	5/10/96	100%
Correct all items on punch list	* 5/16/96	* 5/17/96	* 5/16/96	* 5/17/96	100%
Final punch list	* 5/16/96	* 5/16/96	5/16/96	5/16/96	100%
Customer punch list	* 5/17/96	* 5/17/96	5/17/96	5/17/96	100%
Customer acceptance	* 5/20/96	* 5/20/96	5/20/96	5/20/96	100%
Prepare "as-built" package	* 5/20/96	* 5/24/96	5/20/96	5/24/96	100%
Provide all project documents to customer	* 5/27/96	* 5/27/96	5/27/96	5/27/96	100%
Return surplus materials to distributor for storage	* 5/20/96	* 5/22/96	5/20/96	5/22/96	100%
Complete billing to customer	* 5/28/96	* 5/28/96	5/28/96	5/28/96	100%
Review billing with customer	* 5/29/96	* 5/29/96	5/29/96	5/29/96	100%
Receive final payment	* 5/30/96	* 5/30/96	5/30/96	5/30/96	100%
Close project	* 5/31/96	* 5/31/96	5/31/96	5/31/96	100%
Clear punch list	* 5/10/96	* 5/13/96	5/10/96	* 5/13/96	100%
Clear customer punch list	* 5/17/96		5/17/96	* 5/20/96	100%

* Indicates completion

Attachment F Site Survey Checklist

Checklist for Site Survey—New Construction

Item	Description of Operation	Date Checked	Y	N
1	Is the general contractor responsible for construction and finish of the closets and pathways?			
2	Is there an electrical contractor on the project and are they responsible for the pathways?			
3	Is the electrical contractor responsible for the telecommunications grounding and bonding within the building?			
4	Are the closets completed and ready for use by the telecommunications vendor? If not, when will they be ready?			
5	Are the pathways completed and ready for use by the telecommunications vendor? If not, when will they be ready?			
6	Is the grounding and bonding system installed and ready for use by the telecommunications vendor? If not, when will they be ready?			
7	Is there space on the job site available for storage and staging of materials and tools? Will it be secured?			
8	Will the space be under the control of the telecommunications vendor? If not, you must work out responsibility for loss or damage.			
9	Is the space inside or outside?			
10	Does the general contractor conduct construction progress meetings? When? Can the telecommunications vendor attend?			
11	Does the general contractor have a posted safety plan?			
12	What is the schedule for inspections by local code authorities?			
13	Is the building equipped with suspended ceilings? Are they used to handle environmental air?			
14	Where can the telecommunications vendor set up field operations?			
15	Is there a way to get large cable reels and other heavy materials to the top floor of the building?			
16	Are lifts required on the job site?			
17	Is the building a hardhat area?			
18	When will the telecommunications vendor be allowed access to all spaces in the building requiring telecommuncations work?			
19	When will the walls receive final finishes?			
20	Are elevators and loading docks available and appropriate for the job?			
21	Is rigging required?			
22	Does any equipment require floor loading modifications?			
23	Verify position of sprinklers and water pipes?			
24	Are any alarm systems activated?			

Attachment G Job Change Orders

Company Name
Company Address

Job Change Order
Project Name
Project Location
Project Number

Change Order No._____

Initiated by:_____ Date:_____

Details of request or nature of change:

Labor $_____   Materials $_____   TOTAL $_____

_____   _____
Verified by (Signature)   Date

The undersigned hereby accepts the prices quoted on this Job Change Order and agrees to pay_____ the amount stipulated at the TOTAL above upon satisfactory delivery of the goods and services described above.
Acceptance also authorizes the materials to be ordered and/or the work to be performed.

_____
Authorized signature

_____
Title

_____
Date

Date work complete:_____   Verified by:_____

Attachment H   Scope of Work (Example)

Anywhere School District One
Anywhere Elementary School

General

This project provides for the installation of a Category 5 structured cabling system network at Anywhere Elementary School. The primary medium is 4-pair Category 5 CMP horizontal cable and a 6-strand optical fiber backbone cable between closets. A telecommunications closet will be located in the storage area adjacent to the stage. An equipment room will be located in the office of the media center. Each of these spaces will serve their respective areas as indicated by the zone division line.

Copper Horizontal Cable

The cable to be used on this project is manufactured by Manufacturer C and is 4-pair, 24 AWG, unshielded twisted-pair, CMP-rated. It is available in 305 m (1000 ft) boxes. This cable meets or exceeds the ANSI/TIA/EIA-568-A requirements.

Patch Panels

Category 5 patch panels wired to T568A, will be provided at each closet location. These patch panels will be mounted on a 483 mm (19 in) relay rack along with the hubs and wire management panels. The patch panels will be provided in two sizes: 96-port and 24-port. At both closets, two 96-port and two 24-port patch panels will be required.

Relay Racks

Two 483 mm (19 in) × 2.1 m (84 in) relay racks will be installed as part of this project. One of the racks will be installed at the telecommunications closet and the second will be installed at the media center. Manufacturer A manufactures both racks, catalog number 55053-703. They are black in color. Both racks will be equipped with vertical and horizontal wire management systems to ensure orderly installation of patch cords between various components on the relay rack. These racks provide enough rack space for all of the equipment to be mounted on them at their respective locations. They will be bolted to the floor, and a ladder rack will be provided for additional support and routing of cables. The equipment will then be mounted on the racks. Each rack will be grounded to its respective telecommunications main grounding busbar or telecommunications grounding busbar with 6 AWG, stranded, green-insulated wire.

Faceplates and Modular Jacks

Faceplates will be provided in two different port sizes. The dual-jack installation will be a 2-port faceplate while the classrooms will have a 4-port faceplate, each equipped with four modular jacks. The faceplates are manufactured by Manufacturer A, catalog number XXXX-XXX. They will be office white in color.

Modular jacks will be provided at each location shown on the attached illustration. The modular jack will be Category 5, T568A pinout, and will be one of four colors (i.e., orange, black, red, or yellow).

Raceways and Associated Equipment

Where wire or cable cannot be concealed within the walls at a device location, surface raceway will be installed. This raceway will be 19 mm (0.75 in) EMT conduit. The conduit will originate above the suspended ceiling and continue down to a surface-mounted box. The surface-mounted box will house the faceplate and modular jack(s) as well as provide mechanical protection for the terminations.

Between the main building and the portable classroom buildings, a two-inch EMT conduit will be installed. Junction boxes will connect the 50 mm (2 in) conduit to the 19 mm (0.75 in) conduits that will be installed into each portable classroom building. These conduits will provide mechanical protection as well as support for installation of the wires to the portable classroom buildings.

Optical Fiber Backbone Cable

A 6-strand optical fiber cable will be installed from the equipment room to the telecommunications closet. Manufacturer C manufactures the cable, which is UL listed as OFNR to comply with all applicable building codes. The cable will be installed in the suspended ceiling area between the telecommunications closet at the stage area and the media center.

Optical Fiber Connecting Hardware

Rack-mounted optical fiber distribution centers manufactured by Manufacturer A will be installed at each closet location. They are catalog number XXXX-XXXX and are equipped with catalog number YYYYY-YYY connector panels.

The fiber distribution centers will be equipped with sufficient connecting hardware to accommodate all six strands of glass that originate or terminate at their respective locations. SC-type optical fiber connectors manufactured by Manufacturer B Corporation, catalog number ZZ-ZZZ-ZZZ, will be used to terminate the strands of glass.

Data Patch Cords

A Category 5 data patch cord will be provided for equipment interconnection. Data patch cords will be installed in the length required to provide proper wire management between components installed on the relay racks. They are orange in color.

A Category 5 data line cord will be provided for each Category 5 horizontal wire installed. These data line cords will be provided to Anywhere Elementary School for installation when the actual computer installation occurs. They are 3 m (10 ft) in length and orange in color.

10BASE-T Hubs

We are providing pricing for Manufacturer D Ethernet 10BASE-T hubs to allow for the networking of each Category 5 horizontal cable installed. As per your direction, we are sizing the hubs to facilitate a maximum of two devices in each classroom and in the administrative area. The hubs will be equipped to allow connection to the optical fiber backbone as well as to each other. Each hub will be mounted in its respective relay rack. Patching from the hubs to the patch panels will be accomplished using the data patch cords and optical fiber patch cords provided.

Optical Fiber Patch Cords

Duplex, SC/SC, 3 m (10 ft) optical fiber jumpers will be provided to allow connection of all hubs.

Firestopping

In each classroom and firewall in the hallways, a horizontal penetration must be made to facilitate installation of the horizontal wires and optical fiber cables. Each penetration will be made and then restored using approved materials and methods. Manufacturer E Model No. AAAAA firestop compound will be used to restore penetrations. Where sleeves are required by code, they will be installed.

Installation Methodology

All work will be completed so as to conform to the:

- Manufacturer's specifications.
- 1999 edition of the *National Electrical Code®*.
- 1999 edition of the *National Electrical Safety Code®*.
- ANSI/TIA/EIA-568-A.
- ANSI/TIA/EIA-569-A.
- ANSI/TIA/EIA-606.
- ANSI/TIA/EIA-607.
- TIA/EIA TSB-67.
- Latest edition of the BICSI *Telecommunications Distribution Methods Manual*.
- Latest edition of the BICSI *LAN Design Manual*.
- Latest edition of the BICSI *Telecommunications Cabling Installation Manual*.
- All local codes and ordinances.

> **Note.** Where a conflict exists, local codes and ordinances will supersede all other requirements.

Testing

Each horizontal cable will be tested for compliance with TIA/EIA TSB-67 requirements using a Manufacturer E tester. The tests will be conducted on the basic link configuration. A hard copy and 3.5-inch floppy disk containing the test results will be provided to the District.

Each strand of the optical fiber cable will be tested using a Manufacturer B power meter and light source. The results will be recorded and provided to the District in hard copy and, if possible, on a 3.5-inch floppy disk on completion of the project.

"As-Built" Package

Upon completion of the project, Datawireinstaller, Inc. will provide an "as-built" package to the District that will contain the following items:

- Updated floor plans
- Wire/cable routing schematic
- Facility assignment records
- Horizontal cable test results
- Optical fiber cable test results
- Photographs/videotape

Warranty

The project will be warranted for a period of fifteen (15) years from completion of installation.

Schedule

We anticipate this project will take approximately two weeks to complete. This is contingent upon no unforeseen delays during installation. We plan to perform the work with minimum impact on the staff by working the hours most advantageous to both the school and Datawireinstaller, Inc. This is contingent upon the school providing access to the building at these times. If weekend work is required, we are prepared to perform the necessary work during these times. We would, however, prefer to work during normal working hours.

Most of the materials required for the project are available from our distributors as stock and the other items have short intervals.

Delays to Project

Delays caused by Anywhere School District One, their employees, students, agents, assignees, contractors, subcontractors, or any other person(s) not directly employed by Datawireinstaller, Inc. will result in proportional delays in the completion of the project. All costs incurred by Datawireinstaller, Inc. as a result of said delay, will be passed along to Anywhere School District One on a cost-plus basis.

Damage to Installed Physical Plant

Damage caused to physical plant installed by Datawireinstaller, Inc., by Anywhere School District One, their employees, students, agents, assignees, contractors, subcontractors, or any other person(s) not directly employed by Datawireinstaller, Inc. will be repaired or replaced at the District's option. All cost associated with said repairs or replacement work will be passed along to Anywhere School District One on a cost-plus basis. Any delays caused by said work will result in a proportional delay in the completion of the project.

The information contained within this proposal is considered business proprietary and is not to be duplicated by any manner, by hand, electrically, mechanically, or any other means, without the written permission of Datawireinstaller, Inc. Information Systems Sales and Services. It is not to be disseminated outside of Anywhere School District One without the written permission of Datawireinstaller, Inc. Information Systems Sales and Services.

Chapter 3

Installing Supporting Structures

Setting Up Telecommunications Closets	190
Installing Backbone Pathways	209
Installing Horizontal Pathways	228
Installing Grounding Infrastructure	231
Installing Cable Support Systems	242

Overview

Supporting structures are necessary to allow installation of wire, cable, connecting hardware, and associated apparatus. They are comprised of relay racks, cabinets, D-rings, bridle rings, J-hooks, plywood backboard, cable trays, conduits, slots, sleeves, and their associated hardware. Even though the appropriate cable pathways may be available, one must have facilities to terminate them.

Particular emphasis must be paid to the requirements of American National Standards Institute (ANSI), Telecommunications Industry Association (TIA) and Electronic Industries Alliance (EIA) Standards 568-A, 569-A, 606, and 607 when planning and installing this infrastructure. Proper design for the components described in these standards leads to a successful initial installation and future additions or rearrangements. Failure to adhere to such requirements may cost hours or even days of delay or additional work to complete the job. Responsible and successful cabling installers will obtain as much information as possible when planning for an installation and will implement the latest methods to ensure the migration of that installation into a future project.

Most of the hardware associated with this chapter vary in size, color, shape, manufacturer, etc. The cabling installer will become familiar with the various manufacturers' products and how they adapt to the installation requirements. Unfortunately, we cannot include every item in this chapter. The products discussed are representative of those available in the marketplace.

Setting Up Telecommunications Closets

Overview

There are two types of telecommunications spaces in which the termination of cabling permits cross-connection and interconnection.

- Equipment rooms
- Telecommunications closets

These two spaces share some of the same basic purposes. They both support the installation of cables, connecting hardware, cross-connects, and electronic equipment. For instance, while a telecommunications closet is generally floor serving, an equipment room is generally building serving. The primary difference is in the nature of what they serve.

Equipment rooms are designed to house large equipment items such as telephone cabinets, data processing mainframe computers, Uninterruptible Power Supplies (UPS), or video head-end equipment. The floor loading of equipment rooms must be rated higher than that for telecommunications closets because of the anticipated high concentration of equipment in a confined space. The heating, ventilation, and air conditioning (HVAC) requirements for these spaces are also greater.

Telecommunications closets are designed for only limited equipment and floor loading. They may house splice cases, termination hardware, and relay racks. Small items of equipment such as hubs, multiplexers, and key telephone systems can also be found in telecommunications closets.

Every building is served by at least one telecommunications closet or equipment room, and each building should be provisioned with a minimum of one telecommunications closet on each floor.

The types of cabling facilities which may be serviced by telecommunications closets are:

- Horizontal cables and their connecting hardware
- Backbone cables and their connecting hardware
- Building entrance cables and their connecting hardware
- Telecommunications equipment
- Data processing equipment

- Public network equipment
- Video head-end equipment
- Paging systems
- Intelligent building systems

In smaller buildings, these may be housed in a single equipment room.

Cross-connects between horizontal and backbone systems will be found in telecommunications closets. Work area outlets must be cabled to the connecting hardware in telecommunications closets, thereby providing connection between them and the backbone system. Horizontal cabling should be terminated in a telecommunications closet on the same floor as the area being served. Cabling between telecommunications closets is considered to be a backbone cable.

Additional space may be available in a building that is currently being employed as a makeshift telecommunications closet. The proper equipment must be installed and provisions made for the cable and connecting hardware to upgrade the space to a true telecommunications closet. If the existing closet is a closet shared with electrical, plumbing, or HVAC facilities, then this choice for a telecommunications closet must be reconsidered. Never house telecommunications facilities in a space which handles other building utilities.

It is recommended that telecommunications closets have 0.75-in-thick plywood backboards installed on at least two walls of the closet. The backboard should be painted with two coats of nonconductive, fire-retardant paint of a light color. The plywood provides a space for wall-mounting connecting hardware. A 300 mm (12 in) wide cable tray or ladder rack should be mounted on the same wall(s) as the backboard. Good planning dictates that all walls of a telecommunications closet be equipped with plywood and cable tray/ladder rack to facilitate future growth.

The telecommunications designer's documents will indicate the size, location, quantity, and nomenclature of the equipment to be installed in the closet, along with a routing diagram of the cables to be installed or those that pass through the closet. They will also show the location of the pathways entering or leaving the space and who is responsible for installing them. If equipment racks are to be installed, a plan view of the space will indicate where their respective footprints are located and how they relate to other equipment being installed. In most instances, the telecommunications designer's documents will indicate where voice, data, and video cables are to be terminated. The telecommunications designer may be an architect, consulting engineer, or a Registered Communications Distribution Designer (RCDD®).

Always ensure that appropriate clearances are maintained around all pieces of equipment. In the absence of a specified distance, plan for a minimum of 1 m (3.3 ft) work and aisle space. If a telecommunications closet layout is not provided, prepare one. Details for laying out a telecommunications closet may be found in the most current edition of the BICSI *Telecommunications Distribution Methods Manual*.

Plywood backboards

Plywood backboards are employed on walls in telecommunications closets. Plywood is available in two types—interior and exterior—and in four grades—A, B, C, and D.

Sheets of plywood are normally sized 4-ft wide by 8-ft high. The thickness is variable, but, for the purposes of this document only two will be considered: 0.75 in and 1 in. Plywood that is too thin allows the screws used by cabling installers to penetrate completely through the plywood and sometimes does not offer enough strength to ensure that mounted hardware is securely anchored. Plywood sheets thicker than one inch are not usually required. This, of course, is contingent on the sheet of plywood being properly attached to the building structure.

The finishing grade of plywood (A, B, C, or D) describes the quality of the surface, i.e., degree of knotholes or blemishes. Grade A is the highest grade and is without any surface blemishes. Grade B has the knotholes cut out and replaced with a patch of clean wood. Grade C contains some blemishes and an occasional small knothole. Grade D contains knotholes without any repair or corrective action by the manufacturer. Grading of a sheet of plywood may result in a different grade for each of the two sides. For instance, a sheet of plywood could be graded A/B—one side is A and the reverse side is B.

Plywood should also be void free. This means that the space in each layer inside the plywood where the knotholes are removed is completely filled with replacement wood patches. Voids inside the sheet of plywood may create a weak spot to which the attachment hardware (i.e., screws, toggle bolts, etc.) cannot hold fast.

For telecommunications use, grade A/C should be used. The A side is exposed to the interior of the closet and the C side placed against the building structure or cabinet wall.

If the plywood is to be painted, do not use treated (fire-retardant) plywood. This type of plywood is usually pressure treated with a saline solution. The effects of the treatment will cause the paint to crack, deteriorate, and peal off the plywood backboard. In addition, the saline solution can cause the metal hardware mounted to the plywood to corrode. Untreated plywood, when painted, should be painted with two coats of fire-retardant paint in the color specified by the designer, as stated earlier.

Installing plywood

Plywood sheets used for backboards should be installed with the longest dimension reaching from the floor level up toward the ceiling to its eight-foot height. In the case of a nine- or ten-foot ceiling, do not succumb to the temptation to raise the bottom of the plywood to split the difference. This will raise the working height level such that a ladder may be required to work on equipment mounted at the top of the plywood.

Plywood should be installed in such a manner that there is no separation between adjacent sheets. When installing plywood in a corner, the plywood backboard can be installed plumb and adjacent to the edge of one side of the wall at the corner with the sheet on the intersecting wall butted up against the first sheet to form a smooth, ninety-degree corner.

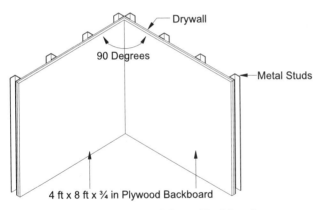

Figure 3.1 Corner installation of plywood backboards.

The plywood backboard must be secured on top of the existing drywall or to the studs in the perimeter walls of the room. When installing the plywood on bare studs (no drywall), drywall screws a minimum of one-half inch longer than the depth of the plywood backboard must be used.

When installing plywood on drywall that has already been installed on the studs, always verify the load rating of the wall prior to installing the plywood. If the load rating will permit this type of installation, use toggle bolts (butterfly bolts) to ensure the stability of the installation. These toggle bolts should be a minimum of one-quarter inch in diameter and must be sufficient in length to allow the bolt to seat behind the drywall after installation.

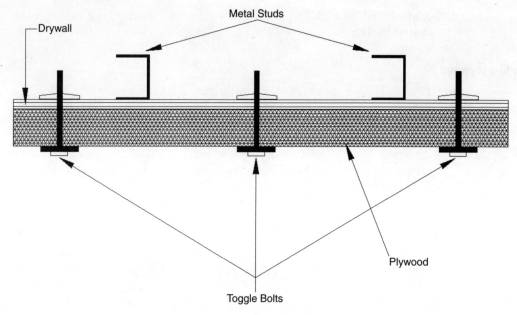

Figure 3.2 Installation using toggle bolts in drywall construction.

For this drywall application, toggle bolts should be installed at approximately 609 mm (24 in) spacing around the entire perimeter of the plywood board. If desired, recess the bolt heads to allow for use of the entire area on the plywood. Never recess the bolt head on any plywood less than one inch in thickness. This will affect the ability of the

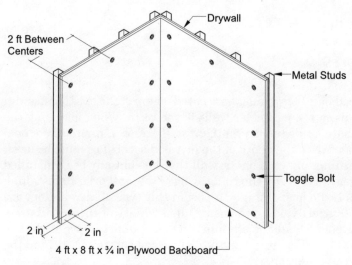

Figure 3.3 Plywood installed using toggle bolts.

plywood to hold the desired load of equipment and termination hardware. The toggle bolts should be installed 50 mm (2 in) from the edges of the sheet of plywood on 609 mm (24 in) centers. Locate the toggle bolts using care to avoid the studs when drilling. The studs will prevent the toggle bolt's wings from opening behind the sheet rock.

If possible, install plywood backboards around the entire perimeter of the closet. This will enhance the use of the wall space in the closet and allow cables to be installed around the walls to where terminal equipment will be located, now or in the future. It will also facilitate attaching cables which pass through vertically to closets above or below.

Cable trays

Cable trays are available in many sizes and configurations. Some are solid-bar stock; others are tubular; while some are manufactured from rod stock. Also, there are enclosed and open cable trays. Cable trays are manufactured from steel and aluminum. Some cable trays are called ladder racks since they resemble a ladder when viewed from above or below. Cable trays provide a pathway to house and support cables installed between telecommunications closets and from telecommunications closets to work areas. They are also used to support cables from one wall to another within a telecommunications closet or equipment room.

This chapter will specifically address the installation of a cable tray within a telecommunications closet. However, the principles involved can be applied to installing cable trays between telecommunications closets as horizontal pathways. The following are examples of open tubular and rod cable trays. For both types, a metal bottom can be installed along with a metal cover to convert it to an enclosed system.

Tubular construction

In this type of construction, the cable tray is constructed of tubular stock. That is, the inside of each side rail and cross rung of the tray is hollow. This type is lighter than the solid-bar stock construction and easier to handle in elevated work areas. Solid stock may offer a stronger installation, but in most cases, the tubular stock will provide the weight-loading factor required for most installations. The tubular cable tray is manufactured with a depth of 38 mm (1.5 in); the width varies from 152 mm (6 in) to a maximum of 0.91 m (3 ft).

A tubular cable tray can be wall mounted or supported by the building structure from above using threaded rods and appropriate attachments. The threaded rods must be installed using properly sized anchors and correct hardware. Threaded rods are available in various lengths and in a variety of diameters. Select the proper diameter rod required to support the maximum load for which the cable tray is designed.

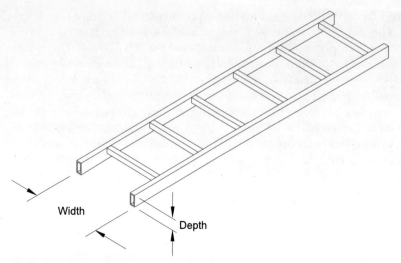

Figure 3.4 Tubular cable tray.

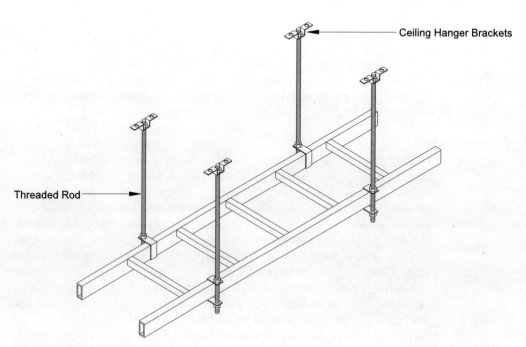

Figure 3.5 Suspended cable tray.

A tubular cable tray can, in addition, be supported by wall brackets. Wall brackets are first installed on a wall along the route of the cable tray. The number of brackets and specific spacing interval is dependent upon the load the cable tray must support. ANSI/TIA/EIA-569-A requires that a supporting attachment be made on a cable tray not more than 609 mm (24 in) from the end, or at a joint between two sections. Additional supports are also required every 1.5 m (5 ft) thereafter. Prior to installing the wall brackets, install properly sized anchors. It is important to select the proper-sized anchors or they will not be able to support the load of the cable tray and the installed cables.

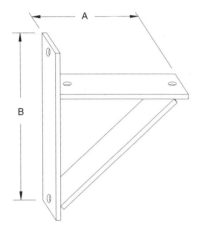

Figure 3.6 Wall bracket (tubular).

Tubular cable trays can be installed on one level or on several levels. Corners and changes in horizontal plane are accomplished by using sections of cable tray cut from standard stock and connected together through the use of proper hardware. Connecting hardware permits all angles, both vertical and horizontal, to be accommodated over the entire route of the cable tray.

Cable retaining posts are available in 152 to 300 mm (6 to 12 in) lengths to allow additional cables to be installed to a depth exceeding that of the cable tray. Without these devices, the cable would not be confined by the edges of the cable tray and may fall from the tray.

Rod stock

Rod-stock cable trays are manufactured by using either ¼-in or ⅜-in diameter steel rods welded together to form a section of cable tray in the form of a wire mesh. It is normally manufactured in two depths: one and one-half inches and two inches. However, it is available in

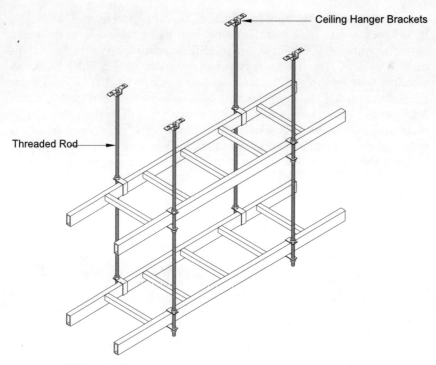

Figure 3.7 Multilevel cable tray.

depths of 100 mm (4 in) and 152 mm (6 in) by special order. It may be ordered in painted, electroplated zinc galvanizing, hot dip galvanizing, and stainless steel sections. They are available in standard lengths of 3 m (10 ft). Its light weight is especially helpful when installing it in confined areas.

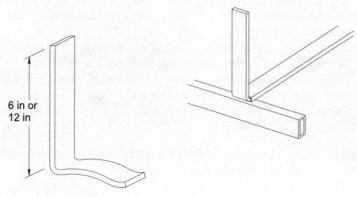

Figure 3.8 Cable retaining posts.

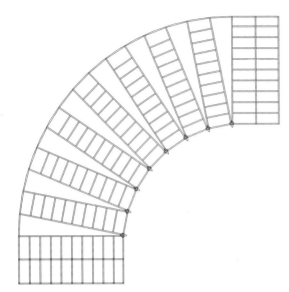

Figure 3.9 Rod-stock cable tray.

Installation is accomplished by using light hand tools, power drills, and an offset blade bolt cutter. Unlike standard cable trays, this system can be formed to adjust to changes in elevation and horizontal direction without the need for specialized adapters. The list of accessories required to assemble this type is not as numerous as for a standard cable tray.

The rod-stock cable tray can be supported using wall brackets or all-thread rods. Wall brackets are first installed on a wall along the route of the cable tray. They are installed at intervals, depending on the load the cable tray must support. ANSI/TIA/EIA-569-A requires that a supporting attachment be made on a cable tray not more than 609 mm (24 in) from the end, or a joint between two sections. Additional supports are also required every 1.5 m (5 ft) thereafter. Prior to installing the wall brackets, you must first install properly sized anchors. It is important to select the proper-sized anchors or they will not support the load of the cable tray and the installed cables.

Rod-stock cable trays can be installed on one level or several different levels. Corners and changes in horizontal level are accomplished by selective cutting of the rods in the cable tray at specific points using bolt cutters and connecting together the points cut by means of the appropriate hardware. By using connecting hardware, all angles (both vertical and horizontal) can be accommodated over the entire route of the cable tray without having to sever the entire cable tray and use specialized hardware.

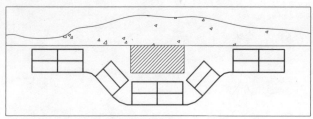

Easily avoid obstacles in the vertical plane

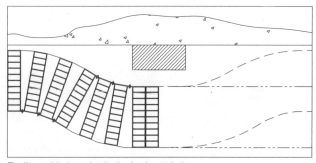

Easily avoid obstacles in the horizontal plane

Figure 3.10 Directional transition.

Installation

In the example that follows, a cable tray is installed from one wall to another in a telecommunications closet. The closet is 6 m (20 ft) wide by 3 m (10 ft) deep with a door in the corner of one wall.

As shown in the following figure, the cable tray is installed from one wall to the opposite wall using wall angles to secure the cable tray at

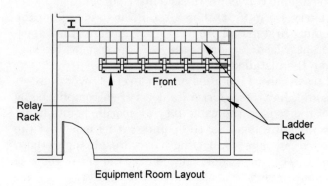

Figure 3.11 Plan view of a typical telecommunications closet with cable tray installed on two walls.

each end and by wall brackets to support the cable tray in the middle. By using both of these attachments, sufficient support for the cable tray and the installed cables is provided. The cabling installer may, instead, choose to support the cable tray from the ceiling using channel stock (see Figure 3.16) and threaded rods.

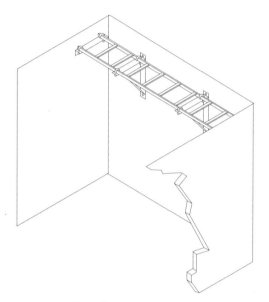

Figure 3.12 Elevation view of cable tray installed on the rear wall of a telecommunications closet.

It is suggested that a cable tray be installed on every wall in the telecommunications closet. This will ensure that a proper pathway exists for cables that route from a point on the backboard of one wall to a location on a different wall. D-rings (see D-ring Installation) should not be used as a substitute for a cable tray.

Additional information is found in the attachments to this chapter. They provide a graphic view of the hardware required to accomplish this and other installations. Refer to them for guidance in different situations. These references are not all-encompassing; however, the manufacturer of the tray selected for installation should be consulted for specifics in planning the installation. This will ensure that all parts are accounted for and that the proper method of supporting the cable tray is implemented. Many manufacturers will provide a computer-aided design (CAD) drawing and a list of materials required.

The telecommunications cabling designer's documents may already include the details referenced above; if so, they must be used for the installation.

D-ring installation

D-rings are used to support small bundles of cables as they route from one termination point on the plywood backboard to another. However, they should not be substituted for a cable tray.

D-rings are available in many sizes, shapes, materials, and colors. Traditionally, D-rings are manufactured in metal (aluminum) and are formed in the shape of the letter D. Plastic versions of D-rings are available but are not as sturdy as metal. A half-D-ring is available in both metal and plastic. The half-D-ring is used to provide support and management primarily where cross-connect jumpers turn at a 90-degree angle. The following figure gives an example of standard metal D-rings.

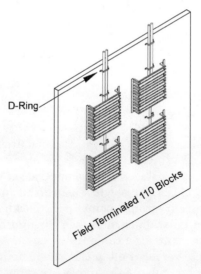

Figure 3.13 Typical backboard layout for D-ring installation.

An alternate to the half-D-ring is the mushroom. Mushrooms are constructed of plastic and contain a center-mounted screw for attachment to plywood backboards. They are also available with threaded bolts for installation on equipment racks.

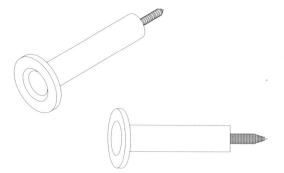

Figure 3.14 Mushroom.

Conduits

Conduits are pathways that house backbone cables and horizontal cables. Conduits that enter a closet should be terminated at specific locations on the wall to enable orderly routing of the cables to termination equipment in the closet or routed through the closet to another location. Ream conduit ends and install bushings. It is especially important to use sweep elbows when dealing with optical fiber cables or large backbone cables. These elbows restrict the cable bend radius. Severe bending can result in cable damage. It is especially important to adhere to a minimum cable bend radius of not less than 10 times the outer diameter of the sheath.

Conduit ends should be positioned adjacent to a corner of the backboard (in the case of a single piece of plywood) or in the corner of the room (if multiple sheets of plywood are installed around the perimeter walls of the room). If conduits cannot be located in these positions, regardless of the reason, cable trays (ladder racks) should be used to route the cables from one location in the room to another. This avoids encroaching on wall termination space.

Where several large backbone cables are passing vertically through the room, a vertical cable tray should be appropriately positioned to support the cables from ceiling to floor level. The cables should be tie-wrapped to the cable tray in an orderly fashion to ensure that they are properly supported and the entire weight of the cable is equally distributed over several cable supports.

Conduits should be physically attached to the top of the plywood backboard using channel stock and conduit brackets when entering from overhead. When entering from below grade, the floor slab or building structure is usually enough to provide for a fixed installation without the use of channel stock. The telecommunications designer's

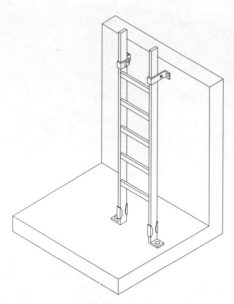

Figure 3.15 Vertical cable tray.

drawings should indicate the location of each conduit and where they terminate in the closet.

The conduit brackets may be of the type that provides electrical continuity between the conduit and the channel stock, to provide a bonding path between conduits and a bonding conductor to be installed from the channel stock. This should then be bonded to either the telecommunications main grounding busbar (TMGB) or to a telecommunications grounding busbar (TGB). Where a small number of conduits are installed, a grounding bushing should be installed at the end of each conduit for the attachment of a 6 AWG copper wire which connects to the TMGB or TGB. Conduits in close proximity to each other can be bonded together. One conduit in the string must be bonded to the TMGB or TGB.

Ground wires

Ground wires and a ground bus should be installed in such a manner that they are not obstructed by cable trays, cables, or terminating equipment. Ground wires and bonding wires should always be installed in the straightest and shortest route between the origination and termination point. A sharp bend may interfere with the effectiveness of the grounding system, since it will modify the characteristics of the grounding path. Ground wires and bonding conductors play an

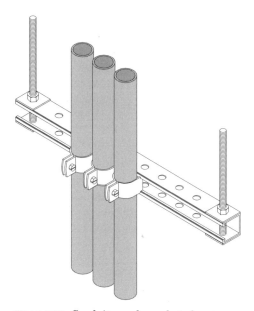

Figure 3.16 Conduits on channel stock.

Figure 3.17 Grounding bushing.

important role in ensuring a safe and reliable building communications system. Additional details regarding their installation are found in Section 8 of Chapter 1 and in the latest edition of the BICSI *Telecommunications Distribution Methods Manual*.

Relay racks

Rack-mounted equipment will first require the installation of standard relay racks, which are metal frames usually providing a large number of vertically arranged, closely spaced, mounting holes. Normal height for these racks is 1.83 m (6 ft) or 2.1 m (7 ft). Custom-sized racks are also available. The common widths are nineteen inches and twenty-three inches. For building telecommunications systems, most installations will require the 483 mm (19 in) wide by 2.1 m (7 ft) high racks. Always use double-sided racks, with mounting holes on the front and rear, when installing ANSI/TIA/EIA-568-A projects. Racks should be accessible from the front and rear. A rack line up (consisting of two or more racks) will require accessibility on one side of the line up for walk space. This will facilitate the installation of cable management devices used to support the cables as they are routed to the rear of the patch panels for termination.

Note. Refer to EIA-310C for hole spacing requirements for racks.

Racks are available in two common depths: 100 mm (4 in) and 152 mm (6 in). Depending on the requirements for an installation, deeper racks may be needed to support a larger number of cables routed to the patch panels. These deeper racks may also offer additional physical stability when several pieces of electronic equipment are attached to the racks.

Cables should be routed on the rear sides of the rack using cable management accessories attached to the rear of the rack's vertical

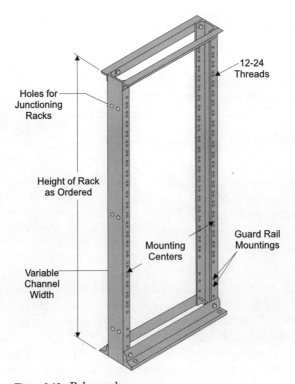

Figure 3.18 Relay rack.

channels or in cable management channels mounted vertically on the sides of the rack. Some manufacturers offer racks with built-in cable management that have deeper and wider vertical channels to allow cable management inside the vertical channels. These racks offer a clean installation with easy rear access to equipment.

> **Note.** Equipment can later obstruct access to cables inside the channels.

Where relay racks are separated some distance from a wall, cable trays should be installed from the wall to the top of the relay racks. This will provide a pathway for cables to be routed to and from the relay racks. Without a cable tray, no means of support will be available and the cables may incur damage. At a minimum, degradation of signal will result. Conduits do not offer sufficient cross-sectional area for large numbers of cables and are not easily attached to the tops of the racks. This is especially true when cabling from telephone PBX cabinets to relay racks or from mainframe computers to adjacent cabinets.

The telecommunications designer's drawings should indicate the method of supporting cables routed to the racks and the means for attaching such supports.

Floor-mounted cabinets

Cabinet enclosures are available to house connecting hardware in lieu of relay racks. They are basically enclosed relay racks with additional specialized hardware installed. Cabinets are available with standard or specialized components. Most cabinets can be ordered with clear-paneled doors at front and rear. Side panels are available in solid or vented versions; top sections can also be solid or vented. The vertical mounting hole spacing on the racks inside of the cabinets is the same as for relay racks. Sometimes the designer will specify cabinets containing power strips, fans, or other electrical apparatus. When these are required, it is especially important that the grounding and bonding instructions be carefully followed.

Cable access to the cabinet is normally provided by knockouts (pre-punched holes) in the cabinet side, top, or bottom. Cable trays and ladder racks provide the means of routing cables to the cabinet. Verify the installation methods specified by the manufacturer prior to attempting to install cabinets. Ensure that the cabinets will fit the footprint allocated for them prior to installation. If there is a question regarding the space allocated, again review the designer's drawings or contact the designer for further clarification.

Wall-mounted racks and cabinets

When floor space is at a premium or when other equipment is occupying all of the available floor space, relay racks and cabinets may not fit in the room. Wall-mounted racks and cabinets should then be considered for the installation. Wall-mounted racks are available in several depths, with or without easy rear access (hinged brackets). If electronic equipment is to be installed on the wall-mounted rack, allow for equipment depth when ordering and installing them.

For example, LAN hubs may require a depth of as much as 355 mm (14 in) when installed on a rack. In this instance, a rack of at least 457 mm (18 in) in depth would be specified.

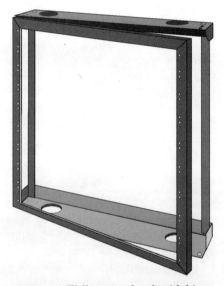

Figure 3.19 Wall-mounted rack with hinge.

Wall-mounted racks and wall-mounted cabinets are available in various heights up to 72 inches and in widths of 21 and 26 inches with a depth of 6, 12, or 18 inches.

Wall-mounted cabinets may be either vented or unvented and can be equipped with clear plastic or metal doors. Some manufacturers also offer cabinets featuring power strips and fans. Fans are required only when the equipment installed in the cabinets generates a significant amount of heat. Power strips should provide both overvoltage and overcurrent protection and should provide a minimum of six positions.

Many electronic equipment manufacturers use bulky external ac transformers to power their equipment. Use power strips designed with adequate spacing for these transformers or every other outlet will be obstructed.

Always ground the cabinets using a minimum of 6 AWG, green-insulated copper wire. Install the ground wire from the special ground lug on the cabinet chassis to the ground busbar located in the telecommunications closet.

Installation checklist

TABLE 3.1 Checklist for a Telecommunications Closet Installation

Items	Tasks
1	Obtain blueprints/specifications.
2	Determine equipment layout.
3	Verify type of drywall construction and load capacity of wall.
4	Install plywood backboard per designer's drawings.
5	Install vertical and horizontal pathways.
6	Install relay racks.
7	Install cable trays/ladder racks horizontally and, if necessary, vertically.
8	Install terminal blocks or patch panels.
9	Install D-rings and other wall-mounted hardware.
10	Install TGB (or TMGB) at the location shown on the designer's drawings.
11	Install grounding and bonding conductors.
12	Install optical fiber connecting hardware.
13	Label all pathways and hardware in accordance with the designer's drawings.
14	Label all terminal blocks, patch panels, and other connecting hardware in accordance with the designer's drawings.
15	Label all grounding and bonding conductors in accordance with their originating and terminating locations.
16	Update design documents to reflect any changes required by field conditions.
17	Verify that the electrical panels, outlets, and lighting fixtures are adequate for the telecommunications closet requirements.
18	After the work is complete, remove all trash, wire clippings, boxes, packing crates, excess materials, and any other equipment or cable that does not permanently reside in the telecommunications closet. Make the space ready for installation of the electronic equipment.

Installing Backbone Pathways

Overview

When installing backbone pathways, it is important to ensure that the route for the pathways is verified prior to actually installing the support mechanisms for the pathways. Fire- and smoke-rated barriers have to be penetrated. Obstructions, such as HVAC ducts, large pipes, and structural beams within the building, have to be overcome. Be sure that the chosen route will provide a clear path.

There are three types of conduit used inside commercial buildings today:

- Electrical Metallic Tubing (EMT)
- Intermediate Metallic Conduit (IMC)
- Galvanized Rigid Conduit (GRC)

EMT is available in 13 mm (0.5 in) trade size through 100 mm (4 in) trade size; IMC and GRC are available in 13 mm (0.5 in) trade size through 152 mm (6 in) trade size. Both the OD and ID of steel conduit are larger than the trade size designation; e.g., 13 mm (0.5 in) trade size EMT has an OD of 18 mm (0.706 in) and the ID is about 17 mm (0.664 in). The metric designators were developed to aid in using design specifications for government jobs. They are designators only and not actual millimeters. The metric trade size designators for electrical conduit differ from pipe sizes to differentiate between conduit and pipe/tubes for other purposes.

PVC conduit should not be used inside commercial buildings. When subjected to extreme temperatures during fires, this type of conduit emits toxic smoke and gases which can cause extensive injury or death to those persons attempting to escape the fire. The products of combustion can also be severely corrosive to telecommunications equipment and can disrupt service to critical telecommunications equipment. PVC conduit can be used under slab-on-grade construction (where codes permit).

Electrical metallic tubing (EMT)

EMT is thin wall metal tubing. The ends are not threaded. It is used widely today in electrical distribution systems and as a pathway for telecommunications wiring. EMT is supplied in standard 3 m (10 ft) lengths. Each length of tubing is referred to as a stick.

Advantages. The advantages of EMT over other types of conduit are that it costs less, weighs less, and is easier to install. It is recognized for physical protection of wiring except where subject to severe physical damage.

Disadvantages. EMT is thinner and thus does not provide the same level of physical protection as IMC and GRC. Since EMT is not threaded, the joints are not of the same strength in applications in which that may be a concern. Some EMT die-cast couplings and connectors have been known to melt when exposed to fire; although the EMT itself has been tested to withstand fire conditions up to almost 2000 degrees for a 4-hour period in full-scale testing. (This was the standard Underwriter Laboratories (UL) E119 time–temperature curve.) For this reason, it is recommended that listed steel connectors and couplings be used in environmental air spaces. Listed die-cast fittings are likely to withstand most fires for sufficient evacuation time. The problem appears to be with unlisted die-cast connectors and couplings.

Couplings. EMT uses two types of couplings to connect sections of conduit together. Compression couplings are installed by first slipping the fitting's nut and then its compression ring onto the conduit. The conduit is then slipped into the main coupling body and the compression ring is compressed as the nut is threaded and then tightened onto the coupler body. This is accomplished using two pipe wrenches or slip joint pliers. The other type of coupling is a set-screw coupling. It is installed by inserting the conduit into the fitting and tightening the set screws with a screwdriver. Both couplings are used extensively throughout the industry.

Be sure that conduit ends are securely placed in the coupling or connector. This is important to ensure that the set screws have room to bite into the conduit for a secure joint. Correct measurement, including length of the elbow legs, is necessary to have sufficient conduit length for secure joints.

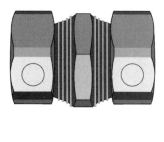
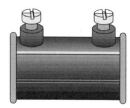

Figure 3.20 EMT couplings.

IMC conduit

IMC has a heavier wall than EMT. Although the wall is less than that of GRC, it is even stronger than EMT. Like EMT, it is available in 3 m (10 ft) lengths (sticks). Each stick of IMC is factory threaded on both ends. When a stick of this conduit is cut to length, the cabling installer must then install new threads on the cut end (see instructions in "Cutting and Threading"). The threads are required to have a 19-mm- (0.75-in) per-foot taper and must comply with ANSI/ASME B.20.1. This provides a secure joint when made up with the standard threaded couplings, which have straight-tapped threads. These conduits can be purchased with an attached coupling, which can be secured without turning the pipe, making installation easier. Elbows of this integral type are also available and are very useful in tight applications. Compression couplings are also available.

Figure 3.21 Intermediate metallic conduit.

Advantages. IMC conduit offers more mechanical protection for the cable and can be used in locations in which it is subject to severe physical impact. It is usually more stable than EMT; the coupled joint is tighter because it is threaded. When subjected to a building fire, IMC remains intact when properly supported and installed. Installed cost is about 30 percent less than GRC.

Disadvantages. IMC is heavier than EMT. It is more labor intensive because of its weight and the requirements for screw-together couplings. It costs more than EMT but less than GRC.

Couplings. Straight-tapped screw-on couplings are used to connect sections of IMC together. When standard couplings are used, the conduit, not the coupling, is turned. This generally means that the conduits on the right and left of the coupling are turned in opposite directions. Pipe wrenches are required to tighten the conduit into the couplings. If they are not sufficiently tightened, the connection will not be airtight and will continue to loosen over time. This could result in the separation of the conduit section and the coupling. Integral coupling type IMC is available in 50 mm (2 in) trade size through 100 mm (4 in) trade size; for this type only, the coupling must be turned to make up the joint. This is especially useful in tight spaces. Couplings used on IMC are the same as those used on GRC. There is nothing to prohibit joining a stick of IMC to a stick of GRC.

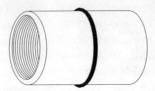

Figure 3.22 Intermediate metallic conduit coupling.

Galvanized rigid conduit (GRC)

GRC is thicker and heavier than IMC. As previously explained, IMC is actually stronger. Like IMC, it is factory threaded and joints are screwed together. In fact, the very same couplings and thread forms are used on IMC and GRC. GRC is available in 3 m (10 ft) lengths (sticks). The sticks of conduit are already threaded on both ends. As with IMC, when a stick of GRC is cut to length, the cabling installer must then install new threads on the cut end. See instructions under "Cutting and Threading." As with IMC, threads are required to have a 19-mm- (0.75-in) per-foot taper and must comply with ANSI/ASME B.20.1.

Figure 3.23 Galvanized rigid conduit.

Advantages. GRC offers maximum mechanical protection for the cables installed inside them. It is the most durable of all conduits. The couplings are airtight and more robust than IMC or EMT. When heavy loads are required, this is the conduit of choice because the threads are slightly heavier due to the thicker conduit wall.

Disadvantages. GRC is heavier than IMC. It is more labor intensive because of its increased weight. It is more labor intensive than EMT because of the requirements for screw-together couplings. It costs more than EMT or IMC.

Couplings. Straight-tapped screw-on couplings are used to connect sections of this conduit together. The threads of GRC are deeper than IMC. This is made possible by the thickness of the conduit wall. When standard couplings are used, the conduit, not the coupling, is turned. This means that the conduits on the right and left of the coupling are turned in opposite directions. Pipe wrenches are required to tighten the conduit into the couplings. If they are not sufficiently tightened, the connection will not be airtight and can continue to loosen over time. This could result in the separation of the conduit section and the coupling. Integral coupling type GRC is available in 50 mm (2 in) trade sizes through 100 mm (4 in) trade size. For this type only, the coupling must be turned to make up the joint. This is especially useful in tight spaces.

Proper support, good tight joints, and bonding for an assured ground are the important elements of a good installation of EMT, IMC, and GRC. Grounding and bonding shall comply with Chapter 8 of the *National Electrical Code®*. Good workmanship in these areas will provide a backbone pathway that will serve well for many years.

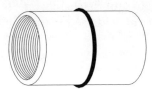

Figure 3.24 Galvanized rigid conduit coupling.

Supports. The support requirements of National Electrical Code Article 348 for EMT, Article 345 for IMC, and Article 346 for GRC shall be followed. Where out-of-the-ordinary heavy loads or abuse are anticipated, cabling installers may choose to add extra supports to assure joints remain secure. EMT, IMC, and GRC shall not be supported by the ceiling grid nor by the ceiling support wires. Separate support shall be provided. This is particularly important when the ceiling is fire rated, as any extra load could compromise the rating.

All exposed raceways are to be run as near to parallel or perpendicular to walls and ceilings as the cabling installer can achieve.

Do not use raceways as support for equipment. Provide separate support, unless otherwise permitted by the NEC.

Install the complete raceway before installing the cables. Be sure joints are tight and the raceway is securely terminated and held firmly in place.

Cutting and Threading IMC and GRC

A roll-type cutter should not be used on EMT. Using a hack saw or band saw will permit reaming of the EMT without flaring the ends. Be sure to make a square cut. It is best to ream EMT with a tool that is designed for that purpose. This will make fittings install easier and better. If other tools must be used, such as pliers, special care must be taken not to flare the ends.

Be sure to measure the exact length of conduit needed. If it is too short, good thread engagement cannot be made with IMC and GRC, or set screws and glands of compression fittings will not engage with EMT.

Use a standard 0.75-inch-per-foot taper die that is sharp. Cut full, clean threads. A worn die or poor threading practices can result in ragged and torn threads.

Adjust the threading dies and use a factory threaded piece to set the die; lock the dies so they are firmly held in the head by tightening the screws or locking collar. A proper thread will usually be one thread short of flush with the thread gauge. This is within permitted tolerances.

IMC and GRC can be cut with a saw or roller cutter. It is very important to make a straight cut and to start the die on the pipe squarely. Wheel-and-roll cutters must be revolved completely around the pipe, tightening the handle about one quarter turn each time it is rotated.

Ream interior edges after cutting and before threading.

One of the most important aspects of good threading is to use cutting oil very freely. Apply it for the first time right after the die has taken hold. Keep the conduit well lubricated throughout the entire threading process.

It is a good practice to thread one thread short to prevent butting of conduit in a coupling and allow the coupling to cover all of the threads on the conduit when wrench tight.

After the die is backed off, clean the chips and lubricant from the thread.

Joint make-up

Always read and follow packaging instructions for any fittings. Use fittings specifically for the raceway type and size being installed. This information is generally found on the container.

Expansion fittings are seldom needed for steel conduit installed in buildings. Expansion joints for the raceway must be evaluated if large temperature extremes are expected, or building expansion joints are in the pathway. The coefficient of expansion to be used is 2.83×10^{-4} mm/°C (6.5×10^{-6} in/°F).

The need for square cut ends has been discussed where threadless fittings are used. These ends are to be clean and assembled flush against the fittings end stop. Threadless fittings for IMC or GRC should not be used unless the manufacturer specifically recommends that application.

All threaded joints have to be made up wrenchtight. Engage at least five full threads, but be careful not to overtighten.

EMT is joined by set-screw or compression fittings. There are a variety of designs and there is no one method of tightening which applies to all. It is important not to over-torque or overtighten screws while assuring they are firmly secured. Most bad joints are due to lack of attention to workmanship and failure to set the screws or compression glands.

For good joints and terminations, it is important to measure the length of conduit needed and select the right elbows.

Conduit elbows and bends

Factory-manufactured bends are recommended for conduit installation. Bends can be manufactured in the field, but specialized equipment must be used to accomplish the required angle of bend. Field-manufactured bends are typically elliptical (oval) in shape rather than completely round. This is caused by the action of the tools used to bend the conduit. A hand bender can be used in the field to bend 16 mm (0.5 in) trade size, 21 mm (0.75 in) trade size and 27 mm (1 in) trade size EMT. Conduits are specified by their trade size. Occasionally, hand benders can be used to bend 35 mm (1.25 in) and 41 mm (1.5 in) trade size EMT conduit.

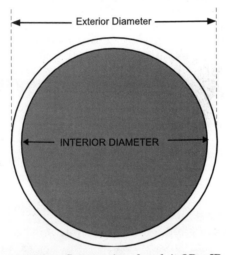

Figure 3.25 Cross section of conduit OD v. ID.

TABLE 3.2 Bend Radius

If the conduit has an internal diameter of . . .	The bend radius must be at least . . .
50 mm (2 in) or less	6 times the internal conduit diameter.

Bends are available in standard radii. Conduit bends are commercially available beginning at 16 mm (0.5 in) trade size.

Caution. Care should be taken when obtaining pre-bent elbows to ensure that they meet the bend radius requirements of the ANSI/TIA/EIA-569-A standard, as "standard" electrical elbows may not comply.

Conduit bends are readily available in various bend configurations. Elbows and bends of 11.25 degree, 15 degree, 22.5 degree, 30 degree, 45 degree, and 90 degree bends are available through distributors. These bends are available in all conduit trade sizes. Special radius and long sweeps are available, but may have to be special ordered.

If odd bend degrees are needed, various configurations of factory-manufactured bends can be used to accomplish the specific requirements. For example, one 45-degree bend and one 15-degree bend can be coupled together to form a 60-degree bend; two 30-degree bends can also be used to accomplish this. However, it is preferable to use one elbow of the proper degree for easier pulling of the conductors. Long sweeps are also available, but may need to be special ordered. Conduit bends are commercially available with varying leg lengths on each side. A leg is the extension of the conduit bend past the point where the bend stops and straightens out.

Most elbows are furnished with a standard leg length for each size. Nonstandard leg lengths are also available, but may have to be special ordered. Leg lengths may vary from one manufacturer to another. The cabling installer should confirm these lengths in order to maintain a uniform installation with accurate termination points on the conduits.

Whenever possible, conduit sweeps should be used rather than bending the conduit in the field using mechanical methods.

Hangers

Many different types of hangers are available to support the installation of this type of conduit. Because of this, a limited number of options will be discussed.

Pipe hangers are used to support conduits. The hanger is a pear-shaped device that is attached to an all-thread rod (ATR). An all-thread rod is a length of rod stock that has been threaded for its entire length at manufacture. ATR is available in various lengths and thickness.

A pipe hanger is suspended from the building structure by an anchor and a section of ATR. The anchor is installed in the concrete structure of the floor or beam. When concrete is not available and steel trusses are installed in the building, beam clamps can be used to support the ATR and hanger. The selection of anchors and beam clamps should be determined by the load weight of the conduit and cable to be supported. The ATR is attached to the pipe hanger with nuts and lock washers. The assembly is then capable of supporting the conduit that is installed through the pipe hanger.

Another type of conduit support is a trapeze. This is a device that is made by using two ATRs and a section of channel stock. The ATR is suspended from the building structure and attached to each end of the channel stock using appropriate nuts and washers. The conduits are then attached to the channel stock with pipe clamps and locked in place.

A third type of conduit support is a one-piece conduit hanger. This is a device that is manufactured in such a way that the compression bolt is part of the hanger itself and, when loosened, it will not come off the

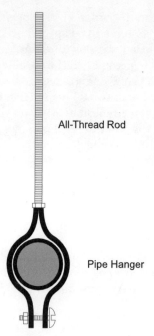

Figure 3.26 Pipe hanger.

hanger. This type of hanger can be mounted directly to the building structure, to red iron, or to a specialized hanger mount.

Other types of conduit hangers are available and should be researched by the cabling installer prior to performing the work.

Conduit Terminations

Backbone conduits should be terminated where they enter or leave a telecommunications closet or equipment room. The recommended location for terminating the conduit is in a horizontal plane where the conduit penetrates the wall of the TC or ER. Allow no more than 50 mm (2 in) of conduit and bushing to extend into the room. Conduits that enter a telecommunications closet should terminate near the corners to allow for proper cable racking. Terminate these conduits as closely as possible to the wall where the backboard is mounted (to minimize the cable route inside the closet).

Conduits should be reamed in accordance with procedures outlined in the "Cutting and Threading" section of this text. Conduits should have a chase nipple installed to reduce cable sheath damage during the pulling operation. A chase nipple can be a plastic insert within the end connector or a plastic ring that is threaded onto the sharp threads of the conduit or fitting.

Conduits shall be equipped with grounding bushings. Grounding bushings are installed on the end of the conduit. With EMT conduit, a set-screw grounding bushing is placed on the end of the conduit and tightened using the appropriate tool. On IMC and GRC conduit, a grounding bushing is screwed onto the threaded end of the conduit. Tighten the bushing until it is secure. A threadless grounding bushing is also available which can provide a more convenient means for locating the ground lug. These are installed with set screws.

Another method for securing conduits is employed when conduits are turned down the wall and terminate at the top of the plywood or turned up and terminate at the bottom of the plywood.

Terminate conduits that protrude through the structural floor 25–76 mm (1–3 in) above the surface. This prevents cleaning solvents or other fluids from flowing into the conduit.

When conduits are turned down in a closet, terminate them above the plywood backboard. This allows full usage of the plywood backboard for termination and routing of cabling. If this method is used, channel stock can be used to attach the conduits in a fixed manner to the closet wall. Each conduit can be attached to the channel stock with a pipe clamp. If the pipe clamps are equipped with "teeth" that bite into the conduit, a grounding bushing is not required on every conduit. A single conduit in each run of channel stock can be equipped with a grounding bushing. The entire section of channel stock can effectively be grounded using a single bushing and ground wire. An alternative method is to install a grounding compression lug onto the channel stock and then route the ground wire to the appropriate ground bar (TMGB or TGB—see Chapter 1: Background Information, Grounding and Bonding).

Note. Backbone conduits should never be turned down in a closet, as it adds an extra bend to the conduit and increases the coefficient of friction during the pulling operation.

The following table provides guidelines for adapting designs to conduits with bends.

Note. Consider an offset as equivalent to a 90° bend.

TABLE 3.3 Adapting Designs

If a conduit run requires . . .	Then . . .
More than two 90° bends	Provide a pull box between sections with two bends or less.
A reverse bend (between 100° and 180°)	Insert a pull point or pull box at each bend having an angle from 100° to 180°.
More than two 90° bends between pull points or pull boxes	For each additional bend, derate the design capacity by 15 percent, or use the next larger size of conduit.

A third bend may be acceptable in a pull section without derating the conduit's capacity if:

- The run is not longer than 10 m (33 ft),

or

- The conduit size is increased to the next trade size,

or

- One of the bends is located within 300 mm (12 in) of the cable feed end. (This exception only applies to placing operations in which cable is pushed around the first bend.)

In practice, a third bend should not be used unless there is no way to avoid it.

Securing conduit formations

Backbone conduits must be secured on each end and throughout the entire route to prevent swinging and swaying during the cable placement. When large, high pair-count cables are installed in backbone conduits, the pulling of the cable places significant tension on the conduit. Tuggers are employed in this placing operation. Sometimes the tugger is anchored to the building or attached directly to the conduit. This tension causes swinging and swaying of the conduit and its hangers. Excessive movement of the conduit can cause the hangers to loosen and possibly come free. If this happens, the conduit and its cable could fall to the floor resulting in damage to the cable, the building, and possibly cabling installers.

Conduits must be secured at each end in such a manner that they do not move. Cross braces can be used throughout the route to stabilize the conduit(s) and prevent movement. This can be accomplished by the use of conduit clamps, channel stock, or ATR placed at opposing angles (180 degrees opposite from each other). This helps prevent lateral movement of the conduit during placing operations. The same anchoring mechanisms can be used to secure the clamps and cross braces as used to hang the conduits from the building structure.

The 1996 edition of the *National Electrical Code®*, Section 300-11, dictates conduit securing and supporting requirements. Chapter 3, "Wiring Methods and Materials," is not referenced from Chapter 8 except for a reference to Section 300-22.c. While Chapter 8 does not directly reference this chapter and its articles, it provides useful information in regards to properly installing backbone pathways.

Pathway preparation

It is important to determine the entire route of a backbone pathway prior to installation of the supporting hangers, ATR, or other support mechanisms. The entire route should be planned ahead of time to ensure that the conduit can be installed without unforeseen obstacles. This is especially true when having to penetrate fire- or smoke-rated walls and floors. If the penetration cannot be established, then all the work done to install the support hardware may have to be repeated at another location.

Always make penetrations through fire- or smoke-rated walls and floors prior to installing the hangers, clamps, and trapezes. Once the conduit is installed, firestop the penetrations using approved methods (see Chapter 5: Firestopping).

Anchors

The types of anchors available are too numerous to mention. The most common anchors are addressed in this chapter.

Plastic anchors. Anchors made of plastic represent the most common anchor used throughout the industry. This type of anchor is used to attach just about everything to any type of surface. These anchors are manufactured by screw or bolt size. The diameter of the screw, as well as its length, determines which anchor to use.

When used in masonry or concrete formations, these anchors provide a secure attachment for light loads.

Expansion anchors. This type of anchor requires that a hole be drilled into the wall or floor where the attachment is to be made. This can be in drywall construction, masonry, or concrete construction.

Drywall installation. The types of anchors used in drywall are sometimes referred to as butterfly anchors (toggle bolts). This is because, once installed, they flare out to resemble the wings of the butterfly.

To install a butterfly anchor, first drill a hole in the drywall sheet, taking care not to blow out the gypsum board. When gypsum board is penetrated improperly, the backside of the sheet flares out, causing the area around the penetration to be thinner and thus weaker. If an anchor is installed in this situation, it will not hold the attachment properly.

After drilling the hole, insert the anchor with the screw end towards the exterior side of the drywall. Once the anchor is firmly seated against the drywall, the locking studs on the anchor base will engage the drywall to prevent the base from turning when the bolt is rotated.

Rotate the bolt in a clockwise direction until the anchor expands and locks over the rear of the gypsum board. This can be determined by tightening the bolt in the anchor. Care should be exercised not to overtighten the bolt since the anchor can rupture the drywall and penetrate the gypsum board. Repairs to the gypsum board will then have to be made and a different location will have to be selected for the anchor.

Gypsum board is not a high load-bearing structure and care should be exercised to prevent overloading.

Masonry and concrete block installation. Masonry walls are generally made of brick or concrete block. Bricks can be solid or with interior cavities. Concrete blocks are similar in that they have interior cavities; however, their cavities are much more expansive than those found in brick.

Bricks are normally 203 mm (8 in) long by 57 mm (2.25 in) high by 89 mm (3.5 in) deep. They can vary in size depending on manufacturer and use. They usually have three circular holes in the flat side of the brick. These holes are designed to reduce the weight of the brick and also to provide cavities for the mortar mix to bond to.

Concrete blocks are normally 457 mm (18 in) long by 203 mm (8 in) high by 203 mm (8 in) deep. These too can vary in size depending on the manufacturer and use. They usually have two oval-shaped holes in one side of the block. These holes are designed to reduce the weight of the block. Occasionally, a concrete block wall will have these cavities filled with poured insulation.

The types of anchors used on masonry and concrete block walls are metal anchors that expand when the bolt is installed inside the anchor. This causes the walls of the anchor to expand, forcing them against the wall of the hole and securing the anchor in place.

First, drill a hole for the anchor in the masonry or concrete block wall. Select a properly sized masonry bit. The bit should be slightly larger than the diameter of the anchor. Most manufacturers provide recommendations on which size bit to use for a specific anchor. The bit will then be attached to a power drill via a chuck.

Conventional power drills, high-torque drills, or hammer drills can be used for this process. The cabling installer should be thoroughly trained in the use of this type of power drill equipment. Failure to receive proper training could result in injury to the cabling installer and damage to the building structure and equipment.

Identify the exact spot where the anchor needs to be installed. Wearing the proper personal protective equipment, install the drill bit in the chuck of the drill. Plug in the drill to a proper power source. Position the drill bit at the location where the hole is to be drilled and engage the drive mechanism of the drill. During the drilling process, be sure to occasionally clean out the hole by pulling the bit out of the hole and discarding the masonry dust. This allows the bit to drill a clean hole and reduce the friction against the bit.

Once the hole is drilled into the formation, use an air source to blow the hole clean. Insert the anchor into the hole. Usually, a hammer is required to completely insert the anchor into the hole.

Place the device to be mounted to the anchor in proper position. Insert the bolt into the anchor. Screw the bolt into position using the proper hand tool until the bolt tightens against the anchor. This can be determined by encountering significant resistance to the bolt rotation. Do not overtighten the bolt.

Epoxy resin anchors. When high loads are going to be supported by anchors, it may be necessary to install special epoxy resin anchors. These anchors use a chemical bonding agent to prevent them from coming loose under most conditions.

The same installation methodology should be employed for epoxy resin anchors as was used for the expansion anchors in masonry or concrete block walls. However, the addition of a chemical bonding agent will be used to ensure the load rating of the attachment. It is especially important to follow the manufacturer's instruction to the letter. Failure to do so could result in an installation that will not support the designed load.

Epoxy glues require a chemical reaction in order for the glue to set up. Most of these glues are a permanent bond. Reaction time for the glues may vary, depending on the chemical makeup of the agents and the accelerator used to cause the chemical reaction.

It is important to verify the size of the hole required for each anchor and to dry fit each anchor into its hole. Keep in mind that the epoxy agent must be applied prior to insertion of the anchor.

Drill the proper size hole in the structure for the anchor to be installed. Clean out the hole with compressed air to ensure all dust is removed. Failure to ensure a clean penetration causes the epoxy to bond to the dust and possibly create a weak anchor.

Dry fit the anchor into the hole to ensure that it will fit prior to installing the epoxy. Once the epoxy is in the hole, there is a limited amount of time before it cures.

Install the epoxy glue in the penetration according to the manufacturer's instructions.

Insert the anchor in the penetration. Immediately install the bolt, with the supporting device being attached to the anchor. Do not place a load on the anchor until the epoxy is allowed to fully cure. Observe the manufacturer's specified time limits.

Once the bonding agents have cured, a load can now be safely applied to the device installed along with the bolt.

Fasteners

A number of fasteners are available today. They are available in many sizes and shapes and have different uses. They fall into the following categories:

- Screws
- Bolts, nuts, and washers
- Specialty fasteners

These fasteners can be used in conjunction with many different clamping mechanisms to affix conduit, cables, or other types of hardware to building structures. A brief description of each and its primary uses will be covered.

Screws. Screws are available in three basic styles: wood screws, metal screws, and drywall screws. All three styles are available with straight slot, star, or hexagonal heads.

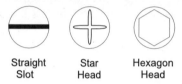

Straight Slot   Star Head   Hexagon Head

Figure 3.27 Screw heads.

Wood screws. Wood screws are used primarily for insertion into wood products. Examples are wood building studs, plywood, wood sheathing, etc. They can be installed directly into the wood by hand or with electrical screwdrivers. Some types of wood products split when the screw is installed. A pilot hole must be drilled into this type of wood product prior to installation of the screw.

When drilling a pilot hole, ensure that the drill bit employed is smaller than the screw. If the drill bit is the same size or larger than the screw, the screw will not secure to the wood product and will be loose in the pilot hole. This will cause the attachment to be loose or fall off.

Wood screws are available in various sizes. The smaller sizes are 6, 4, 2, 0, etc. Larger sizes are 8, 10, 12, etc.

The size of the screw indicates its capabilities in fastening power. The larger the load, the larger the screw must be to handle the load. Consequently, a smaller number screw size will be used.

If the hole in which a wood screw has previously been installed has reamed out, a bigger screw can be used to ensure a secure fastener.

To install a wood screw, first obtain the proper size screw for the load to be fastened to the wood structure. Determine whether the screw has a straight slot, star head, or hexagonal head.

Obtain a screwdriver that matches the screw head. This may be a manual screwdriver or an electric-powered screwdriver (battery powered or ac powered). Screwdriver bits are available for use in power drills. Care should be exercised when using power tools.

Using a pencil or other mechanism, mark the exact location where the screw is to be inserted into the wood. Place the screw against the point marked on the wood through the device being mounted to the wood.

Insert the head of the screwdriver into the head of the screw. Make sure that the screwdriver matches the type of head, both in type and size. An undersized screwdriver blade will cause the screw head to ream out and become unusable.

Turn the screwdriver in a clockwise direction until the screw is fully inserted and the object is secured to the wood. Do not overtighten the screw. Overtightening will cause the screw to ream out the wood and become loose, derating the holding power of the screw.

Continue this process until all of the items are secured to the wood.

Metal screws. There are two basic types of metal screws. They are self-tapping and machine screws.

Self-tapping screws are those types that have very sharp threads that go all the way from the head to the sharp (pointed) tip of the screw. These sharp threads bite their way into the metal when the screw is inserted and twisted into a hole. This process causes the screw to tap the hole with threads to match the screw.

Machine screws have threads that are usually finer than self-tapping screws. In addition, they usually have a blunt tip rather than a sharp tip.

Both of these types of screws require a predrilled hole to be prepared in the metal surface prior to insertion of the screw into the hole. If a machine screw is used, the hole must be tapped to match the threading on the screws.

The cabling installer must select a tapping tool that matches the threads and the size of the hole. After selecting the proper tapping tool, insert the tapping tool into its driving device. Place the pointed end of the tapping tool into the hole and turn it in a clockwise direction until the hole is fully tapped. The tool, turning easier than during the initial insertion and rotation process, can usually determine that the tapping tool has begun to exit the opposite side of the hole. Turn the tapping tool in a counterclockwise direction, removing it from the hole. Visually inspect the hole to ensure the threads extend completely through the metal. Also, look for debris in the hole. Remove any excess metal and debris.

The screw can now be installed into the prepared hole along with the device it is fastening to the metal.

Self-tapping screws can be used on thinner types of metal. The thicker the metal to be penetrated, the more difficult it will be to install a self-tapping screw. Ensure that the predrilled hole is not larger than the self-tapping screw. Place the self-tapping screw into the hole. Using a screwdriver with the proper size and type of blade, turn the screw in a clockwise direction, driving it into the hole until the screw is fully inserted and bottoms out with the screw head against the metal.

This driving and turning causes the tapping threads of the screw to tap the hole.

Turn the screw in a counterclockwise direction, removing it from the hole. Visually inspect the hole to ensure the threads extend completely through the metal. Also, look for debris in the hole. Remove any excess metal and debris. Place the screw through the hole in the object to be fastened. After this, insert the tip of the screw into the hole and rotate in a clockwise direction until the object is securely mounted to the metal surface.

Drywall screws. Drywall screws are similar to self-tapping metal screws. They have very sharp threads and a very sharp point. The threads and points are durable enough to allow the screws to penetrate galvanized metal studs without predrilling. Drywall screws are available in galvanized and untreated finishes. Galvanized screws can be used outside or in areas where there is a likely presence of moisture. Untreated screws should be used inside only. Most drywall screws are available with a star head. They can be easily installed with a star driver bit and a power drill or power screwdriver. Drywall screws can be used on gypsum board, wood, or thin sheet metal.

Using a pencil or other mechanism, mark the exact location where the screw is to be inserted into the structure. Place the screw against the point marked on the structure through the mounting hole in the device being fastened.

Insert the head of the screwdriver bit into the head of the screw. Make sure that the screwdriver bit matches the type of head, both in type and size. An undersized screwdriver bit will cause the screw head to ream out and become unusable.

Turn the screwdriver bit in a clockwise direction until the screw is fully inserted and has secured the object to the structure. Do not overtighten the screw. Overtightening will cause the screw to ream out and become loose, derating the holding power of the screw.

Continue this process until all of the items are secured.

Bolts, nuts, and washers. Using bolts, nuts, and washers requires careful planning. The bolt, nut, and washer must match in size.

Bolts have different thread patterns. Select the nut that exactly matches the bolt diameter and thread pattern. Using a nut that is too large may cause the thread pattern inside of the nut to strip out, rendering it useless. Using a nut that has the wrong thread pattern may cause cross threading. Cross threading occurs when a thread pattern on a bolt and a nut do not correspond, causing the threads to mismatch when the nut is rotated onto the bolt. While the nut may appear to seat onto the bolt, it is stripping out the thread pattern on the bolt, inside the nut, or both. This will derate the holding power of the assembly and may cause it to fail.

Bolts usually are available with hexagonal heads. Nuts also are available with a hexagonal shape. This allows conventional hand tools to be used to install them.

Washers are available in too many types to be discussed here. As a general rule, they fall into two categories—lock washers and flat washers. Lock washers are available in two versions: star washers and split washers. Star washers are available with internal stars or external stars.

A split washer is a round washer that has one point on the washer which has been interrupted, causing it to raise up on each side of the split. In addition, the washer is under tension, so that the split is separated, forming jagged edges. When the washer is installed and compressed against the bolt head or nut, the jagged edges bite into the bolt head or nut and resist rotation of either in the opposite direction. This locks the nut or bolt head into place.

Star washers have manufactured tines. These tines are sharp and very durable. They can appear on the inside or outside edges of the washer. When the washer is installed and compressed against the bolt head or nut, the tines bite into the bolt head or nut and resist rotation of either in the opposite direction. This secures the nut or bolt head into place.

Flat washers may be round or square. Usually round washers are used since they match the circular pattern of the hole and bolts. The size of the washer depends on the load rating of the assembly. The load rating of the assembly also depends on the size of the bolt. Together they must match to ensure the stability of the installation.

The area (size) of the washer will vary depending on the job to be performed. When used with wood products, a bigger area prevents the wood from deteriorating around the area where the bolt is installed. When using flat washers with a smaller area on wood, the area immediately under the washer may become compressed so much that the wood is weakened, causing the assembly to become loose or fail completely. Do not overtighten these assemblies on wood installations.

Specialty fasteners. Specialty fasteners make up a large group of fasteners that are used for a single specific purpose and may require special tools to install them. The most common specialty fastener used in telecommunications installation is the drive ring. Drive rings are metal nails with metal loops affixed to them. The loop is used to support cable after it has been installed. Drive rings are used with wood structures and may be used with masonry structures, along with an anchor.

Another type of fastener that falls into this category is the bridle ring. Bridle rings are metal rings formed out of galvanized wire stock with one side open. The bridle ring can be obtained with wood screw threads for installation on wood structures or with machine threads for use with beam clamps or pretapped holes in metal structures.

> **Note.** Use care when using specialty fasteners. They are not always manufactured to maintain the minimum bending radius of high-performance cables.

Installing Horizontal Pathways

Stub-up/stub-out conduits

These terms imply that a section of conduit is used to provide a pathway in a vertical and then horizontal direction from a point of termination. While similar in many ways, they are significantly different from an installation perspective.

Stub-up installation. Stub-ups are usually single sections of small diameter metallic conduit. They originate at a single- or double-gang box installed in drywall or paneling. The stub-up continues vertically through the wall cavity, where it penetrates the wall cap and stubs up into the suspended ceiling area. It terminates at that point and is usually equipped with a conduit bushing and a pull string. Sometimes the stub-up is equipped with a 90-degree bend that is turned back into the room, especially when installed in fire- or smoke-rated walls.

Stub-out installation. Stub-outs are usually short runs of small diameter metallic conduit. They originate at a single- or double-gang box installed in drywall or paneling. The stub-out continues vertically through the wall cavity, where it penetrates the wall cap, and continues into the suspended ceiling area. In a typical installation, the conduit continues out of the room area and into an adjacent hallway. The conduit may terminate as it exits the wall of the hallway or may continue to another type of supporting structure such as a cable tray or ladder rack. It terminates at that point and is usually equipped with a conduit bushing and a pull string.

> **Note.** Conduit sizes larger than 33 mm (1.25 in) ID are not generally employed in this type of installation because the knock outs on receptacle boxes do not accept box adapters for larger conduits.

Surface-mounted raceway

Surface-mounted raceway is installed on the surface of walls, ceilings, floors, modular furniture panels, and modular furniture. It is available in both metallic and nonmetallic versions. The metallic version is commonly used to conceal power. The nonmetallic version is most often used to conceal low-voltage wiring. It can, however, be used to conceal power as well. Metallic surface raceway should be bonded and grounded when used for power or low-voltage wiring.

Metallic surface raceway. This type of raceway is available in many sizes and configurations. Do not confuse it with wireway or hinged-cover duct systems.

This raceway is available in two sections: a base and a cover. The base is installed on the wall surface using fasteners. The wiring is installed and then the cover is installed over the top of the wiring. The cover is held in place by snapping it over the base. The entire assembly is secured to the surface by the use of two-hole clamps that are specifically designed to fit over the installed assembly. The two-hole clamps are anchored to the wall by fasteners.

Nonmetallic surface raceway. Like metallic raceway, nonmetallic raceway is available in many sizes and configurations. Two types of nonmetallic raceway are available: noncategorized installation raceway and Category 5-type installation raceway. The primary differences between the two are the elbows and fittings used to couple sections of the raceway together. Nonmetallic raceways, being nonconductive, prevent most of the problems associated with bonding and grounding of pathway components.

Nonmetallic surface raceway is available with a single channel or divided channels. When divided channels are required, they are available in two- and three-channel versions. The dividers used to create the channels are removable. They can be removed by breaking them out of the raceway section. By flipping the divider back and forth in a 180-degree arc, the attachment to the base will be weakened, allowing the divider to be pulled out of the section of raceway.

This type of raceway is available in two component sections or a single-component raceway. Some manufacturers refer to the single component versions as a latching duct. This version is manufactured as a single assembly and, once anchored in place, the cover simply snaps onto the opposite side wall latching the cover to the base. Other manufacturers offer a hinged version. The hinged version is a two-piece assembly. One side of the cover is equipped with a round projection that is snapped into a cavity on one side of the base. When closed, the opposite side latches onto the other side of the base.

These raceways are available in a wide range of colors. Most of the colors employed by the manufacturers match the NEMA colors (Ivory, Black (E), Grey, Red, White, Brown, and Almond). Most manufacturers offer special colors at additional cost. In most instances, these special orders require additional lead time for delivery. Some manufacturers also allow their raceways to be painted after installation. This increases the difficulty of reentering the raceway after the paint has cured and may not be aesthetically desirable.

These raceways are also available with or without adhesive backing. If adhesive backs are used on a project, be aware that some of the adhesive backing will not fully adhere to all types of wall finishes. The backing will hold for a limited time, after which the raceway will fall off the structure. This may necessitate use of an additional adhesive, anchors and screws, or some other type of fastener to ensure the stability of the installation.

Various connectors are also available for surface raceway. Splice connectors are used to join two sections of the raceway in a straight line. Internal, external, and flat elbows are used to change direction with the raceway. In locations where joining three sections of the raceway together creates a T, a T cover is used to conceal the joint. End caps are used at the termination of a raceway where a surface-mount box is not used. Conduit adapters, reducers, and ceiling fittings are also available for surface-mount raceway.

Three hand tools are essential for the proper installation of surface raceway. They are a level, a straight edge, and a ruler. A PVC cutter is used to cut sections of the raceway to the desired length. A compound power miter saw is helpful in cutting difficult angles. With the proper blade, it provides an exact fit between the two sections of raceway. By using this power tool, any angle of cut is obtainable to exacting measurements creating an almost imperceptible joint after installation of the raceway.

Surface-mount boxes. Surface-mount boxes are available in single-gang and double-gang versions. They are also available in shallow, standard, and deep versions. The single-gang box measures approximately 50 mm (2 in) wide by 100 mm (4 in) tall by 54 mm (2.125 in) deep. The double-gang version measures approximately 100 mm (4 in) wide by 100 mm (4 in) tall by 54 mm (2.125 in) deep.

These surface-mount boxes are available in a wide range of colors. Most of the colors employed by the manufacturers match the NEMA colors (Ivory, Black (E), Grey, Red, White, Brown and Almond). Most manufacturers offer special colors at additional cost. In most instances, these special orders require additional lead time for delivery. Some manufacturers also allow their boxes to be painted after installation.

These surface-mount boxes are also available with or without adhesive backing. If adhesive backs are used on a project, be aware that some of the adhesive backing will not fully adhere to all types of wall finishes. The backing will hold for a limited time, after which the raceway will fall off the structure. This may necessitate use of an additional adhesive, anchors and screws, or some other type of fastener to ensure the stability of the installation.

Installing Grounding Infrastructure

Overview

Grounding and *bonding* are terms used to define the practice of connecting all components of a system together to a main building ground electrode inside a building, for the purposes of reducing or eliminating the differences of potential between all of the utilities inside of the building structure. A grounding conductor is defined by the *National Electrical Code®* (*NEC*) as:

> A conductor used to connect equipment or the grounded circuit of a wiring system to a grounding electrode or electrodes.

Bonding is defined by the NEC as:

> The permanent joining of metallic parts to form an electrically conductive path that will ensure electrical continuity and the capacity to conduct safely any current likely to be imposed.

The combination of grounding and bonding culminates in a system that equalizes the difference of potentials between all components to as close to zero volts as possible.

Grounding and bonding provides additional safety factors where equipment and people are involved. It protects people from being shocked by voltage potentials and provides a point of discharge for static electricity from technicians prior to working on electronic equipment. In telecommunications installations, grounding and bonding reduces or eliminates stray voltage and current that might interfere with signals traveling through telecommunications cables.

Grounding and bonding reduces the effects caused by lightning, static electricity, and ground faults in electrical equipment. Properly grounding the shields of cables can help reduce noise and crosstalk from adjacent cables.

Electrical grounding and bonding are covered throughout the NEC in several Articles. For telecommunications, consult Article 800. This article will reference other articles in the NEC that should be reviewed as the need arises.

The NEC is written primarily for equipment and personal safety. Manufacturers may also require additional bonding and grounding. Always review the manufacturer's specifications for equipment grounding and bonding. When a conflict exists between a code and a manufacturer's specification, request an interpretation from the local governing agency or a variance.

The building may be grounded at different points depending on its size, age, and location. Smaller buildings are typically grounded at the ac service meter base. A ground conductor is installed from the power neutral bus in the meter base to a man-made electrode (typically a driven ground rod). Larger buildings will be grounded at the main distribution panel (MDP) in the building. A grounding electrode conductor is installed from the equipment grounding bus in the MDP to a man-made electrode called the grounding electrode system. The different derived electrodes are described in detail in NEC Sections 800-40, 250-80, and 250-81.

Multistory buildings also should have a Telecommunications Bonding Backbone (TBB) that appears on each floor of the building. The TBB is connected to the Telecommunications Main Grounding Busbar (TMGB), which is connected to the grounding electrode conductor at the main electrical distribution panel via the Bonding Conductor for Telecommunications (BCT). The TBB is then connected to a Telecommunications Grounding Busbar (TGB) in each Telecommunications Closet (TC) via a pigtail which has been spliced onto the TBB. This splice must be accomplished using an irreversible mechanical connector or an exothermic welded connection. An example of an irreversible mechanical connector is an H or C connector.

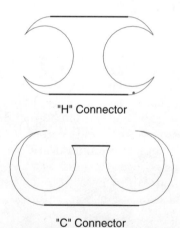

Figure 3.28 Example of an H and a C connector.

The specific grounding and bonding needs of a particular installation will vary due to:

- Building size.
- Equipment designs.
- Special manufacturer's requirements.
- Local code requirements.

Local code information is obtained from the local government agency charged with the responsibility for code enforcement. Some local government agencies may not have adopted a code. In this situation, state government agencies may be charged with implementing code restrictions or enforcing the NEC. Remember that the governing authority is the only entity that can interpret the code and make exceptions or interpretations to the code. Do not rely on code information from other contractors or people who claim to know the local restrictions. They may not be totally familiar with the local requirements.

Other things to consider when planning for bonding and grounding are:

- Multiple closets on each floor of the building.
- Multiple vendors' products to be installed.
- Special electronics equipment.

There can be as few as one and a maximum of four telecommunications grounding and bonding points within a building structure. They are the:

- Main building ground electrode.
- Main electrical distribution panel.
- Telecommunications Main Grounding Busbar (TMGB).
- Telecommunications Grounding Busbar (TGB).

Refer to the following figure for the building grounding infrastructure. Connection of the Bonding Conductor for Telecommunications (BCT) to the bus inside of the main electrical distribution panel requires the services of a licensed electrician. Only a licensed electrician should install ground conductors to the safety ground busbar inside the main electrical distribution panel. This is the only point inside a building where the neutral (current-carrying conductor) and the equipment grounding conductor (green-wire ground) are connected together. If they are connected at any other place in the building, a ground loop will exist, resulting in stray current.

A schematic representation of a typical commercial multistory building is shown in the following figure.

The size of the BCT provided by an electrician from the main electrical distribution panel to the TMGB will vary depending on the service requirements of the building and the complexity of the telecommunications distribution system. For the BCT, a minimum size of 6 AWG, insulated (solid or stranded) wire with green insulation should be installed for this purpose with consideration being given to a BCT of

234 Chapter Three

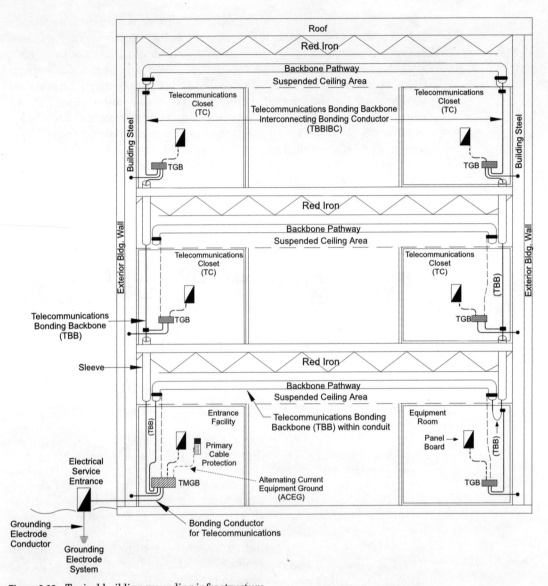

Figure 3.29 Typical building grounding infrastructure.

3/0 AWG. If this wire is installed in metallic conduit, the wire should be bonded to the conduit where it exists, using conduit grounding bushings (see Figure 3.17) and a conductor that equals its size. Any grounding busbar should contain enough multiple points of connections to facilitate the connection of ground and bond wires from the different components inside the closet to the TMGB or TGB.

The TBBs are connected to the TMGB using a two-hole connector that is connected to the TBB via an irreversible mechanical connection or an exothermic weld.

To determine the effectiveness of a ground wire, measure the resistance between the ground source and a reference ground that is isolated from the other ground system. The closer the cabling installer can get to one ohm (preferable less), the better the system.

Building entrance protectors must be connected to the ground system to protect circuits from lightning and power faults. Only building entrance protectors installed on customer-owned physical plant should be grounded/bonded by telecommunications technicians employed by the customer. Physical plant installed by the local regulated telephone company, cable TV company, or other regulated entity can only be maintained, modified, or otherwise changed by their employees. Work by technicians other than regulated company employees may result in service outages or billing rendered to the customer for unauthorized tampering with regulated company facilities.

Avoid splicing a ground or bond conductor. Ground and bonding conductors should be installed in the shortest, straightest route between the equipment being grounded/bonded and the points being connected. If the cabling installer must splice a ground or bond conductor, use an irreversible mechanical connector or an exothermic weld.

Local code requirements

Always review the local code requirements before proceeding. This includes what issue of the NEC® is adopted and what, if any, exceptions to the code are adopted by the governing authority. Most of the code requirements for the job should be included in the telecommunications cabling designer's documents. The cabling installer should never take this information for granted, since the telecommunications contractor is fully responsible for all work done on the project.

If no code has been adopted locally, consult with the state fire marshall's office to determine what state agency is responsible for that geographic area and what codes are in effect. Do not depend on other cabling installers, contractors, or even company personnel in making these determinations.

Standards and codes

Except when local codes are in conflict, follow the national codes and standards. Familiarize yourself with the NEC and ANSI/TIA/EIA-607. Ensure that all work performed complies with these standards. Determine whether manufacturers have requirements which exceed the NEC, ANSI/TIA/EIA-607, or local code requirements. If a conflict exists, obtain an interpretation from the code governing body.

Ground source

If architectural blueprints which contain electrical drawings are available, refer to them to determine what has been provided and installed by the electrical contractor. Refer to the electrical riser diagram and the electrical engineer's notes that accompany it to determine what grounding electrode system has been designed and installed. Consult with the electrical contractor on the job to inquire whether they followed the design or whether changes have been made since the drawings were issued.

If the electrical contractor is still on the job, consult with the contractor to determine where the ground electrode is and whether they have provided a Bonding Conductor for Telecommunications (BCT) to the main telecommunications closet for use by telecommunications technicians. If the electrical contractor is not on the job, consult with the general contractor.

The size of the BCT to the main TC will vary, but should be not less than 6 AWG. ANSI/TIA/EIA-607 recommends consideration of a 3/0 AWG. The BCT must originate at the main distribution panel for the building electrical system. This is sometimes referred to as the main ac switch. This location can be identified because it is the only location in a building where the neutral bus and the equipment ground bus are tied together. The BCT will then be routed in the shortest straightest route to the telecommunications entrance facility. This is the location that houses the building entrance protectors. If the TEF is equipped with a BCT, determine whether a TMGB is provided.

Grounding hardware

If a TMGB is not provided, determine the maximum number of ground/bond wires to be installed in the closet and install a TMGB that will handle them. The minimum number of positions on a TMGB should be in accordance with the immediate requirements of the application and with consideration of future growth (ANSI/TIA/EIA-607). It is necessary to be familiar with the manufacturer's requirements for each piece of equipment requiring grounding/bonding. Always follow local codes and ordinances as well as national codes and standards.

The recommended connector is made of a copper alloy with a crimp connection for the ground/bond wire. It is equipped with a flat plate, featuring a hole(s), so that it can be screwed or bolted to the ground/busbar. A second type of connector is the H or C connector used to join two sections of bond wire together to form a connection. This type of connector can only be used on a bond wire. Ground wires should be continuous and without splices or connectors from the point of origination to the point of termination. Bonding conductors for telecommunications to the TMGB shall use listed 2-hole compression connectors, exothermic welded type connections, or equivalent. A single-hole connector can be used when connecting only one device to the busbar. However, 2-hole connectors are preferable. When dissimilar metals are bonded to a busbar, a conductive paste or grease should be used to minimize electrolysis.

When grounding or bonding specific telecommunications hardware, always review the manufacturer's specifications.

Telecommunications main grounding busbar (TMGB)

Determine whether the electrical contractor has provided and installed a BCT from the main electrical distribution panel to the TEF. If the BCT is provided and installed, proceed with installing the TMGB. If not, determine the location and makeup of the building's grounding electrode system. The building's grounding electrode system may consist of:

- Building structural steel, where effectively grounded.
- Exposed metal service entrance raceway.
- Ground rods, pipes, conduits and plates which have at least 2.5 m (8 ft) of contact with surrounding moist soil.
- Concrete-encased electrodes.
- Metallic cold water pipe—must be supplemented by one of the four types listed above. (It is necessary for this metal piping to be in contact with moist soil for a minimum of 3 m [10 ft]. The interior cold-water piping of a facility is acceptable, providing it is at the building's street side of the water meter. Connections to metal drain pipes are not an acceptable means of providing a grounding electrode connection, due to the high probability of polyvinyl chloride [PVC] piping used in these systems.)

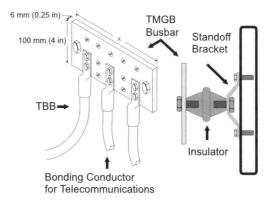

Figure 3.30 TMGB.

If the number of ground/bond wires exceeds the number of positions on the busbar employed at the site, a TGB must be installed to allow connection of the additional ground/bond wires. The TGB should be as close to the TMGB as feasible.

Use a minimum of 6 AWG, green-insulated, solid/stranded wire to connect the TMGB to the TGB. Mount the TGB at the same height as the TMGB. (The recommended height is 152 mm (6 in) below the top of the plywood backboard.)

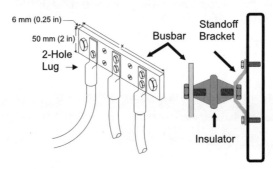

Figure 3.31 TGB.

Steps—Installing TMGB

Step	Installing TMGB
1	Select the length of busbar to be installed. The TMGB should be 100 mm (4 in) high by 6 mm (0.25 in) thick by the length necessary to terminate all of the conductors to it.
2	Determine the exact position of the busbar on the backboard.
3	Prelocate the mounting hardware that comes with the TMGB. Mark to location where the two mounting holes for each bracket will be attached to the backboard.
4	Predrill holes for the screws using a hand drill and an appropriately sized drill bit for the leg screws being used. Install mounting brackets using the lag screws contained in the kit.
5	Install the insulators on the mounting brackets using the bolts contained in the kit.
6	Install the busbar on the insulators using the bolts contained in the kit.
7	Always install the BCT to the TMGB before proceeding with the additional grounding.

8. Identify the BCT installed from the main electrical distribution panel. Determine how much wire is required to position it onto the busbar. Cut off the excess wire using cable cutters.

9. Using an insulation-removal tool, remove the correct amount of jacket from the wire.

Note. The correct amount is determined by the size of the connector and the gauge of the wire.

10. Insert the bare end of the copper wire into the end of the connector. Obtain a compression tool equipped with a die that matches the connector body.

11. Place the round connector body into the die of the compression tool. Close the jaws of the tool fully. This action may have to be repeated on another section of the connector to fully ensure contact with the wire. Leave approximately 6 mm (0.25 in) of copper showing between the end of the insulation on the wire and the end of the connector that the wire is installed into.

12. Position the ground wire and connector by routing them to the busbar. Ensure that a minimum bend radius of eight times the conductor outside diameter is used where the wire changes direction. Use smooth bends in the wire. Attach the ground wire to the backboard using wire clamps installed at intervals of not more than 457 mm (18 in).

13. Label the ground wire as the main ground source for the room and where it originates. The label should contain where it came from, where it is going to, and its unique identifier as described in ANSI/TIA/EIA-606. If this is the BCT, the cabling installer must also install a warning label on each end of the wire. Refer to ANSI/TIA/EIA-607, Chapter 5.

14. Upon completion of the installation of the TMGB, update all installation documents with the necessary information that will allow a cabling installer to identify any ground source and the methods employed in providing the ground/bond.

The procedure outlined above is also used for connections between the TMGB and the other components of the telecommunications grounding system within the building. Refer to ANSI/TIA/EIA-607.

Other components that will have to be installed in a multicloset building will be:

- Telecommunications Bonding Backbone (TBBs).
- Telecommunications Grounding Busbars (TGBs).
- Telecommunications Equipment Bonding Conductors (TEBCs).

- Telecommunications Bonding Backbone Interconnecting Bonding Conductors (TBBIBCs).
- Alternating Current Equipment Ground (ACEG).

Telecommunications bonding backbone (TBB)

A TBB must be installed from the TMGB through each set of closets in a multistory building or in a series of closets in a low, wide building. The TBB must be a minimum of 6 AWG, green-insulated conductor. Consideration should be given for sizing the TBB at 3/0 AWG.

TBBs should always be terminated at the TMGB using a two-hole connector. In each of the other closets, a short length of conductor, sized at the same size of the TBB, must be tapped onto the TBB using an H or a C connector or an exothermic weld. This conductor will be extended to the TGB and terminated using a two-hole connector.

Telecommunications grounding busbar (TGB)

In each additional closet, a TGB must be installed. The TGB must be 50 mm (2 in) high by 6 mm (0.25 in) thick and as long as necessary to terminate all of the conductors onto it. It must also be insulated from its mounting (i.e., wall, plywood, etc.) by a 50 mm (2 in), or greater, insulator. The TGB must be installed in a manner similar to the TMGB.

Telecommunications equipment bonding conductor (TEBC)

Each piece of telecommunications equipment should be bonded to the TGB to equalize potential within the grounding mesh. To do so, install a TEBC from each piece of equipment to the TGB. When equipment is mounted on a relay rack, a rack grounding kit can be installed on the rack and the various pieces of equipment bonded to it. Then a TEBC can be installed from the relay rack to the TGB. The minimum size conductors must be a 6 AWG, green-insulated wire. Normally a TEBC can be terminated using a single-hole connector. However, if it connects multiple pieces of equipment (i.e., from a relay rack), it should be terminated using a two-hole connector.

Telecommunications bonding backbone interconnecting bonding conductor (TBBIBC)

Whenever two or more TBBs are run vertically in a building, on each third floor and the top floor of the building, a TBBIBC must be installed between the TGBs in the closets. The TBBIBC must be a minimum of a 6

AWG, green-insulated copper wire. Like all ground/bonding conductors, it should be routed in the shortest, straightest manner between its two points of termination. It should be terminated using a two-hole connector.

Alternating current equipment ground (ACEG)

Many times a telecommunications closet will be equipped with an electrical distribution panel (EDP). This panel will serve all of the electrical service contained within the walls of the closet. It is usually on a dedicated feed from the MDP or from some other EDP. If such an electrical panel is located within the room, an ACEG must be installed from the equipment grounding bus inside the panel to the TGB. This conductor must be a minimum of 6 AWG, green-insulated copper wire. It should be terminated at the TGB using a two-hole connector.

Steps—ground test

Measure the resistance between the TMGB and the remaining ground points of the building grounding system with a megger. The maximum system resistance allowed per NEC 250-84 is 25 Ω, however, resistance of 1-5Ω maximum is generally found suitable in large commercial buildings. The ideal resistance is 1Ω or less.

Step	Ground test
1	A short jumper wire should be installed between the P- and C-terminals of the earth ground resistance tester. This allows the meter to be used in the two-point configuration.
2	Terminate one cable at the E-terminal of the tester.
3	Terminate the other cable at the P- or the C-terminal of the tester.
4	To compensate for the resistance of the instrument cables, connect the open end of the instrument cables together. Depress the measurement button of the test instrument and record the indicated value. This is the resistance of the meter leads and should be subtracted from each reading during the bonding resistance measurements.
5	Once the lead resistance has been obtained, measurements should then be made between selected points in the grounding system. Resistances that can be checked are between the following: • Metallic water pipe main and the structural steel • Metallic water pipe main and any driven electrode

- Service entrance conduit and the structural steel
- Electrical system grounding electrode and the building lightning rod grounds
- Power transformer feeding the electronic equipment and the structural steel
- TMGB and the ac grounding electrode system
- TMGB and the TGB
- TGBs located in separate equipment rooms

Other combinations may exist depending on the size of the facility or the type of customer operation that resides in the building.

Note. To make this bonding verification complete, the cabling installer should verify the tightness of connections to the different grounding electrode systems. Corroded connections and loose connections comprise the majority of deficiencies in the resistance between grounding electrode systems. At a minimum, the system should be checked annually.

Installing Cable Support Systems

Overview

Proper support structures are critically important for the implementation of a cabling system. Cable support systems include the following:

- Cable trays and associated support hardware
- Ladder racks and associated support hardware
- Enclosed wire pathways and associated support hardware
- Plywood backboards, clamps, rings, or hangers
- Conduits and associated hardware

In most cases, a cabling installer will install these support structures. Support systems provide a pathway for the cable, thus reducing and eliminating stress which could damage the copper pairs or glass strands inside a cable sheath. The following steps are general in nature. Additional steps may be required, depending on the size, quantity, and load rating of the individual supporting structures. Telecommunications designers should provide all the information in their design documents that will allow the cabling installer to select the necessary hardware and employ proper methods to correctly install these structures.

Steps—install cable support systems

Step	Install cable support systems
1	Obtain blueprint/specifications/designer's documents. ■ Determine the size, type, and quantity of pathways to be installed. ■ Determine the proposed route of the pathways between closets and from closets to work areas. ■ Identify any obstructions along the proposed route and determine how to overcome them. ■ If it is necessary to pass through any walls, be prepared to make the required penetration—concrete, concrete block, drywall, or other wall construction. ■ When installing cables in a suspended ceiling without permanent pathways, avoid cable paths which introduce obstacles. ■ Identify the plan to support cable in suspended ceilings and what type of hardware will be installed.
2	Verify load capacity of the cable support system. Verify by: ■ Reviewing the telecommunications designer's documents. ■ Determining the weight of individual types and sizes of cable to be installed. ■ Identifying the building structure to ensure a proper attachment of the support system.
3	Verify the adequacy of existing cable support structures. Where existing supports are to be used, visually examine them to determine if they are firmly attached, not worn or broken, and are capable of bearing the extra weight.
4	Verify cable support installation accessibility. Examine the following: ■ Area where the building beams are located ■ Area where the concrete floor slab above is accessible and usable for installing anchors or other cable supports directly to them ■ Paths in all directions from the point of observation to see if the pathway is clear for cable supports
5	Cable separation. ■ Maintain specified distances from possible sources of EMI, per ANSI/TIA/EIA-569-A, and the latest edition of the BICSI *Telecommunications Distribution Methods Manual*. ■ For both safety and performance purposes, keep power cables physically separated from data and voice cables.

6   Verify materials and tools availability.

Before beginning an installation of a cable support system, be sure that the required materials, hand tools, and power tools are available. This should also include safety tools such as safety glasses and, where required, hardhats.

7   Mount D-rings.
- Review the designer's drawing and specifications to determine exact placement, quantity, size, and type of D-rings.
- Install the D-rings, working from the top left side of the plywood backboard to the bottom right side of the plywood backboard:
  - Measure the location of the first D-ring to be installed according to the designer's documents.
  - Position the D-ring on the backboard and mark the location of the holes in the D-ring, using a pencil.
  - Using a drill and a bit, predrill the location of the D-ring holes.
  - Position the D-ring on the backboard and install a No. 6 × ¾ inch, sheet-metal screw using a screwdriver.
  - Install the second No. 6 × ¾ inch, sheet-metal screw.
  - Repeat the above steps for each additional D-ring to be installed.

8   Mount J-hooks in structures above the suspended ceiling level.

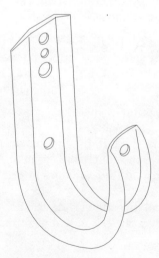

Figure 3.32 J-hook.

- Determine the J-hook size required in each cable path and lay them out along the cable route.

- Identify the location of the first J-hook to be installed.
- Position the J-hook at its proper location and mark the holes for the anchors, using a pencil.
- Predrill the holes and install the correct anchors (depending on the type of structure, i.e., a masonry structure will require one type of anchor while metal structures or drywall will require others).
- Install the anchors.
- Reposition the J-hook at the desired location.
- Using the correct screwdriver, install the appropriate screw through the hole in the J-hook and into the anchor, securing the J-hook to the anchor.
- Repeat the above for the second screw.
- Install the remaining J-hooks using the same procedures, until all J-hooks are installed.

9   Install cable trays or ladder racks.

The process of installing cable trays and ladder racks is manufacturer specific, due to the many different sizes, types, and configurations. The cabling installer must consult the manufacturer's specifications and installation guidelines prior to attempting to install these pathways.

Be aware that both trays and ladder racks can be installed parallel to and against a wall or can be suspended from the building structure using threaded rods. They can also be supported using channel stock or special hanger brackets.

Since these pathways are used to support both backbone cables and horizontal cables, they are found in most areas of the buildings. It may be the case that other utilities (i.e., HVAC, plumbing, electrical) or structural obstructions will require that the pathway change elevation and direction. Careful planning is important to minimize changes in direction which can be costly from a material as well as a labor perspective.

Penetrations through four-hour, fire-rated masonry walls are especially challenging to the cabling installer. Firestopping these penetrations requires special knowledge and materials to comply with the NEC. The manufacturers of firestop materials provide detailed instruction on how to install their products to comply with special circumstances. If a situation is identified where the selected firestop manufacturer's technical manual does not contain a certified solution, contact the manufacturer. Provide all the pertinent information regarding the penetration. In most cases, the manufacturer can provide an engineered solution. This engineered solution moves the liability for its design and use to the manufacturer from the contractor. Do not attempt to field design firestop solutions.

10. Install conduits.

   Within buildings, the cabling installer may be faced with only three types of conduit. All three are metallic and require similar methods of supporting them. The three types are:
   - Rigid galvanized conduit (GRC).
   - Intermediate metallic conduit (IMC).
   - Electrical metallic tubing (EMT).

   The most common is EMT. Rigid conduit and IMC conduit are primarily installed for backbone cables. Though EMT may also be installed for backbone cables, large cables may pose special loading problems for the hangers and supports normally used to install EMT.

   The types of fittings used to connect EMT conduit are:
   - Compression couplings

   Compression couplings are designed to reduce in diameter to that of the conduit outside diameter when tightened with a wrench. This process increases the friction between the outer wall of the conduit and the inner wall of the coupling to a point where the conduit does not slip or move under normal conditions.

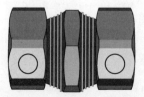

Figure 3.33 Compression coupling.

   - Set-screw couplings.

   These couplings are placed over the end of the conduit with little or no friction. Once the end of the conduit is seated into

Figure 3.34 Set-screw coupling.

the coupling, two screws are tightened on the outside of the coupling, causing the coupling to become firmly attached to the conduit. The main disadvantage of this coupling is that over time, it may become loose and allow the conduit to pull away from the coupling.

11. Conduit hangers.

   Conduits are suspended from the building structure using a variety of hangers. These hangers are available in many sizes and types depending on the type of conduit and method used to install the hangers. Hangers may be attached to threaded rods, building steel, masonry walls, and other hangers. The cabling installer should refer to the manufacturer's

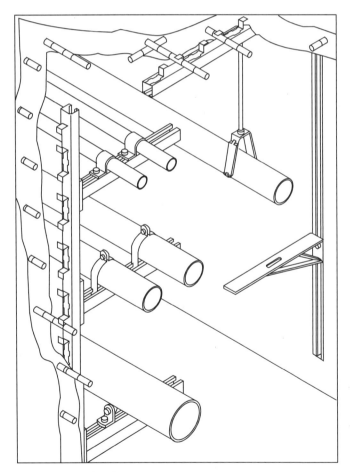

Figure 3.35 Conduit hangers.

specifications and use only the correct type and size for the conduit to be installed.

12. Installing wire ducts.

   Wire ducts are different from cable trays and ladder racks in that they are completely enclosed pathways. They usually feature a hinged cover or one which snaps in place. They are available in a number of sizes and shapes, depending on the manufacturer. As with cable trays and ladder racks, the installation methodology is directly related to the manufacturer's guidelines. Proper planning and the use of the correct hangers and attachments for wire ducts is important to ensure a satisfactory installation.

13. Pathway documentation.

   It is just as important to properly document pathway installations as it is to identify the cables they house and support. Not knowing where a pathway originates or terminates is as necessary as knowing where the cable originates and terminates.

   Proper identification is especially important for pathways left vacant at the completion of the project. Not only may a great deal of time elapse before they may be used, the team that performed the original installation may not be the same group which performs the additional work. The pathways must be identified not only on the drawings, but each pathway should also be physically equipped with a label that identifies it and states where it originates and terminates.

   If the pathways are installed by the electrical contractor on a project, in most cases, the electrical contractor is required to provide an as-built print depicting the routes of the various pathways. The cabling installers should obtain the information contained on these blueprints and transfer it to their work documents. Upon job completion, this information should be contained in the telecommunications as-built package.

14. General housekeeping.

   It is important to clean up behind a work operation when installing pathways. Larger equipment is required to install these pathways and the materials themselves are bulkier, taking up more space in the hallways and passageways of the building. This causes congestion and, if left in place, can impact the work efforts of other trades.

   The hangers, their associated hardware, anchors, screws, and other materials present a safety hazard to workers and should be properly stored during work operations and upon completion of the day's activities. If a work area cannot be completely cleaned, safety cones and barricades should be placed to prevent accidental intrusion into the work area by others until the work is complete and the area is cleaned.

Keeping a clean work area also reflects on the quality of the work which the installation team performs and on the company it represents.

Chapter 4

Pulling Cable

Cable Pulling Setup	251
Pulling Horizontal Cable in Conduit with Fishtape	259
Pulling Horizontal Cable in Open Ceiling	266
Pulling Backbone in Vertical Pathway—from Top Down	269
Pulling Backbone in Vertical Pathway—from Bottom Up	276
Pulling Backbone—Horizontal	283
Pulling Optical Fiber Cable	287

Overview

Pulling cable for telecommunications installations inside of buildings requires using many techniques.

Some cables are very large and heavy, disguising the fact that the pairs inside the cable are actually very fragile. Be prepared to learn the correct way to handle the cable so that the installation will meet specifications.

Setting up, or getting ready, is the proper way to ensure a smooth job of pulling cable. These precable pulling tasks are essential:

- The work area must be secured (cleared of pedestrian traffic) for safe conditions to exist for everyone.

- Equipment for supporting the cable reels must be in place.

- Equipment for pulling the cable must be available.

- Equipment for temporarily holding the cable in place must be available.

- Correct lengths of cable must be on hand.

- Before the pull begins, locations for accessing and pulling the cable should be identified.
- Adequate manpower must be available.
- Equipment allowing cabling installers to communicate must be on hand.

Horizontal cabling begins at the telecommunications closet on each floor and ends at the work area outlets at the workers' desks. Types of support structures include:

- Conduit
- Cable trays
- J-hooks
- Modified types of bridle rings
- Beam clamp
- Innerduct
- Building red iron or roof support structures

Conduit provides a good pathway for the cable. Conduit may be made of:

- Electrometallic tubing (EMT).
- Rigid metal.
- Rigid polyvinyl chloride (PVC).
- Fiberglass.

Conduit made of flexible metallic tubing should not be used unless it is the only practical alternative, such as connection to modular furniture. Check local codes to ensure flexible tubing is permitted.

Conduit provides a safe environment that prevents cable from being accidentally cut or damaged. Normally, conduit is installed from the equipment room to each of the telecommunications closets in what is usually called a home run.

Moving the cable through the conduit from one end to the other requires some type of object (pull string or rope) to precede the cable through the conduit. This is known as fishing a conduit.

Methods for installing the pull string/rope through the conduit include:

- Using the manual fishtape method.
- Using air-propelled or vacuum methods.

A fishtape is a steel or fiberglass wire rigid enough to be pushed all the way through the conduit, or in longer runs, to a pull box. It is used to retrieve the pull string from the far end.

Air-propelled methods include:

- A vacuum on one end and a foam rat or ball attached to a pull string on the other end.
- A compressed air bottle or mechanical blower used to propel a pull string attached to a foam ball or rat.
- Other devices either purchased or made up on the job.

Horizontal cabling in an open ceiling uses a different pulling method from those described above. Various ways are used to pass the pull string through the trusses or other structural elements. Cables should have a straight and smooth path; this is essential for a good cable pull. There are various methods used for placing pull string, including attaching:

- A ball to a pull string and throwing it through the open ceiling space.
- A fiberglass pole to the pull string.
- Remote-controlled appliances to the pull string.

Cables should always be installed parallel or perpendicular to exterior and interior walls.

Backbone cabling provides interconnections between the telecommunications closets, equipment rooms, and entrance facilities. This cable has higher pair counts (25 to 2400 pairs) than horizontal cables and weighs more than horizontal cables. A 2400-pair, 24 AWG backbone cable could weigh as much as 31.3 kg per meter (21 lb per foot). For vertical pathways, cabling is more easily pulled from the top down than from the bottom up because gravity helps with the pull. It is the preferred method if the reels of cable can be moved to the top floor. Sometimes this cannot happen because the reels are too large to fit through closet doors or on a freight elevator. When necessity demands it, backbone cable is pulled from the bottom up. In both cases, special cable-handling devices are required, including:

- Tuggers.
- Cable reel brake.
- Temporary take-up devices.
- Bullwheels.
- Pulleys.

- Mesh grips.
- Bull lines (pull ropes).

The type of equipment needed depends on the direction of the pull. If pulling from the bottom up, a tugger, similar to an electric winch, may be needed. If pulling from top down, a reel brake may be needed. A reel brake is a mechanical device used to stop or slow a freewheeling reel, thus keeping it from unreeling too fast due to gravity.

Communication with coworkers is essential in every cable pull. Everyone should be prepared to alert the person pulling the cable, with or without power equipment, to ensure the cable is traversing the route smoothly without twisting, kinking, or getting bound up in some way. Two-way communication is normally employed to monitor the job progression.

Adhering to fire and building codes and standards gives cabling installers the ability to minimize the likelihood of the spread of fire and smoke in a building. All penetrations through fire- or smoke-rated walls and floors need to be firestopped with approved methods to reduce the chance of spreading fire and smoke. Firestop materials are available as:

- Putty.
- Pillows.
- Caulking compounds.
- Cementitious compounds.
- Blankets.
- Wrap strips.
- Collar devices.
- Composite sheets.
- Sprays.
- Mechanical systems.

All firestopping solutions are a combination of firestopping materials, holding devices, packing materials, and other devices that make up a listed (approved) method. Do not use unapproved methods to firestop a penetration. Always use an engineered and approved system to properly firestop a penetration. Contact an appropriate firestop manufacturer for any situations that are not addressed by the manufacturer's listed methods.

Cable Pulling Setup

Overview

There are many preparation tasks which affect safety. Once the job of pulling cable begins, everything and everyone should be in the right place. This can only occur with proper planning. A good cabling installer is efficient and organizes the work.

A good cable setup means all materials are in place so the cables can be handled properly. A cabling contractor may need specialized equipment capable of holding large reels. Smaller equipment, such as cable trees, may be brought in to handle the many rolls containing low pair-count cable. If all preparations are made correctly, time will be saved once pulling begins.

The job site should be secured from office workers or other occupants of the building. Other contractors as well as building occupants need to know that an installation is going on. The safety of everyone involved is important. Securing the area is a way to let other people in the building be aware that an installation is underway. The equipment in use may create a hazard where people walk.

Prepare starting the cable into the conduit by:

- Securing the area.
- Setting up the cable.
- Setting up the pull string.
- Selecting and identifying the cable labeling system to be used.
- Identifying pull points.

Steps—pulling setup

Step	Cable pulling setup
1	Secure the area. - Set up cones, signs, barricades, or caution tape in work areas to alert everyone of danger in the area. - Place caution tape across the entrance to the area to restrict access to anyone other than cabling installers or other authorized personnel. - Notify appropriate personnel that work is beginning.
2	For large cable reels, set up a jackstand or reel dolly. A reel dolly is basically a jackstand with wheels that allows easy relocation of the assembly. - Find out if the cable reel being installed is so large and heavy that a jackstand or reel dolly is required. - Many contractors use jackstands for holding the reels of cable off the floor. Sometimes this equipment is made from homemade supports and a pipe is used to appropriately suspend the reel.

256   Chapter Four

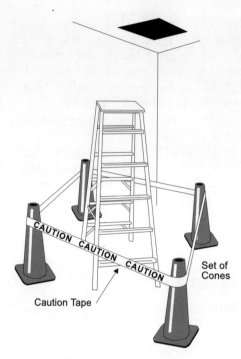

Figure 4.1 Secured area with safety cones and caution tape.

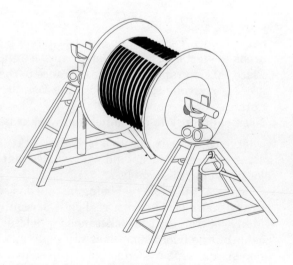

Figure 4.2 Large reel and adjustable jackstands.

- Select a location that is large enough for the number of spools needed.
- Set up the jackstand inside the telecommunications closet if there is room for it, or set it up in the area just outside the closet.
- Place a pipe or crossbar (known as a mandrel) through the center hole in the reel. Get help from coworkers to lift the spool onto the jackstand. Jackstands used for placing larger cables usually have a mechanical ratcheting or hydraulic lift mechanism to assist in lifting the reels into place for pulling.
- Make sure the mandrel can support the weight of the cable reel(s).
- In some pulling operations, reel brakes may be needed to control the pay out of the cable.

3   For smaller reels, set up a cable tree.
- Cable trees are used when there are multiple small spools of cable being pulled at the same time.
- Select the spools of cable that will be needed for the job. Cable can be ordered with various put-up lengths. Usually put-ups are 305 m (1000 ft), 457 m (1500 ft), and 1524 m (5000 ft).
- Select a location near the cable feed point that is large enough for the cable tree.
- Bring in the cable tree and set it up inside the closet if there is room, or in an area outside the telecommunications closet, such as in the hallway.

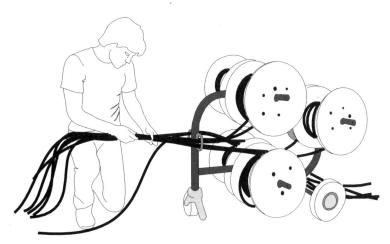

Figure 4.3 Cable tree.

- Mount the spools on the cable tree.
- Find the end of each cable reel and bring them together, through the guide hole on the pay-out end of the cable tree.
- Select one of the cables in the group as a guide cable.
- Take the next two cables and tape them to the guide cable approximately 76 mm (3 in) behind the guide cable. Continue attaching cable to the bundle in groups of two until all cables are attached to the bundle.
- Prepare the guide cable for attachment to the pull string.

4. For even smaller amounts of cable, set up a cable pay-out pack.
   - Knock a hole in the top of the pack or remove the plastic insert to expose the cable inside.
   - Cable is shipped in several different containers:
     * Pay-out-pack
     * Reel-in-a-box
     * Spool
   - The pay-out packs and the reel-in-a-box are convenient and require no rack to hold them. For pulling multiple strands of cable, the packs can be stacked and numbered sequentially for proper labeling later.
   - Find the end of the cable and thread it out through the hole in the top of the pack.
   - Keep the wire free from any obstruction that would crimp or bend it, causing damage.

5. Set up pull string or pull rope. If no pull string or pull rope exists, and it is required in the conduit system, one must be installed.

6. Identify pull points for each horizontal cable run.
   - Determine the distance of the complete cable run. Ensure the distance does not exceed 90 m (295 ft). Take care to

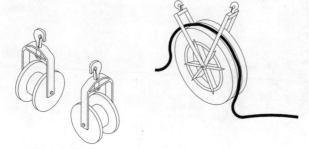

Figure 4.4 Sheave and pulley hangers

identify the changes in direction, vertically or horizontally. These changes in direction may cause the run to exceed the 90 m (295 ft) limit.
- Take note of the number and locations of bends and identify where to feed the cable around sharp bends or turns. Pulley hangers may be used at some points to save manpower.
- Allow no more than two 90-degree bends per 30 m (98 ft) cable pull.
- Divide the total cable run distance into segments where each segment is less than the maximum of 30 m (98 ft), including end measurements.
- Identify a pull point, where necessary, for accessing and handling the cable.
- Where possible, minimize the number of pull points, since each pull point requires an additional person, a pulley, or a separate pull.

Note. When personnel are not available to be stationed at all the pull points, the cable will have to be pulled to the furthest manned pull point and then placed in a figure eight on the floor. To continue the pull, workers are relocated to the next vacant pull points until the cable pull is complete.

Pulling Horizontal Cable in Conduit with Fishtape

Overview

Horizontal cable is installed between the telecommunications closet and the work area outlets for workstations. It supports many different kinds of information for its users:

- Voice communications
- Data communications
- Other building information systems (such as CATV, alarms, security, and audio)

Pulling horizontal cabling is one of the most important jobs of a successful installation. Because the cable seems bulky and well protected, some cabling installers incorrectly believe that it is almost indestructible. Cable must never be bent or kinked excessively.

Conduit installations are usually designed to be parallel or perpendicular to the external walls of the building. No one conduit segment will be more than 30 m (98 ft) in length and no more than two 90-

degree bends are allowed in one segment. Exceeding these limits increases the coefficient of friction and can possibly stretch the cable and damage it.

A pull string is a thin cord used to pull cable through the conduit. The pull string must extend through the entire length of the conduit before cable can be pulled. This can be done by a manual method using fishtape or by pressurized-air methods using a foam rat or a foam ball attached to a pull string. Fishtape passes through the conduit to reach the pull string attached to the cable at the far end. Some fishtapes have swivel-type clips on the end for attaching to objects that need to be pulled back through the conduit. Fishtape is used to pull the string through the conduit, not for pulling the actual cable. Some pull string is designed to break if excessive tension is placed on the cable.

Horizontal cabling in conduit may require the use of a lubricant. This is necessary only for high-pair-count cable. Many products are available for lubricating the cable as it enters the conduit to assure a smooth pull.

Always label cables and their reels prior to pulling cable into place. It is easier to identify and label the cables before they are pulled through the conduit.

The building plans or blueprints are the record of what is in the building. After the installation is complete, the building owner has the as-built plans to refer to for future work. It is a lasting record of cable information that documents the placing of cables.

Steps—conduit pull

Step	Pulling horizontal cable in conduit with fishtape
1	Estimate length of run by walking off the distance or using a measuring wheel. ■ Find the area on the floor that is just below the location in the ceiling where the conduit run is located. ■ Walk the distance of the cable run, noting where the conduit bends and pull boxes are located. ■ Estimate the distance walked and allow for all changes in elevation (e.g., include the distance from conduits to floors or ceilings). **Note.** A common general practice is to count ceiling tiles of known lengths, such as 0.6 m (2 ft) by 0.6 m (2 ft) lay-in or 0.6 m (2 ft) by 0.6 m (2 ft) decorator tile. ■ Ensure that each cable dispenser has enough cable to reach the full length of the run.

- Place an identification label on each cable end.
- Label each cable dispenser and cable a few feet back from the cable's bitter end. Do not put the label too close to the end or the tape will cover the label.

2 Determine the length of fishtape required. After estimating the length of the run, add a few feet to each end to be sure enough fishtape is available.

3 Feed fishtape through conduit.
- Put on all necessary personal safety equipment.
- Feed the fishtape into the end of the nearest conduit.
- Do not force the fishtape if the conduit contains existing cables.
- If there is a metallic tab on the end of the fishtape, cover it with electrical tape to prevent snagging other cable already in the conduit.
- Never feed a fishtape into a conduit without knowing where it may go. It could accidentally be fed into a live electrical junction box.
- Push the fishtape through the conduit until it comes out the far end of the conduit.

4 Attach pull string to fishtape.
- Go to the far end of the conduit where the end of the fishtape is now located.
- The end of the fishtape will have a hook, a ring, or some type of attachment device.
- Attach the end of the fishtape to the pull string to be pulled through the conduit securely to prevent it from coming off during the pulling operation.
- If the clip at the end of the tape has been removed, secure the attachment by wrapping a piece of electrical tape around the pull string and the fishtape.

5 Pull the fishtape out of the conduit.
- Go back to the other end of the conduit.
- Slowly rewind the fishtape reel to retrieve the pull string from the far end of the conduit, storing the fishtape in the holder as it is taken up.
- Detach the fishtape from the pull string.
- If there will be a delay prior to cable-pulling efforts, tie off the pull string at both ends. This will prevent accidentally pulling the string back into the conduit and the need to re-fish the conduit.
- Put the fishtape away.

262   Chapter Four

6. Overcoming pulling problems.
   - Feed out a few meters (feet) of cable to prepare for entering the conduit.
   - Arrange the cable(s) to form a smooth transition from a single cable to a larger bundle. Tape cables as necessary to hold them in place while the pull string is attached to the cable bundle.
   - Attach the pull string using a rolling hitch (three half hitches and a clove hitch). Each half-hitch will bite into the cables and prevent them from slipping. The clove hitch is of little importance as long as a tight knot is placed around the cables after each half-hitch. Place electrical tape over the knots to prevent them from slipping or becoming caught during the pull.

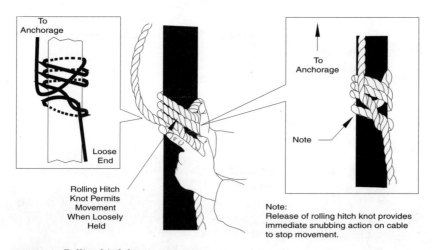

Figure 4.5 Rolling hitch knot.

   - Attach a trailer string to the cables to provide a string for future pulls.
   - As the cable is being pulled through the conduit, keep slack available at the entrance so the cable can flow freely into the conduit.
   - Guard against developing excessive bends or kinks as cable is led to the conduit.
   - Some conduits may require the use of pulling lubricant to prevent excessive pulling tension on the cable(s).

7. Pull cable.
   - Maintain proper bend radius of four times the diameter of the 4-pair cable.

- Monitor pull force closely during installation to be sure that the manufacturer's specifications are never exceeded.
- Maximum pull force for one 4 pair, 24 AWG cable is 110 N (25 lbf). Bends and lubricants also affect performance.
- Avoid excessive tension and deforming of the cable when going around corners or bends.
- Consider using a 11 kg (25 lb) breakaway swivel attached between the pull rope and the unshielded twisted-pair (UTP) cable to ensure that no more than 11 kg (25 lb) of tension is exerted on the pulled cable.
- Leave enough slack in the TC to reach the furthest corner and add the distance from floor to ceiling. This allows service slack to be stored within the closet for any possible mistakes or future reconfigurations.
- Most optical fiber distribution cabinet manufacturers require an additional 3 m (10 ft) of slack within their cabinets in addition to the service slack described above.
- Allow enough slack to ensure a minimum of 0.3 m (1 ft) of slack for copper conductor cables and 1 m (3.3 ft) of slack for optical fiber cables at the work area after termination.
- Tie off both ends of the trailer string.

8   Identify cables.
- Once the amount of required service slack has been determined, mark the cable prior to cutting it off the cable dispenser.
- The cable label shall be a unique identifier that is clearly visible on each end of the cable after the pull is complete. If this is a temporary label, ensure that the person who follows can understand it.

9   Document cable information.
- Obtain a copy of the plans or blueprints.
- On the plan show clearly what type of cables were installed.
- Document the origination and termination point of each cable.
- On the plans show clearly which conduit was used.
- Describe the application of the installed cables.
- Install a label on the cable at each end that conforms to the labeling scheme to be used on the project. The label should identify the cable(s) so that the cabling installers will be able to identify the cables for termination. When terminating the cable, permanent labels should be installed.

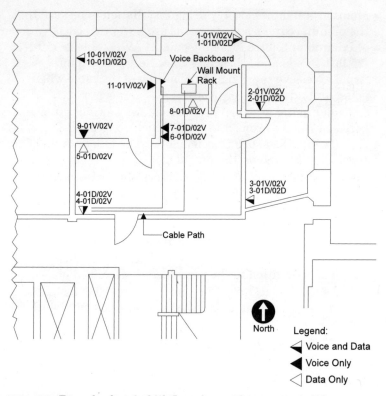

Figure 4.6 Example of marked-job floor plans with common symbols.

10 Job site clean up.
    Keeping a job site clean:
    - Prevents development of safety hazards.
    - Reflects on the professionalism of the cabling installer and the cabling installation company.

    General housekeeping specifics include but are not limited to:
    - Picking up used pull strings and pull ropes immediately after use.
    - Disposing of removed sheath and wire scrap from cables terminated in closets.
    - Storing cable reels and boxes when pulling function is complete.
    - Placing termination scrap in an appropriate container while terminating cables at work area outlets.
    - Vacuuming residue from cutouts in gypsum walls.

- Disposing of all personal-use items (i.e., luncheon materials, coffee cups), and used cleaning supplies.
- Storing tools and equipment properly at the end of the work day.

When the fishtape method is not practical, i.e., excessive conduit lengths, the following optional methods may be employed.

Optional Method 1—Blow string through conduit. Alternative methods to using fishtape include a vacuum cleaner, or air bottle and foam ball. These methods produce the intended result of retrieving the pull string at the far end.

- Select a propellant object that is lightweight, such as a plastic bag crumpled up or a piece of foam rubber designed for the purpose.
- Tie a pull string to the object.
- Place the propellant object in the conduit.
- Place the grommet on the conduit and attach the air hose.
- Using pressurized air, blow the ball attached to the string through the conduit until it reaches the other end.

Warning. Never look into the end of a conduit when a pull line is being blown through it.

- Go to the opposite end of the conduit and tie off the pull string, separating it from the ball.

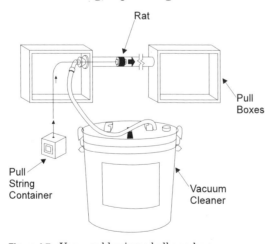

Figure 4.7 Vacuum blowing a ball or a bag.

Optional Method 2—Vacuum string through conduit.
- Attach a vacuum hose to the conduit to pull the object forward from one pull point to the next.

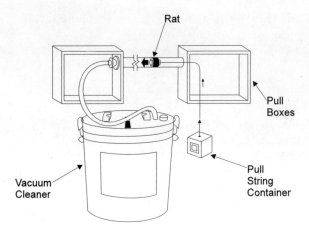

Figure 4.8 Vacuuming a ball.

- Retrieve the foam ball with the pull string attached.
- Secure the string at the receiving end of the conduit.

If it is necessary to pull heavy cable, attach the pull string to a heavier string or rope strong enough to pull the weight of the cable in the conduit.

Pulling Horizontal Cable in Open Ceiling

Overview

The procedure for cable installations in open ceilings is different from that in conduits. Cables may be supported by beam clamps which hold bridle rings of various sizes to accept the cable. J-hooks may also provide cable support. Beam clamps and J-hooks are mounted every few meters (feet) and at each change of direction; the specific distance is such that each J-hook supports less than 11 kg (25 lb). Remember, the minimum bending radius for high performance cables is four times outside diameter of the cable. High performance cables are not normally supported by bridle rings, D-rings, or wire hooks. Always use supports that provide support for minimum bending radii.

Since there is no conduit, the weight of the cable is supported by mounts above the ceiling. The cable path must be determined before the support devices or J-hooks are mounted. Alternatively, the cable is manually pulled through the red iron supports.

Similar to conduit pulls, a pull string with adequate tensile strength or pull tape may be used to pull cable into place in open ceiling runs by first threading it through the supports. A gopher pole, which telescopes

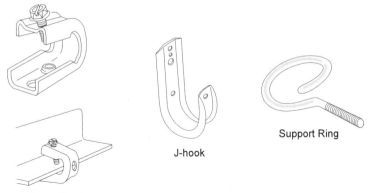

Figure 4.9 Beam clamps, support rings, and J-hooks.

to various lengths, is helpful in threading the string through the supports. Use of the gopher pole limits the number of ceiling tiles that must be moved to provide access to the work area. It also reduces the number of times a ladder has to be moved.

Other devices available for installing pull strings in open ceiling installations include a small crossbow or a remote-controlled toy.

> **Note.** Do not attach tools to pull strings and throw them through the ceiling. This presents a safety hazard and may cause damage to the ceiling system, tiles, or other utilities contained in the ceiling space.

Each 90-degree bend identifies a pull point, which requires a cabling installer at that spot to assist in pulling the cable. Using a pulley can help minimize labor.

Steps—open ceiling

Step	Pulling horizontal cable in open ceiling
1	Tensile strength of string. ■ Pull string has a tensile strength rating which appears on the box. ■ Tensile strength ratings range typically from 330 N to 880 N (75 to 200 lbf).
2	Verify manufacturers' specifications for cable tensile strength limitations. ■ Read manufacturers' guidelines for instructions as to cable strength limitations.

- Excessive tension on the cable stretches the conductors inside the cable, degrading the cable's performance.
- Pay close attention to the requirements for maximum bend radius, which for 4-pair UTP, is equal to four times the diameter of the cable.
- For extensive quantities of cable in a ceiling environment use cable trays or ladder racks.
- Avoid selecting paths that require exceeding the maximum cable bend radius.

3. Route string through the ceiling area.
   - Select a gopher pole that will reach the required distance. Telescoping poles reach from 1.83 to 7.62 m (6 to 25 ft).
   - Attach the pull string to the hook on the end of the gopher pole.

4. Attach string to cable.
   - Tape the pull string to the lead end of the cable, along with an additional pull string for future use.
   - Manually lift the cable into a position to follow the string through the open ceiling area.

5. Place cable in support devices.
   - Lift the cable(s) and place them in the cable supports. Once all cables have been installed into a support, close the support if required.

6. Precautions.
   - Pull cable, being careful not to snag or pull the string against sharp objects, such as ceiling obstructions (HVAC duct, etc.) or ceiling grid stringers.
   - Avoid creating friction that causes burns or tears in the jacket of the cable.

7. Job clean up.
   Keeping a job site clean:
   - Prevents development of safety hazards.
   - Reflects on the quality of cabling installer.

   General housekeeping specifics include but are not limited to:
   - Picking up used pull strings and pull ropes immediately after use.
   - Disposing of removed sheath and termination scrap from cables terminated in closets.
   - Storing cable reels and boxes when pulling function is complete.
   - Placing termination scrap in an appropriate container while terminating cables at work area outlets.

- Vacuuming residue from cutouts in sheetrock walls.
- Disposing of all personal-use items (i.e., lunch materials, coffee cups) and used cleaning supplies.
- Replacing all ceiling tiles after use.
- Storing tools and equipment properly at the end of the work day.

Pulling Backbone in Vertical Pathway—from Top Down

Overview

Backbone cable in the vertical riser shaft may be high-pair-count cable, which is extremely heavy, or high strand count optical fiber cable. Therefore, several considerations must go into deciding the best method for making this kind of installation. The cable may be placed vertically directly in an open riser shaft, through cores, sleeves or slots, or within a large conduit.

The cabling installer must determine the size and type of reel onto which the cable is to be loaded. These reels may be steel which must be returned to the factory, or wood which is disposable. The size of the reel is dependent on the size and quantity of the cable. High-pair-count cable can be loaded onto reels that are up to 2.1 m (7 ft) high and 1.5 m (5 ft) wide.

Cables should be ordered with a factory-equipped pulling eye. If this is not possible, a substitute can be installed prior to placing the cable.

When cable is received at the job site, cable length should be verified. Although cable is shipped from the manufacturer with a cable run label that indicates the length of the cable, do not take this for granted. Inspect both ends of the cable to verify the footage markings on the outer jacket and compute the actual length of the jacket. This is the only accurate method for verifying cable length. When shipped from a distributor, cable is usually taken from a larger reel and spooled onto a smaller reel. Sometimes a distributor has not verified the cable length and it may be too short or too long.

Determine if a loading dock is available to off-load the reel upon delivery to the site. An elevator of sufficient size and load capacity must be available to transport the reel to the top floor. Many commercial buildings have freight elevators that are designed specifically for this type of operation.

Once a reel is brought to the top floor, it must be set up for the placing operation. A reel dolly or jackstand is necessary to lay out the cable. It may be necessary to use a reel brake to help control the descent of the cable as it is pulled off the reel.

During pulling operations, a reel must be situated in a location that provides enough slack for the top floor once the placing operations are complete.

Pulleys may be needed to handle the cable from the reel location to the point where it will be dropped down to lower floors. These pulleys can be attached to the overhead structure in the building and provide a pathway for the cable as it is being placed.

In the closet where the cable will enter the vertical pathway, a bullwheel will be required to ensure that the jacket is not damaged as it enters the pathway. Situate the bullwheel so that it will allow the cable to drop vertically into the pathway. Bullwheels must be attached to the building structure, thereby preventing any lateral movement. Any lateral movement will cause the cable to become misaligned with the vertical pathway and will damage the outer jacket.

Cabling installers should be located on each floor through which the cable will pass as well as the floor where it will terminate. They must be equipped with the proper tools to perform critical tasks during the placing operation. Some of the tools include:

- Personal protection equipment such as safety glasses, hardhats, gloves, etc.

- Communication equipment such as portable, handheld radios.

- Temporary restraining devices for the cable. Cable type and size will determine the necessary restraining devices. One of the most common methods of temporarily restraining cable is the use of a large hemp or manila rope equipped with an eyelet. The eyelet is attached to some secure structure in the closet. A half hitch is made in the rope, and the rope is pulled on by the cabling installer. The amount of tension placed on the rope will control how quickly the cable slips through the half hitch. This method allows the cable to be halted and restarted during cabling installation without losing control of the cable as it descends (with or without a reel brake).

A guide rope should be installed throughout the cable route. This rope is used to guide the cable through the various spaces rather than pull the cable.

If a pulling eye was not factory installed, install a device to substitute for it. An alternate method of pulling is the use of a pulling grip which is a flexible wire mesh device that clamps down on the cable as tension is applied. These devices are available in various sizes based on cable diameter. Ensure that it is properly sized for the cable. Place it over the cable jacket and then tape it to the jacket. This will prevent it from coming off should the pull be halted and the tension relaxed on the grip.

If a wire mesh grip is not available and the pathway is large enough, the pull rope can be attached over the jacket in a series of half hitches. Vinyl tape is placed over the guide rope, securing it to the cable.

If the pathway is too small to use half hitches, and a pulling grip is not available, a pulling eye can be created on the cable using the conductors. To do this, remove a section of the outer jacket, shield, and inner jacket. Separate the conductors into two equal groups. Weave the two groups back onto themselves, forming an eye on the end of the cable. Once the eye is formed, vinyl tape should be installed over its entire length, including a section of the outer jacket. By using this device, tension will be evenly dispersed over each wire in the cable as well as the jacket.

The coefficient of friction may be one of the obstacles to installation of cables, especially in conduit pathways. The use of cable lubricants can significantly reduce friction and speed cable installation. Select the lubricant to be used based on the type of pull, type of conduit, cable jacket, and size of the cable. Install the lubricant as the cable is placed.

Once everyone and everything is in place, the cable can be installed. Communicate with each member of the team, and advise them that the pull is starting. Begin to lower the cable and, monitor its progress throughout the pull. Continue the placing operation until the cable is completely placed.

When the cable reaches the bottom floor, the cabling installer must temporarily restrain the cable. Beginning on the bottom floor, the cabling installer should pay out the additional slack needed in each closet. Once the slack is brought into the closet, it should be secured in its final location.

Securing the cable on each floor ensures that the weight of the cable will not become excessive at any one point. There are many different devices that can be used to secure the cable. The most common one used today is the split grip, which is a wire mesh grip that is open on one side. The wire mesh grip is placed around the jacket of the cable. A pin is then weaved through the open sides of the wire mesh grip, securing them around the cable jacket.

A section of steel rod, a minimum of 13 mm (0.5 in) thick, is inserted through the loop in the end of the wire mesh grip. The cable is then allowed to slide through the opening in the vertical pathway until the steel rod lays on the top of the pathway.

Another method of securing the wire mesh grip is to attach it to the building or a plywood backboard with an anchoring device.

The cable should then be identified, labeled, and the documents updated to reflect the work operation.

After completing the work operation, the cable reel and any remaining cable can be returned to storage or prepared for salvage. If the cable was shipped on a wooden reel, the reel can be dismantled and disposed of. If the cable was shipped on a metal reel, it should be returned to the manufacturer for credit.

Step	Pulling backbone in vertical pathway—from top down
1	Verify proper cable length and path. ■ Examine the vertical backbone pathway for the installation and check for a clear pathway. Avoid, if possible, a cable path with obstacles or transitions. Check vertical pathways visually and, if necessary, with a flashlight. ■ Determine the actual length of the cable run, based on information from the blueprint or by installing a pull tape equipped with sequential footage markings. This pull tape, if sufficiently sized, can also be used as the guide rope.
2	Move cable reels to the upper floor of the vertical feed point.
3	Set up the cable pulling area and put reels on a jackstand or reel dolly. Secure the work area.
4	Attach a reel brake to the cable reel.

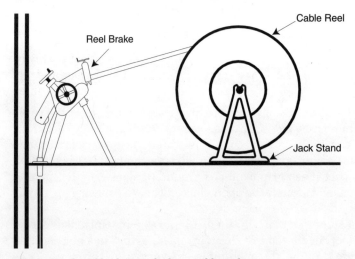

Figure 4.10 A reel brake attached to a cable reel.

5	Install and attach guide rope to cable lead. This pull rope helps to guide the cable as it descends.
6	Attach pulleys, if required. ■ Decide whether the weight (based on size and length of cable) requires the use of pulleys to help control the gravitational pull.

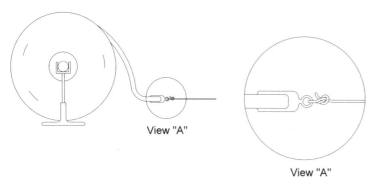

Figure 4.11 A pull rope attached to cable lead.

- Feed the cable off the reel and onto the pulleys mounted above the route to the pathway so as to support the cable to the point of entry into the vertical pathway.
- Mount the pulley on any appropriate superstructure available at the location or on overhead steel. If none is available in the place where needed, set up a swing-set type of pulley mechanism.
- Use a bullwheel, if necessary, to provide a sweeping arc of the cable down into the vertical shaft.
- Use a bullwheel to direct the cable into the vertical pathway keeping the cable from scraping the edge of the floor and chafing the jacket, which damages the integrity of the cable.

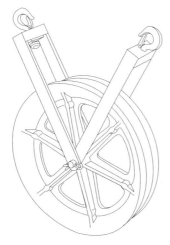

Figure 4.12 Bullwheel.

274　Chapter Four

    7   Communicate with coworkers for routing.
- Establish locations for coworkers to clearly view the progress of the cable descending.
- Have constant communication between floors to notify the person controlling the reel brake of the progress being made.
- Listen carefully for a report from a coworker who may spot a problem in the cable path that no one else can see.

    8   Pull (or drop) until finished.
- Watch and observe the functions of any pulley in use.
- Maintain control of all pull strings or guide ropes in use.
- Stay in position as a communicator with coworkers until the pull or drop is finished.
- Be prepared to temporarily take up the cable should a problem develop in the placing operation and prior to permanently securing the cable on each floor.
- All cables must be secured at each floor level. Cables that are extremely heavy may fall if they are not properly installed. Free-hanging cables create a serious hazard.

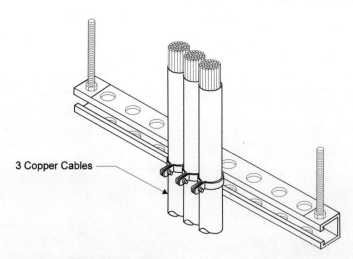

Figure 4.13　Channel with straps.

    9   Route and secure cables in vertical pathway.

    10   When the vertical pathway enters a telecommunications closet:
- Make a sweep of the cable toward the termination point.
- Secure the cable in D-rings, with cable ties, straps, split, wire mesh grips, or clamps and place in provided cable trays.

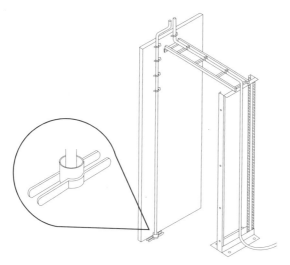

Figure 4.14 Cable on tray from vertical pathway.

- Mount the cable to the wall using D-rings, cable clamps, or special-purpose cable ties, for cables of 25-pair or less.

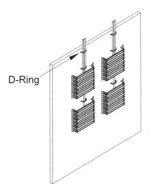

Figure 4.15 Backboard layout with D-rings.

11   Identify cable.
- Place a unique identification number (from the installation drawings) at each end of the cable showing the point of origin and destination. Also, place a label on the cable where it enters or leaves closets on each floor.

12   Document cable information.
- Obtain a copy of the plans or blueprints.
- Show clearly what kind of cable was installed.

- Show clearly which conduit was used.
- Describe the purpose of the installed cables.

Pulling Backbone in Vertical Pathway— from Bottom Up

Overview

Conditions that exist when pulling backbone cable from the top down are the reverse of those that exist when pulling from the bottom up. The first choice is to pull the cable from the top down. Sometimes, however, it is not possible to get the large spools of cable up to the top floor of a building. In those situations, specific equipment must be brought in to handle the task of hauling the cable up through the riser to the necessary heights. The heavy weight of the cable, in addition to the gravitational pull, makes a bottom-up installation more difficult.

The cabling installer must determine the size and type of reel onto which the cable is to be loaded. These reels may be steel which must be returned to the factory, or wood which is disposable. The size of the reel is dependent on the size and quantity of the cable. High-pair-count cable can be loaded onto reels that are up to seven feet high and five feet wide.

Cables should be ordered with a factory-equipped pulling eye. If this is not possible, a substitute can be installed prior to placing the cable.

When cable is received at the job site, cable length should be verified. Although cable is shipped from the manufacturer with a cable run label that indicates the length of the cable, do not take this for granted. Inspect both ends of the cable to verify the footage markings on the outer jacket and compute the actual length of the jacket. This is the only accurate method for verifying cable length. When shipped from a distributor, cable is usually taken from a larger reel and spooled onto a smaller reel. Sometimes a distributor has not verified the cable length and it may be too short or too long.

Determine if a loading dock is available to off-load the reel upon delivery to the site.

Before pulling the cable, determine the required cable length and how the weight will be distributed. The number of floors and closets that must be accessed are factors in the decision process.

Once the reel is brought to the bottom floor, it must be set up for the placing operation. A reel dolly or jackstand is necessary to allow for paying out the cable. The reel must be situated in a location that will provide enough slack for the bottom floor once placing operations are completed.

Cable sheaves may be necessary to handle the cable from the reel location to the point where it will be pulled up to upper floors. These cable sheaves can be laid on the floor along the route to the vertical pathway and provide a pathway for the cable as it is being placed. In the closet where the cable will enter the vertical pathway, a bullwheel will be required to ensure that the jacket is not damaged as it enters the pathway. The bullwheel should be located so that the cable ascends vertically into the pathway. A bullwheel must be attached to the building structure so that they do not move laterally. Any lateral movement will cause the cable to become misaligned with the vertical pathway and damage the outer jacket. This operation may require that the bullwheel be attached to the building in three places. The three attachments will support the bullwheel and prevent it from pulling away from the pathway entrance and from moving laterally.

Cabling installers should be located on each floor through which the cable will pass as well as the top floor where it will terminate. These individuals must be equipped with the proper tools to perform critical tasks during the placing operation. Some of the tools include:

- Personal protection equipment such as safety glasses, hardhats, gloves, etc.
- Communication equipment such as portable, handheld radios.
- Temporary restraining devices for the cable. Cable type and size will determine the necessary restraining devices. One of the most common methods of temporarily restraining cable is the use of a large hemp or manila rope equipped with an eyelet. The eyelet is attached to some secure structure in the closet. A half hitch is made in the rope, and the rope is pulled on by the cabling installer. The amount of tension placed on the rope will control how quickly the cable slips through the half hitch. This method will allow the cable to be halted and restarted during placing operations without losing control of the cable if it falls (with or without a reel brake).

The pull rope chosen for this operation is critical. Manila or hemp rope should be used for this operation because it does not stretch when placed under tension. It also does not deteriorate quickly when subjected to the friction (and heat caused by the friction). It does not have a tendency to twist as tension is applied. This will require the use of a swivel between the cable and the pull rope. The swivel will prevent the cable from twisting during the placing operation thus damaging the jacket, the shield, and the conductors.

If a pulling eye was not factory installed, install a device to substitute for it. An alternate method of pulling is the use of a pulling grip which is a flexible wire mesh device that clamps down on the cable as tension is applied. These devices are available in various sizes based on cable diameter. Ensure that it is properly sized for the cable. Place it over the cable jacket and then tape it to the jacket. This will prevent it from coming off should the pull be halted and the tension relaxed on the grip.

If a wire mesh grip is not available and the pathway is large enough, the pull rope can be attached over the jacket in a series of half hitches. Vinyl tape is placed over the guide rope securing it to the cable.

If the pathway is too small to use half hitches, and a pulling grip is not available, a pulling eye can be created by using the conductors. To do this, remove a section of the outer jacket, shield, and inner jacket. Separate the conductors into two equal groups. Weave the two groups back onto themselves, forming an eye on the end of the cable. Once the eye is formed, vinyl tape should be installed over its entire length, including a section of the outer jacket. By using this device, tension will be evenly dispersed over each wire in the cable as well as the jacket.

Attach the swivel to the pull rope, then attach the cable to the swivel.

To lift a cable of significant weight, a tugger is required. A tugger is an electrical motor equipped with a capstan. It is secured to the building structure on the top floor. The pull rope is wound around the capstan. When the cabling installer places tension on the rope, it tightens around the capstan. Friction created by the tightening causes the rope to advance toward the cabling installer. The installer can control the rate at which the rope advances, thus controlling the advance of the cable along its route.

> **Warning.** It is absolutely critical that the tugger be securely attached to the building structure. If the tugger is not properly attached to the building structure and comes loose, it becomes a missile and can cause damage.

The coefficient of friction can be an obstacle when installing cable, especially in conduit pathways. The use of cable lubricants can significantly reduce friction and speed cable installation. A lubricant should be selected based on the type of pull, type of conduit, cable jacket, and cable size. The lubricant should be installed as the cable is placed. When placing the cable from the bottom up, use a gel-type lubricant.

Once everyone and everything is in place, the cable can be installed. Communicate with each member of the team and advise them that the pull is ready to begin. Begin pulling the cable up through the vertical pathway. Monitor its progress throughout the pull. Continue the placing operation until the cable is completely placed. Ensure that enough slack is pulled to the top floor so that it can be lowered back down to the individual floors, where it may be required, before it is permanently secured.

Once the cable reaches the top floor, the cabling installer must temporarily restrain the cable. Beginning on the bottom floor, pay out the additional slack that is needed in each closet. Once the slack is brought into the closet, the cable can be secured in its final location.

Securing the cable on each floor ensures that the weight of the cable will not become excessive at any one point. There are many different devices that can be used to secure the cable. The most common one used today is the split grip, which is a wire mesh grip that is open on one side. The wire mesh grip is placed around the jacket of the cable. A pin is then weaved through the open sides of the wire mesh grip, securing them around the cable jacket.

A section of steel rod, a minimum of 13 mm (0.5 in) thick, is inserted through the loop in the end of the wire mesh grip. The cable is then allowed to slide through the opening in the vertical pathway until the steel rod lays on top of the pathway.

Another method of securing the wire mesh grip is to attach it to the building or a plywood backboard with an anchoring device.

The cable should then be identified, labeled, and the documents updated to reflect the work operation.

After completing the work operation, the cable reel and any remaining cable can be returned to storage or prepared for salvage. If the cable was shipped on a wooden reel, the reel can be dismantled and disposed of. If the cable was shipped on a metal reel, it should be returned to the manufacturer for credit.

Step	Pulling backbone in vertical pathway—from bottom up
1	Verify proper run and cable length. ■ Examine the conduit/vertical pathway for the planned installation and check for a clear pathway. Avoid, if possible, a cable path with obstacles or transitions. ■ Use the blueprints to estimate the actual length of the cable run or verify by installing a pull tape equipped with sequential footage markings. This pull tape, if sufficiently sized, can also be used as the guide rope.
2	Move cable reels into place and secure the pulling area. Place the reel on a reel dolly or jackstands. If necessary, place cable sheaves along the floor to the entrance of the vertical pathway.
3	Install the pull rope. ■ Check to see that the pull rope, which must hoist the cable to the upper floors, is lowered down to the proper building level so it can be attached to the cable. ★ Bring pull rope to upper floor. ★ Secure one end at top. ★ Lower rope to bottom floor or vertical feed point.

4. Determine appropriate rope size/strength.
   - A critical factor in the success of the job depends on the strength of the rope used in the pull.
   - Check manufacturer's specifications about the length and weight of the cable that will be installed to ensure the rope meets the strength requirements.
5. Attach the cable to the rope.
   - Check the pulling end of the cable to see if it is equipped with a factory-installed pulling eye. Attach the pull rope to a swivel. Attach the swivel to the cable.
   - Utilize a wire mesh grip, if a pulling eye is not provided on the cable.

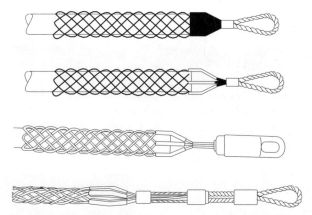

Figure 4.16 Wire mesh grips.

6. Use a bullwheel, if required.
   - Use a bullwheel to maintain the proper bend radius as the cable comes off the reel and is fed up into the opening.
   - The arc formed by pulling the cable under the bullwheel protects the integrity of the conductors.
   - Unlike pulleys, bullwheels do not alter the pulling force required.
7. Attach tugger and pulleys, if required; verify that they are anchored and secured.
   - Confirm that the tugger is at the required location.
   - Verify that the tugger is solidly anchored and firmly secured. Adjust tugger pulley wheels to proper tension to begin tugging.
   - Verify that the short, tugging motions that the tugger makes will not move the equipment due to vibration.
   - Verify that each pulley is positioned in a secure location and that it is firmly anchored.

Pulling Cable 281

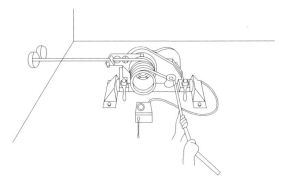

Figure 4.17 Tugger in position and properly secured to a concrete slab.

- Pulling equipment must be anchored to the structure of the building (i.e., concrete anchors, steel cables, bolts, etc.).

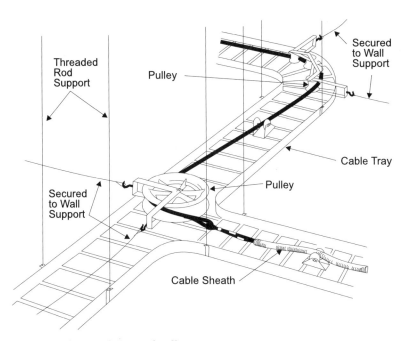

Figure 4.18 A properly secured pulley.

8   Lubricate cable/pipe.
    - Obtain an approved lubricant which is not abrasive or damaging to the cable.

- Apply a liberal amount of the lubricant to the open end of the conduit.
- Apply the lubricant to the end of the cable where it is attached to the pull rope. Some dispensers are commercially available for automatically putting the lubricant on the cable as it enters the conduit.

9. Communicate with coworkers.
   - As the cable pull begins, communicate with coworkers to pass on information about how things are going.
   - Ask all coworkers to call out the job status—if everything is OK or if there is a problem.
   - Encourage everyone to take responsibility for the job and to report even the slightest appearance of a problem. It is better and safer to prevent the problem than to correct or repair one.

10. Start tugger slowly; notify personnel at the other end.
    - Make sure a coworker is at each end of the pull and at all midway positions. Otherwise, workers will have to move between floors. The pulling operation may have to be temporarily halted for them to relocate.
    - Notify coworkers of the conditions at the end where cable is being fed into the riser shaft.
    - Start the tugger slowly, so that the tugs do not come too fast and everyone can observe what is going on.

11. Maintain operation until cable is in place. Keep coworkers in position and watching for problems until the pull is complete.

12. Route and secure the cable, as necessary.
    - Secure cable in riser shaft.
    - Bring cable into telecommunications closet.
    - Bring cable toward the termination point.
    - Continue securing the cable with tie wraps or other appropriate devices to the backboard, tray, or rack.
    - Prepare service slack, leaving slack in the cable for termination. Mount the cable to the wall, using D-rings, clamps, or ladder rack and cable ties.
    - The service slack is used to:
      ★ Help relieve tension on the cable.
      ★ Provide slack for future moves and changes.
    - The service slack should be long enough to reach the other side of the closet in case of relocated or additional equipment. Keep in mind that these cables usually require a minimum bending radius of 10 times the outside diameter of the cable.

13 Identify cable.
- Place a unique identification number (from the installation drawings) at each end of the cable showing the point of origin and destination. Also, place a label on the cable where it enters or leaves closets on each floor.

14 Document cable information.
- Obtain a copy of the plans or blueprints.
- Show clearly what kind of cables were installed.
- Show clearly which conduit was used.
- Describe the purpose of the cables installed.

Pulling Backbone—Horizontal

Overview

Horizontal backbone cable is used to interconnect closets. Horizontal backbone multipair cable may or may not be installed in conduit between closets.

Cable support mechanisms are required to provide whatever protection is necessary to keep the cable free of excessive tension. As noted earlier, excessive tension will often cause damage to the pairs inside the cable.

Horizontal backbone cable runs may be supported by beam clamps, cable trays, lay-in raceway, or trapeze system.

ANSI/TIA/EIA-569-A guidelines require cable pulls to be no longer than 30 m (98 ft) between pull points. Maintaining proper pulling tension is essential. Pulling too hard may damage the cable. High-pair-count backbone cables must be placed individually to manage their weight.

The cabling installer must determine the size and type of reel onto which the cable is to be loaded. These reels may be steel which must be returned to the factory or wood which is disposable. The size of the reel is dependent on the size and quantity of the cable. High-pair-count cable can be loaded onto reels that are up to 2.7 m (9 ft) high and 1.5 m (5 ft) wide.

Cables should be ordered with a factory-equipped pulling eye. If this is not possible, a substitute can be installed prior to placing the cable.

When cable is received at the job site, verify the cable length. Although cable is shipped from the manufacturer with a cable run label that indicates the length of the cable, do not take this for granted. Inspect both ends of the cable to verify the footage markings on the outer jacket and compute the actual length of the jacket. This is the only accurate method for verifying cable length. When shipped from a distributor, cable is usually taken from a larger reel and spooled onto a smaller reel. Sometimes a distributor has not verified the cable length and it may be too short or too long.

Determine if a loading dock is available to off-load the reel upon delivery to the site.

Horizontal backbone installations sometimes require the use of pulling equipment, such as a pulley or a tugger. A tugger is an electrical motor equipped with a capstan. It is securely attached to the building structure on the top floor. The pull rope is wound around the capstan. When the installer places tension on the rope, it tightens around the capstan. Friction created by the tightening causes the rope to advance towards the installer. The installer can control the rate at which the rope advances, thus controlling the advance of the cable along its route. In other installations, the tugging is adequately performed manually by an installer.

The pull rope chosen for this operation is critical. Manila or hemp rope should be used for this operation because it does not stretch when placed under tension, nor does it deteriorate quickly when subjected to friction and heat. However, it does have a tendency to twist as tension is applied. To prevent this, a swivel can be used between the cable and the pull rope. The swivel prevents the cable from twisting during the placing operation and prevents damage to the jacket, the shield, and conductors.

If a pulling eye was not factory installed, install a device to substitute for it. A wire mesh grip that is properly sized can be installed on the cable. The mesh grip should be placed over the cable jacket and then taped to the jacket. This will prevent it from coming off should the pull be halted and the tension relaxed on the grip.

When a wire mesh grip is not available and the pathway is large enough, the pull rope can be attached over the jacket in a series of half hitches. Vinyl tape is placed over the guide rope, securing it to the cable.

If the pathway is too small to use half hitches, a pulling eye can be installed by using the conductors. To do this, remove a section of the outer jacket, shield, and inner jacket. Separate the conductors into two equal groups. Weave the two groups back onto themselves, forming an eye on the end of the cable. Once the eye is formed, vinyl tape should be installed over its entire length, including a section of the outer jacket. By using this device, tension will be evenly dispersed over each wire in the cable as well as the jacket.

Attach the swivel to the pull rope; attach the cable to the swivel.

Installers should be equipped with the proper tools to perform critical tasks during the placing operation. Some of the tools include:

- Personal protection equipment such as safety glasses, hardhats, gloves, etc.

- Communication equipment such as portable, handheld radios.

Placing horizontal, high-pair-count backbone cables presents the same problems as pulling from the bottom up. The probability of additional changes in direction will increase the difficulty. The coefficient of friction increases because the entire weight of the cable is horizontal and creates friction with everything it contacts. Open-type supports, such as cable trays and ladder racks, preclude the use of lubricants. As a result, pulleys may have to be employed to overcome the problem of friction.

Once everyone and everything is in place, the cable can be installed. After informing each member of the team that the pull is starting, begin pulling the cable up through the horizontal pathway. Monitor its progress throughout the pull, and continue the placing operation until complete. Ensure that enough slack is pulled to the last closet.

Cable should be identified, labeled, and the documents updated to reflect the work operation. Documentation is an essential part of the job and includes labeling the cable and updating the building plans to specifically indicate the cable installed by the cabling team.

When the work operation is complete, the cable reel and any remaining cable can be returned to storage or prepared for salvage. If the cable was shipped on a wooden reel, the reel can be dismantled and disposed of. If the cable was shipped on a metal reel, it should be returned to the manufacturer for credit.

Step	Pulling backbone—horizontal
1	Install and attach pulling mechanism to the cable. - If possible, use manual strength only to pull cable. - When more power is required, use a tugger. - Attach the pull rope to the tugger. - Use a swivel to prevent the cable from twisting as it is pulled forward.

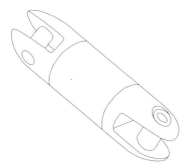

Figure 4.19 Swivel to prevent cable twisting.

2. Attach pulleys, if required.
   - Decide whether pulleys are required to advance the cable during the pull or to stabilize it.
   - Pulleys also help guide cables around obstacles.

3. Communicate with coworkers.
   - As the cable pull begins, communicate with coworkers to pass on information about how things are going.
   - Ask all coworkers to call out the job status—if everything is OK or if there is a problem.
   - Encourage everyone to take responsibility for the job and to report even the slightest appearance of a problem. It is better and safer to prevent a problem than to correct or repair one.

4. Pull cable until it is in place.
   - Carefully monitor the cable's progress to the last location.
   - Maintain control of any pulley that may be in use.
   - Maintain control of all pull ropes that may be in use.
   - Stay in position as a communicator with coworkers until the pull is finished.

5. Route and secure cable in closets, if necessary.
   - Route cable into closet.
   - Route cable to the termination point.
   - Secure the cable with tie wraps or other appropriate devices to the backboard, tray, or rack.
   - Prepare service slack, leaving slack in the cable for termination. Mount it to the wall using D-rings, clamps, or ladder rack and cable ties.
   - The service slack is used to:
     * Help relieve tension on the cable.
     * Provide slack for future moves and changes.
   - Service slack should be long enough to reach the other side of the closet in case new equipment is needed. Keep in mind that these cables usually require a minimum bending radius of 10 times the outside diameter of the cable.

6. Identify cables.
   - Place a unique identification number (from the installation drawings) at each end of the cable showing the point of origin and destination. Also, place a label on the cable where it enters or leaves each closet.

7. Document cable information.
   - Obtain a copy of the plans or blueprints.
   - Show clearly what kind of cables were installed.

- Show clearly which conduit was used.
- Describe the purpose of the installed cables.

Pulling Optical Fiber Cable

Overview

Optical fiber is one of four types of cable suitable for backbone systems. Optical fiber cable is more rugged than generally perceived; however, just as with copper cable, care must be taken when pulling optical fiber so as not to exceed the manufacturer's recommended pulling tension. Most optical fiber cable manufacturers provide maximum pulling tension based on cable construction.

There is an increased use of optical fiber cable to the desktop as the need for increased bandwidth to serve imaging and other services develop in the marketplace. Within buildings, the predominant placement of optical fiber cable is in the backbone between telecommunications closets.

Regardless of the application, whether it be backbone or horizontal, optical fiber cables may be installed in innerducts as an indicator that fiber cable is present. This method of installation also tends to reduce the pulling tension required, especially when multiple innerducts are installed in conduit.

Plenum and nonplenum rated innerducts are available in a variety of colors Though usually purchased with a pull rope preinstalled inside for attaching the fiber cable to be pulled, they are also available without pull ropes. The pathway of a fiber cable must be free of sharp bends and turns. Normally, innerducts are inside conduit, through sleeves, or placed in cable trays. Care should be taken to ensure that the properly rated innerduct is being installed.

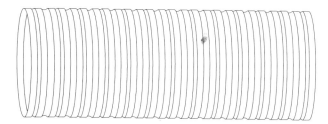

Figure 4.20 Innerduct.

As in copper cable installations, communication among coworkers is essential to ensure the pulling function progresses properly without excessive stress or tension on the cable.

Complete the pull and route the cable to the closets; identify and document cable information.

Steps—optical fiber

Step	Pulling cable—optical fiber
1	Ensure that only tested and accepted lengths of optical fiber cable are installed. • Verify that there is fiber continuity while the cable is on the reel. In order to do so, the cable must be ordered with access to both ends of the fiber on the reel. • Test the optical fiber cable for continuity. This can be done by shining a light source (such as a flashlight) into the fiber cable or using a power meter and light source or optical time domain reflectometer (OTDR). A bare-fiber adapter or mechanical reusable splice is necessary to connect either the power meter and light source or OTDR to the unterminated fiber on the reel. **Warning.** Never look directly at the end of a previously installed optical fiber cable because of the possibility that a laser light source (which is not visible to the human eye) may be present. This can cause permanent damage to the retina of your eye.
2	Install and secure innerducts. • Install innerduct for optical fiber cable. • Verify the application (plenum or nonplenum) and place the innerduct(s), depending on the specific location (vertical shaft, cable tray, or open ceiling) with the appropriate support. If secured with tie wraps in a plenum area, use plenum-rated tie wraps. • Innerduct can be ordered with or without a pull rope already placed inside. If ordered without a rope, follow the procedure for installing a pull string in a conduit. • If the requirement is to place innerducts within a conduit, determine the size and number of innerducts permitted. Innerducts are classified by outside diameter (OD), whereas trade-size conduits are classified by inside diameter (ID). This fact allows a total of four 25 mm (1 in) innerducts to be placed inside a 100 mm (4 in) trade size conduit. • 25 mm (1 in) innerduct is the usual size placed within buildings; however, 33 mm (1.25 in), 38 mm (1.5 in), and 50 mm (2 in) innerducts are available for larger fiber cables.

Figure 4.21 Four-inch conduit with four one-inch innerducts.

- Attach the optical fiber cable(s) to the pull rope for pulling of the optical fiber cable through the innerduct. There are two methods of connection commonly used. The most common is to remove the sheath for about 300 mm (12 in) length from the fiber cable and place the exposed aramid yarn strength member through a loop in the rope. Secure it with tape, as shown in the following figure.

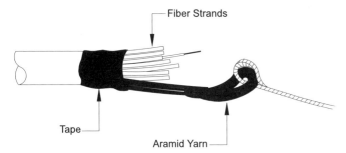

Figure 4.22 Connecting aramid yarn.

- The second method is to place the optical fiber cable(s) in a multiweave, wire mesh grip of the appropriate size which has a swivel pulling eye to which the pull rope is attached.

Figure 4.23 Multiweave wire mesh grip with swivel pulling eye.

3. Install without innerduct.
   - In those instances where optical fiber cable is installed without an innerduct, care must be taken to ensure that the run is as straight as possible.
   - At transitions, be sure that a coworker is placed at such points to relieve excessive tension during the pull. This ensures that the optical fiber cable does not rub against obstructions which can nick or cut the jacket.
4. Pull fiber.
   - Feed the optical fiber cable into the innerduct. Normally the optical fiber cable can be pulled by hand.
   - Do not exceed the recommended pull tension or the recommended bend radius of the optical fiber cable.
5. Always leave a service loop.
   - The service loop is used to relieve all tension on the optical fiber cable and provide slack for future moves or changes.
   - Always bring the optical fiber cable with the innerduct into a point of termination using the grommets provided in the optical fiber cabinet. The optical cable should be secured to the cabinet, ladder rack, cable tray, or backboard. The optical fiber strands should be stored and dressed inside the cabinet.
   - The optical fiber cable service loop should be long enough to reach the other side of the wiring closet in case relocated or additional equipment is encountered in the future and the fiber must be relocated within the closet. The service slack stored inside the cabinet should be 3 m (10 ft).
6. Secure the cable to the cabinet per manufacturer specifications.
   - All fiber cables should be anchored in the termination cabinet by using the aramid yarn of the cable to act as a strain relief, as stated in the manufacturer's instructions.
   - Care should be taken to not overtighten tie wraps around optical fiber cable when dressing. Keep them loose enough that they can be turned from side to side using finger pressure.
   For instance, the bend radius requirements of some optical fiber cables are 10 times the cable's outside diameter when the cables are at rest and 20 times the cable's outside diameter while the cables are under stress. These values may be 15 times and 30 times the cable's outside diameter for some cables.
7. Identify cables.
   - Place a unique identification number (from the installation drawings) at each end of the cable showing the point of origin and destination.

8   Document the job.
- Obtain a copy of the floor plans or blueprints.
- Indicate clearly what type of cables were installed (i.e., 24-strand, plenum-rated multimode fiber).
- Show clearly which conduit was used and which innerduct, if applicable.
- Describe on the floor plan the purpose of the installed cables [i.e., backbone from main cross-connect (MC) to 3rd floor telecommunications closet (TC)].

Chapter 5
Firestopping

Firestopping Systems	298
Testing and Guidelines for Firestops	300
Evaluation of Firestop Systems	302
Categories of Firestop Systems	305
Nonmechanical Firestop Systems	307
Firestopping for Brick, Concrete Block, and Concrete Walls	312
Firestopping for Gypsum Board Walls	315
Firestopping for Floor Assemblies	317
Firestopping for Floor/Ceiling Assemblies	317
Firestopping Considerations (General)	319
Typical Installations	319

Overview

Fire containment is an essential part of a cabling installer's job. Because it is often necessary to create openings in existing walls and floors within buildings, it becomes equally necessary to safely close all openings that were created.

Fire, smoke, and superheated gases are hazards that create great damage in a relatively short amount of time. Once they are started in a building, they can spread rapidly and are difficult to control.

One way that these hazards spread is through openings in walls, floors, and ceilings. The cabling installer must use approved methods to restore any penetrations made during the installation to the original rating of the structure (barrier) being penetrated. The cabling installer is also responsible to firestop any holes created by the removal of existing penetrations.

The term "firestop" refers to the installation of "qualified" firestop materials in holes made through fire-rated floors, walls, or ceilings for the penetration of pipes, cables, or other construction/building, service/utility items. Firestop products are used to restore the opening to the original fire-rated integrity. Most common requirements are for 1, 2, or 3 hours. In rare instances, a 4–6 hour rating must be provided.

Firestopping is necessary for five reasons:

- Save lives
- Buy time
- Protect critical systems
- Protect investments
- Liability

Firestops are expected to stop the spread of fire, but they must also stop the real killers in a burning structure which are as follows:

- Smoke
- Cold gas/noxious fumes
- Superheated toxic gases

There are two basic types of penetrations:

- Through penetration is a hole made in a fire-rated wall or floor to run pipes, cables, or any type of building service, completely through from one side of the fire barrier to the other side.
- Membrane penetration is a hole in one side of a fire-rated wall or floor for a single surface-barrier penetration (i.e., an electrical outlet/switch box).

All firestop materials are independently tested by third parties to ensure functionality. The products are tested as assemblies (i.e., firestop manufacturer A sends its firestop putty to UL [or one of several other] laboratories to test it for a 100-pair communications cable in a 50 mm [2 in] conduit. It passed a two-hour test when 13 mm [0.5 in] of putty was applied over 76 mm [3 in] of mineral wool inside the conduit).

The above information would become a UL-qualified assembly for manufacturer A. This is very important on a professional level as well as on a personal level. Products must always be installed according to the manufacturer's instructions based on independent testing. The cabling

installer and the company may be held legally, criminally, and financially responsible if the manufacturer's instructions are not followed.
This means the cabling installer:

- Cannot substitute products that are not part of the qualified assembly.
- Cannot add more of a product to get a higher rating.
- Cannot add or change the type or location of penetrations.
- Can be held personally responsible for improperly firestopping or not firestopping.
- Can contact the manufacturers to get written assistance for situations not covered in their installation practices for qualified assemblies.

There are ten basic types of firestop products:

- Mechanical systems
- Mortar/compounds
- Composite sheets
- Collar/devices
- Blankets
- Caulks/sealants
- Putty
- Wrap strips
- Pillows/bags
- Sprays

Mechanical systems consist of manufactured elastomeric components presized and shaped to fit around standard cables, tubes, and conduits. These products are made of several substances that resemble rubber and are flexible. These systems use mechanical pressure to hold the elastomeric components in place and to provide a tight seal around the penetrating devices. Whether or not frames are included, some means of applying compression to the modules is required.

Nonmechanical firestop systems are generally pliable. These include items such as putties, caulks, blankets, silicone foam, pillows, and other types of materials that can be molded to fit into an opening to seal it.

Firestop products have unique properties to help contain the fire. Each product may contain one or several properties depending on its design.

- A firestop that is endothermic has the ability to absorb heat. This prevents the temperature from rising on the other side of the fire barrier.

- A firestop that is intumescent will swell or enlarge when under the influence of heat. The material swells around and within the penetration to provide a tight seal. In some applications, it is designed to crush the penetration, such as with combustible conduits.

- A firestop that is ablative develops a hard char that resists erosion from fire and flames. This allows a soft firestop to become rigid to help it withstand the high-pressure gases and the force from a hose stream.

Cable insulation and tie wraps contain materials that may be toxic when heated. Cable and tie wraps with special heat-resistant ratings must be used in plenum environments.

Firestopping

This chapter provides guidelines for reestablishing the integrity of fire-rated architectural structures and assemblies (e.g., walls, floors, ceilings) when these barriers are penetrated by:

- Pipes.
- Ducts.
- Cables.
- Conduits.
- Innerducts.
- Cable trays.

When referring to firestopping systems, the terms "qualified," "tested," "listed," "classified," and "approved" are essentially synonymous. Each refers to firestopping systems that have been tested by an independent laboratory and certified as effective (different laboratories use different terms). For simplicity, this chapter will use the term "qualified" to represent all of these terms.

Some independent testing laboratories for firestopping include:

- FM—Factory Mutual Systems
- OPL—Omega Point Laboratory
- SWRi—Southwest Research, Inc.
- UL—Underwriters Laboratories, Inc.

- ULC—Underwriters Laboratories, Inc. of Canada
- W-H—Inchcape Warnock Hersey (ETL)

Each of the recognized test facilities tests to American Society of Testing Materials standard (ASTM) E814 (UL 1479). This is the standard for "Through Penetration Firestop Products."

The terms "system" and "design" are nearly synonymous; both refer to an arrangement of specific firestopping materials in a specific configuration. The term "system" is used in this context throughout this chapter.

Important. This chapter does not address local authorities' requirements concerning the design, use, and specifications for these structures or assemblies.

Role of Firestopping in Fire Protection

A comprehensive fire protection program must include fire:

- Prevention.
- Detection.
- Suppression.
- Containment.

Firestopping is a special activity under the fire containment category.

The methods, materials, and considerations of this chapter are generally applicable to the following activities within a building.

- Construction
- Renovation
- Rehabilitation

However, firestopping requirements more rigid than those described in this chapter may be set by:

- Local building codes.
- Construction documents.
- BICSI *Telecommunications Distribution Methods Manual,* current edition.
- Reference standards.
- Insurance underwriters.

Secondary Functions of Firestop Seals

In many cases, firestop seals may be required to perform secondary safety or security functions such as:

- Acting as environmental protection seals.
- Special requirements (such as an EMI/EMP barrier).
- Sealing around pipes that reach high temperatures and that may move axially or laterally.
- Resisting effects of explosion.

The methods and considerations for providing firestop seals explained in this chapter may have consequences that extend beyond the primary function of firestopping.

L (air leakage) is a measurement of air to indicate low-temperature emissions of toxic chemicals from plastic through a firestop barrier. A lower number is a better "L Rating." The minimum reportable value is < 1.0 in cubic meters per minute per square meter or cubic feet per minute per square feet of area.

Firestopping Systems

Introduction

Preventing fire from passing through a barrier penetration may depend on a single material which offers a complex balance of:

- Thermal resistance.
- Thermal conductivity.
- Adequate sealing at high temperatures.
- Controlled consumption.
- Durability to survive the:
  - Turbulence of a fire.
  - Rapid cooling and erosive impact of a hose stream.

Many firestop systems combine several materials, each offering specific physical properties that contribute to the success of the overall design. An essential role in surviving the dynamics of a fire test is played by the interrelationship between:

- Component products and dimensions.
- Anchoring and installation techniques.

Appropriate systems

A firestop seal system must provide an appropriate balance between:

- Durability.
- Ease of installation.
- Ease of maintenance.

Some seal systems can be installed from:

- Only one side, when access to the other side of the barrier is impossible.
- The bottom of an opening, because of the characteristics of the material and inaccessibility to the top.

Seal repairs or reinstallations must be:

- Qualified by performance tests or "engineered" judgments.
- Simple to achieve.

Selecting firestop materials/systems

For a construction or renovation project, select firestop systems on the basis of:

- Requirements of local authority having jurisdiction.
- Qualification testing.
- Installation efficiency.
- Maintenance convenience.
- Future cable changes.

Firestop systems consist of the floor or wall assembly and the penetrating item(s) as well as the seal materials. The complete system is qualified, not individual materials.

Note. Substitution of any system component may invalidate the system.

Qualified electrical apparatus

Some electrical apparatus are tested under exposure to fire and qualified for use in fire-rated assemblies. These items are known as "qualified fire-rated assemblies." These include:

- Boxes.
- Junction boxes.
- Breaker panels.

- Fixtures.
- Poke-throughs.

These are designed so that no additional firestopping is needed (other than normal patching where the item penetrates a fire-rated assembly). Installation criteria are listed in the installation instructions for each component.

> **Warning.** Firestop all nonqualified electrical apparatus. When fire rating is unknown, contact a firestop manufacturer for clarification.

Testing and Guidelines for Firestops

The fire resistivity of through-penetration firestops is evaluated under positive pressure, time-versus-temperature furnace conditions. This testing allows the resistivity to be assessed through controlled fire exposure.

There are four basic steps to each test:

- Install product into test barrier.
- Allow the product to dry/cure.
- Expose firestop to controlled fire/temperatures to a time–temperature curve expectation of:

    1 hr @ 977 °C (1700 °F)—2 hr @ 1010 °C (1850 °F)—3 hr @ 1052 °C (1935 °F)—4 hr @ 1093 °C (2000 °F).

- Immediately after burn, the firestop is subjected to a hose stream of 45 PSI.

Underwriters Laboratories, Inc.'s "Fire Resistance Directory" rates successfully tested firestop systems with F and T ratings. Additionally an L rating is available (see Table 5.1).

Fire rating classifications—United States

Two fire-rating classifications in the United States for through-penetration firestops are available under the rating criteria of ASTM E 814. A third fire-rating classification has been added to UL 1479.

> **Note.** A fourth fire-rating classification (H) is considered a separate classification only in Canada. In the United States, it is part of the F rating.

These classifications may be relevant in assessing the fire hazard of a particular application. The classifications are explained as follows.

TABLE 5.1 United States' Fire Rating Classifications

Rating	Description

F (Required number of hours)	The firestop withstands the test fire (direct flame) for the rating period without:
	■ Permitting flames to:
	  ■ Pass through the firestop.
	  ■ Occur on any element of the unexposed side of the firestop (i.e., auto-ignition).
	■ Developing any opening in the firestop during the hose-stream test that permits a projection of water beyond the unexposed side.

T (Required number of hours)	The firestop: ■ Meets the criteria for an F rating. ■ Limits temperature rise during the rating period. The temperature on any unexposed surface can increase no more than 180 °C (325 °F) plus ambient. **Note:** Ambient temperature is the current surround temperature.
L (Air Leakage)	**Note:** This rating is an optional part of UL 1479 and is not included in ASTM E 814. ■ Indicates a firestop's ability to provide an effective smokestop by reporting a number value. ■ Measurements are made at ambient and 204 °C (400 °F). ■ A lower number is a better L rating. The minimum reportable value is < 1.0 in cubic meters per minute per square meter or cubic feet per minute per square feet of area.

Fire rating classifications—Canada

TABLE 5.2 Canada's Fire Rating Classifications

Rating	Description
F (Required number of hours)	Corresponds to the E 814 F rating in the United States, but without the hose stream test.
FH (Required number of hours)	Corresponds to the E 814 F rating in the United States.
FT (Required number of hours)	Corresponds to the E 814 T rating in the United States, but without the hose stream test.
FTH (Required number of hours)	Corresponds to the E 814 T rating in the United States.
H (Hose Stream Rating)	**Note:** This rating is listed separately only in Canada. In the United States, it is a part of the F rating (see ASTM E 814 or UL 1479). ■ Indicates the firestop's ability to withstand hose stream application. ■ The test sponsor can elect, with the advice of the testing body, to conduct the hose stream test on the sample constructed for the fire exposure test. The test must be conducted within 10 minutes of completion of the fire exposure test.

Evaluation of Firestop Systems

Introduction

Firestops are:

- Specific combinations of materials installed and (possibly) supported or anchored in a certain way.
- Qualified by independent agencies, based on the material's performance when tested in a particular configuration.

  Note. The characteristics of a noncombustible material do not qualify that material as a firestop unless it has passed full-scale performance testing.

It is not possible to test every arrangement of firestops. Many test assemblies are based on testing the worst case in a number of variables. This worst-case configuration may extend to less severe configurations. Contact qualified manufacturers with questions regarding engineering judgments in selecting a suitable firestop system for a particular application.

Qualification testing for openings

Qualification testing of firestop systems provides evidence of performance across the range of opening configurations known or expected to exist.

A firestop system must be qualified for use in walls or floors according to its intended use to include the type of penetrating items.

Other qualification information

Additional testing may be required to confirm that the firestop can perform other functions, such as:

- Accommodating pipe movement without violating the integrity of the seal.
- Providing protection against pressure or flooding.
- Providing environmental protection.
- Special requirements (such as an EMI/EMP barrier).
- Resisting the effects of explosion.

Selecting a firestop assembly

The most common types of penetrating items are:

- Metallic pipes.
- Nonmetallic pipes.
- Insulated pipe.

- Cables.
- Raceways.
- Cable trays.
- Electrical box.
- HVAC ducts.
- Blank openings (no penetrating items).
- Grease ducts.

The most common types of building materials penetrated are:

- Gypsum wallboard.
- Concrete block walls.
- Poured concrete floors.
- Poured concrete walls.
- Concrete over steel deck.
- Wood floors.

To ensure the best firestop product/system choice, answer the following questions for each penetration. The time spent to answer these questions will assist in the task of determining the right firestop product and system. This is a noninclusive list and other considerations may apply.

Checklist for penetrating a firestop assembly	Answer
1 Is the type of barrier to be penetrated floor or wall?	
2 Is the barrier construction concrete, gypsum, wood, etc.?	
3 What is the barrier thickness or number of layers?	
4 What is the required F rating?	
5 What are the penetrating item(s)—pipe, cable, cable tray, duct, etc.?	
6 What is the penetrant makeup—steel, PVC, copper, aluminum, etc.?	
7 What is the size and number of penetrating items?	
8 What is the size of opening? Exact size is best.	
9 Is penetrant centered or is there a point of contact?	
10 What is the annular space?	
11 What type and thickness of insulation, if any, is on the penetrating item?	
12 What is the shape of opening (round, square, rectangle, or sledgehammer starburst)?	
13 Will the penetrant regularly be exposed to water?	
14 Will retrofit be required?	
15 Will there be regular movement/vibration of the penetrating item?	

Firestop systems

A firestop system is simply a specific size opening, in a specific type of building material, with a specific type of penetrating item (firestopping material and/or apparatus) which has been third-party tested to achieve a specific fire rating that is measured in hours of predictable protection.

Each firestop manufacturer is issued system numbers, which are particular to them, for each combination of items tested. No two manufacturers will have the same system numbers, unless one of the parties is a private-label manufacturer for the other.

Understanding the UL fire-resistance directory numbering methodology

- Alpha–alphanumeric

or

- Alpha–alpha/alphanumeric

First alpha designator:

- F = Floors
- W = Walls
- C = Combination (floors and walls)

Second alpha designator (floor or wall construction type):

- A = Concrete floors < 127 mm (5 in)
- B = Concrete floors > 127 mm (5 in)
- C = Framed floors
- D = Deck construction
- E–I = Reserved for future use
- J = Concrete or masonry walls < 203 mm (8 in)
- K = Concrete or masonry walls > 203 mm (8 in)
- L = Framed walls
- M = Bulkheads
- N–Z = Reserved for future use

Third alpha designator:

Indicates a combination of two construction material types tested in one classified system.

The numeric designators:

The first digit identifies the type of penetrant. The following three digits are sequential counters that increase with each additional system that successfully passes a burn test.

The numeric designators are:

0000–0999 = No penetrating item (blank openings)

1000–1999 = Metallic pipe, conduit, or tube

2000–2999 = Nonmetallic pipe, conduit, or tube

3000–3999 = Cables

4000–4999 = Cables in cable tray

5000–5999 = Insulated pipes

6000–6999 = Miscellaneous electrical penetrants (bus duct)

7000–7999 = Miscellaneous mechanical penetrants (HVAC)

8000–8999 = Mixed multiple penetrants

9000–9999 = Reserved for future use

System Number Example:

Company A System # CAJ 8049

No one manufacturer can test every possible installation contingency. Therefore, engineering system determinations are made for unusual applications using existing test data of similar situations. Contact qualified manufacturers with questions regarding engineered judgments when selecting a suitable firestop system for a particular application.

Categories of Firestop Systems

Introduction

Firestop systems can be roughly divided into two broad categories—mechanical and nonmechanical.

Mechanical systems

Mechanical systems consist of premanufactured elastomeric components shaped to fit around standard cables, tubes, and conduits. The elastomeric modules are:

- Fitted around penetrating elements.
- Arrayed within a frame (see the following illustration).

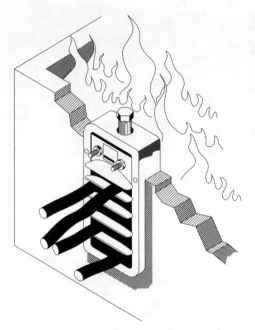

Figure 5.1 Elastomeric modules (within frames).

The most useful systems are highly modular to ensure maximum flexibility in accepting multiple elements of different diameters.

Whether or not frames are included, some means of applying compression to the modules is required to establish a tight seal.

Systems that do not use frames are intended to fit either:

- Standard conduit.
- Sleeves.
- Cored holes.

The elastomer is specially designed to withstand the fire and hose stream test for the rated time. Frame or hardware components must be steel to survive test temperatures.

Mechanical firestop systems are considered highly durable, while providing:

- Reliable pressure and environmental sealing.
- Excellent resistance to shock and seismic vibration.
- Support for penetrating elements.
- The opportunity to reconfigure the penetrating elements as required.

- EMI/EMP capabilities, if required.
- Explosion-proof integrity.

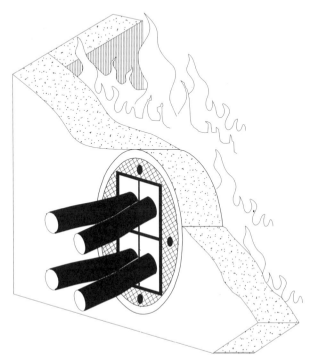

Figure 5.2 Mechanical firestop systems.

Nonmechanical systems

Nonmechanical firestop systems are available in a variety of forms, each having desirable properties for specific situations. They share the common benefit of adapting to irregular openings and off-center penetrating items.

Nonmechanical Firestop Systems

Introduction

Nonmechanical firestop materials are available in the form of:

- Putties.
- Caulks.
- Cementitious materials.
- Intumescent sheets.

- Intumescent wrap strips.
- Silicone foam.
- Premanufactured pillows.
- Prefabricated intumescent collars.
- Sprays.
- Blankets.

Types of putty

Firestop putties are very common because of their wide range of applications, ease of installation, and long shelf life. They may be intumescent and most are endothermic. These putties:

- Have the consistency of glazing putty.
- Remain permanently soft and pliable until exposed to heat.
- Allow easy firestop reentry.
- Come in bulk, bars, or sticks.
- May be installed in conjunction with ceramic fiber, rock wool filler, or other approved fill material.

Systems are available that allow complete installation of a putty seal from one side of a penetration. The consistency of some putty allows it to be installed from the underside of an opening.

Firestop putty is also available in the form of a pad. The pad is used to seal the back of outlet boxes or other electrical fixtures installed in a membrane penetration.

Testing indicates that the putty pad prevents flame-through and undesirable heat buildup on the nonfire side of the wall. The pad:

- Seals knockouts and openings in the fixture.
- Prevents smoke and fire from entering hollow wall cavities.

Collars

Some firestop systems incorporate intumescent putty within a metallic collar, sometimes referred to as a pipe choke system. When the collar is exposed to heat, it forces the intumescent material inward to seal and crush the penetrating item. Collars are usually used for firestopping plastic piping, insulated metal piping, cable/cable bundles, plastic conduit, innerduct, or any other material that may burn away in a fire and leave a significant void.

Types of caulk

Another type of sealant is the caulk type. Several firestop materials are available in caulk form. All of these materials:

- Quickly skin over to form a tight seal, but may require up to two weeks to fully cure.
- Are dispensed either from standard caulk tubes or large pails.

The types of caulk vary somewhat in:

- Their ability to adhere to various surfaces.
- Their flexibility and moisture resistance.
- The quantity required to achieve a rated firestop seal.

Firestop caulks may vary in composition such that they may be:

- Latex based.
- Water based.
- 100 percent solids.
- Solvent based.

   Caution. Do not use solvent-based caulk seals that may give off toxic or noxious fumes in confined areas that are not well ventilated.

- Self-leveling.

   Note. Some of these products are flammable until dry.
   Some of these materials are:

- Intumescent.
- Endothermic.
- Ablative.

   Note. Material Safety Data Sheets (MSDS) of the manufacturers are usually required for submission to the building owner, general contractor, and end user.

Some caulking materials can be installed from the underside of an opening without dripping or slumping. A self-leveling type is available for application as the top side of a firestop.

Some caulking materials may require the removal of the existing caulking firestop and the installation of all new materials when the seal is penetrated to add or remove cables.

Cementitious materials

Some firestop materials are available in a cementitious (cement-like) form. These materials are:

- Premixed or a dry powder to be mixed with water.
- More adaptable to large openings than putty or caulk.
- Easily applied by trowel or pumped for larger installations.
- Easily bored through to add additional penetrations.
- Lighter and more cohesive than cement.

When using cementitious materials, make special provisions to allow for thermal expansion or motion of the penetrating item. Wraps or blankets are relatively soft and permit easy installation around penetrating elements.

Do not substitute ordinary grout, plaster, or cement for a cementitious firestop. Although this substitution is often tried, it is not acceptable. According to tests and field observations, seals made of these materials may:

- Crack.
- Fracture or fall out.
- Be extremely difficult to repenetrate.

Intumescent sheets

Some firestops are available as intumescent sheet materials. Intumescent sheets with a sheet-metal backing can be used to seal large openings or nonsheet-metal sheets may be used with caulk or putty to fabricate a honeycomb-like partitioned opening for cable, conduit, metal, or nonmetallic pipe.

When using intumescent sheets, always follow the manufacturer's instructions for:

- Layering.
- Spacing/size.
- Anchoring.
- Sealing.

Intumescent wrap strips

Some firestop systems incorporate the installation of intumescent wrap strips. Intumescent wrap strips are usually used for firestopping plastic piping, insulated metal piping, cable/cable bundles, plastic conduit, innerduct, or any other material that may burn away in a fire and leave a significant void.

Silicone foams

Early nonmechanical firestop systems were based on two-component silicone foams. Mixed in proper proportions at the right temperature, the two components expanded rapidly to form a cellular structure surrounding the penetrating items.

Silicone foams are mainly used for large openings. The opening must be made structurally adequate and leak-tight to resist and control the expansive forces of the foam.

The disadvantages of two-component foam systems include:

- Special mixing equipment is recommended.
- They do not create a pressure-tight seal (even though foams are resilient in nature).
- Openings are likely to develop along the axis of pipes in cases where there is pipe movement.
- A compatible primer is required to improve adhesion to some of the materials in use.

Warning. Free hydrogen gas is evolved during curing. This requires forced ventilation in confined quarters.

Industry experience indicates that two-component silicone foams can make a low-maintenance seal only if:

- Environmental conditions are maintained.
- Correct installation techniques are employed.

Premanufactured pillows

Firestop pillows are a recent development in firestop sealing. Unlike earlier pillows which contained noncombustible fibers, contemporary pillows contain a specially treated, compressible fiber matrix. When exposed to fire, the matrix:

- Swells to provide further sealing.
- Becomes rigid, allowing the pillow seal to withstand the force of a hose stream as required by fire-test standards.

This approach to firestop sealing is especially attractive when:

- Frequent cable changes are anticipated.
- A fire watch would otherwise be required during construction.

Pillows are often regarded as temporary firestops because of their convenience. However, pillow fire test qualification standards are the same as for any other sealing system.

Note. In some applications, pillows may require wire restraints to pass the hose stream test.

Sprays

Spraying an insulating firestop over critical cables provides additional time to utilize the cable in the event of a fire. When a fire occurs, containment may not be enough. Power, control, and communication cables that contribute to critical functions may need to be protected, to avoid or delay losing services needed under emergency conditions.

The spray covers the cables and can make future cable identification difficult unless good documentation is maintained.

Firestop blankets

Firestop blankets are used primarily for large air handling and exhaust vents. The blankets are wrapped around the vents to contain the fire (e.g., a fire in a grease duct system from a commercial kitchen).

The blankets may be used in situations where a cable tray containing critical power or data cables must remain in operation during a fire. The blanket allows access for future cable plant modifications and inspections.

Firestopping for Brick, Concrete Block, and Concrete Walls

Pipes, cables, conduits, cable trays, and innerducts

Use qualified firestop systems to seal penetrations in brick, concrete block, or concrete walls for pipes, rigid electrical conduits, cable trays, and innerducts.

Verify that penetration conditions fall within the following firestop system parameters:

- Wall thickness
- Hourly rating required
- Opening size
- Annular space (ring outside of the pipe or cable being protected and inside the hole or sleeve)
- Sleeves (present/required/optional)
- Material selection (if subject to movement)

Install the materials according to manufacturer-tested methods.

Pipes, cables, conduits, and innerducts in cored or sleeved openings

Use qualified materials to firestop sleeved penetrations. Install the materials according to manufacturer-tested methods. If necessary, have the seal design evaluated for its ability to tolerate movement.

Secure sleeves in place according to the manufacturer's instructions or construction drawings.

The following illustrations show:

- Conduit penetration through concrete or masonry wall or floor:

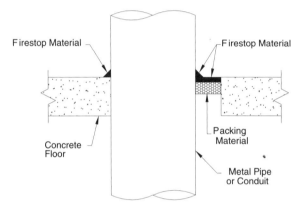

Figure 5.3 Conduit penetration in concrete floor.

- Cable penetration through concrete or masonry wall or floor:

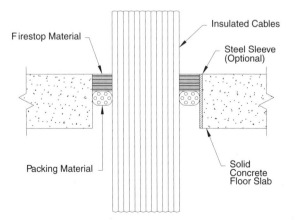

Figure 5.4 Cable penetration in concrete wall.

- Penetration firestop for single optical fiber cable with a PVC innerduct through a concrete wall:

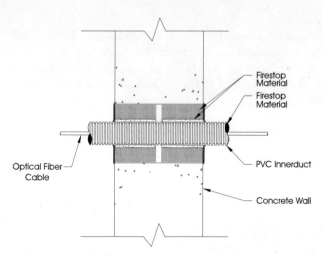

Figure 5.5 PVC innerduct penetration in concrete wall.

Note. Before installing innerduct, make sure that it has an appropriate fire rating (i.e., riser, plenum rated).

- Penetration firestop for a single optical fiber cable with a PVC innerduct through a concrete floor:

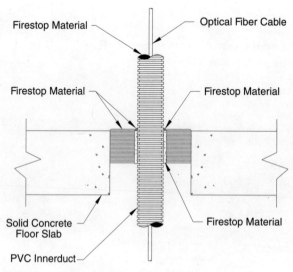

Figure 5.6 PVC innerduct penetration in concrete floor.

Cable trays

Close cable tray penetrations with a qualified firestopping system. Install the system according to the manufacturer's instructions. A qualified seal system is shown in the following illustration.

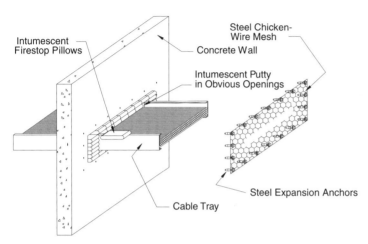

Figure 5.7 Qualified cable tray seal system in concrete wall.

Firestopping for Gypsum Board Walls

Pipes, conduits, and innerducts

Use qualified firestop systems to seal penetrations in gypsum wallboard assemblies. Verify that penetration conditions fall within the following firestop system parameters:

- Hourly rating
- Opening size
- Annular space
- Sleeves (present/required/optional)

The firestop system is typically required to be installed symmetrically on both sides of the wall. Install the materials according to manufacturer-tested methods.

Communications cable

Telephone or data communications cable not enclosed in conduit may penetrate gypsum walls if a qualified firestop system is used.

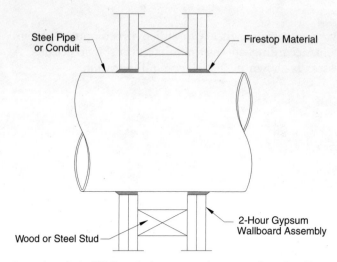

Figure 5.8 A qualified steel pipes system in gypsum board wall.

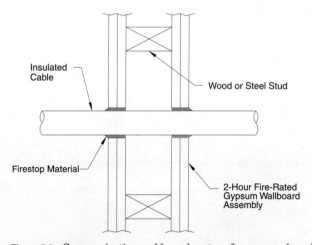

Figure 5.9 Communications cable seal system for gypsum board wall.

PVC innerduct

Optical fiber cable enclosed in a PVC innerduct may penetrate gypsum walls if a qualified firestop system is used.

Firestopping a shaft condition

A penetrating electrical apparatus that is connected to a conduit that penetrates one or more floor/ceiling assemblies may create a shaft condition. In this situation, use an appropriate firestop method inside the conduit.

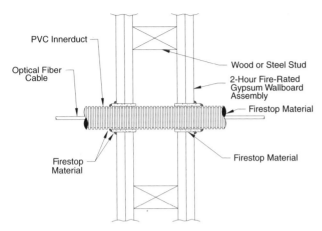

Figure 5.10 PVC innerduct penetration of gypsum board wall.

Cable trays

Through penetrations for cable trays must be:

- Boxed out with gypsum board.
- Sealed with a firestopping system that is:
  - Qualified.
  - Installed according to the manufacturer's instructions.

Firestopping for Floor Assemblies

Making penetrations

Never make penetrations in either new or existing concrete floor systems without the approval of a structural engineer. Penetrations can damage the structural load-bearing capacity of the floor.

Pipes, cables, conduits, ducts, innerduct, and cable trays

Use qualified systems to firestop through penetrations in concrete floor systems for pipes, cables, conduits, ducts, innerducts, and cable trays. The thickness (or depth) of firestop materials must:

- Match the thickness recommended by the manufacturer.
- Be established by formal testing to ASTM E 814, UL 1479, or ULC CAN4-S115.

Firestopping for Floor/Ceiling Assemblies

Introduction

Independent penetrations are the major firestopping concern when dealing with floor/ceiling assemblies.

Effects of fire on ceilings

Warning. Firestopping of ceiling penetrations is especially critical. Once a ceiling is violated by fire, the increase in temperature reduces the load-bearing capacity of the structural support members in the assembly.

Pipe, conduit, innerduct, cable trays, and cable penetrations (in ceilings)

The following describes the firestopping methods used for pipe, conduit, innerduct, cable trays, and cable penetrations in ceilings of plaster, gypsum board, or acoustic materials.

Ceiling material	Firestopping method
Gypsum board and plaster	Seal with qualified systems appropriate for the level of protection required.
Acoustic materials	■ Fully surround with mineral wool or ceramic fiber insulation and flashing. ■ Use a permanently pliable putty or firestop caulk.

Pipe, conduit, innerduct, cable trays, and cable penetrations (in floor/ceilings)

The following describes the firestopping methods used for penetrations for pipes, conduits, innerduct, cable trays, or cables in concrete and wood floor/ceilings.

Floor/ceiling material	Firestopping method
Concrete	Use the same method as specified for concrete floor systems. **Note.** This applies to floors that are a part of a floor/ceiling assembly.
Wood	Use: ■ Sleeves, depending on local job conditions. ■ Mineral wool insulation for the full thickness of the floor element. Top coat with a firestop caulking material. ■ Intumescent materials, depending on the level of protection required.

Note. Make penetrations the minimum size necessary to accommodate the penetrating element.

Choosing a seal system

Use a seal with a qualified firestop system for the barrier and penetrations being sealed. Use good engineering judgment in relating the configuration and features of a seal to the application.

The seal system must:

- Perform successfully in a fire test.
- Demonstrate (in a suitable pressure test) that positive pressure sealing can be achieved, relying on similar substrate materials.
- Be flexible enough to accommodate the independent movement of the penetrating device without losing its seal.

Firestopping Considerations (General)

Match existing conditions

Select a qualified firestop system that matches the actual conditions that exist (i.e., penetrating item(s), material, wall/floor/ceiling configuration). (See "Fire Rating Classifications" in this chapter.) If an exact match cannot be made between firestop systems listed and actual conditions, select a system that is close and seek an engineering judgment by a qualified local authority having jurisdiction.

Selection criteria

When selecting a qualified system, take into account the following:

- Consider the actual conditions tested and covered by the system.
- When substituting one manufacturer's system for another manufacturer's system, ensure the firestop materials selected are appropriate for the actual conditions.
- Never substitute products of a qualified firestop system with another manufacturer's products. Products from one manufacturer's system are not interchangeable with products from another manufacturer's system.
- The difference between the actual floor/ceiling/wall thickness and conditions tested during qualification may affect listed F and T ratings. Generally speaking, the thinner the construction qualified, the better the F and T performance may be if actual construction is thicker. The reverse may not be true.
- A firestop material capable of sealing against the passage of smoke and toxic gases through the penetration is highly desirable.

Typical Installations

Sealing a floor penetration with putty

The following illustrations show an example of installing a putty seal from one side of a floor penetration.

Step	Installing a putty seal
1	Tear off a small portion of the putty. **Note:** Use a drop cloth or other cover to protect the floor surface from the putty and insulation material.

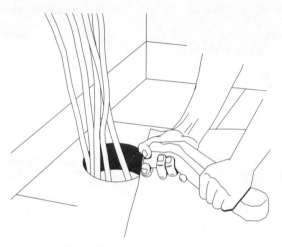

Figure 5.11 Tear off putty.

2	Use the putty to build a bottom in the penetration, according to the manufacturer's instructions.

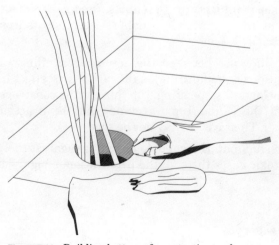

Figure 5.12 Building bottom of penetration seal.

3   Fill the penetration with ceramic fiber or rock wool fill, stopping far enough from the upper rim to allow for a top layer of putty at the manufacturer's recommended thickness.

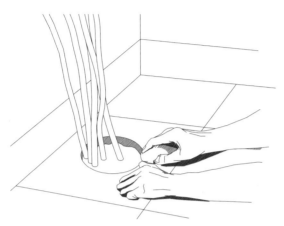

Figure 5.13   Filling the penetration.

4   Use the putty to build a top on the penetration, according to the manufacturer's instructions.

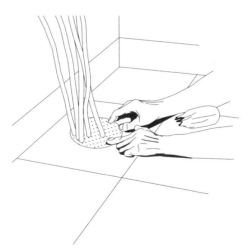

Figure 5.14   Building a top on penetration.

Sealing an outlet box with putty

To seal an outlet box, follow the steps below.

Step	Sealing an outlet box with putty
1	Press a pad of putty into place, covering one side of the outlet box. Ensure that the pad overlaps at the top, bottom, and sides, as shown.

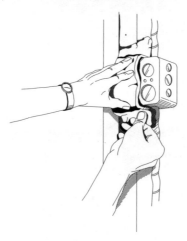

Figure 5.15 Overlap the pad on outlet box.

2	Press a second pad of putty into place on the other side of the box, as shown.

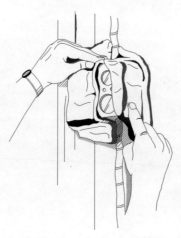

Figure 5.16 Press a second pad of putty on outlet box.

3	Press the seams together to join the two pads, as shown.

Figure 5.17 Joining the pads on outlet box.

Steps—Restore penetrations (general)

Step	Restoring fire-rated penetrations
1	Firestop conduits, pipes, and innerducts in brick, concrete block, or concrete wall. The requirements for firestopping are variable and are directly dependent upon: ■ Fire barrier thickness. ■ Hourly rating required. ■ Opening size. ■ Installing the materials according to manufacturer-tested methods.

324   Chapter Five

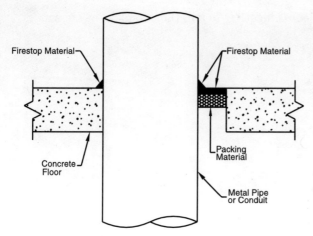

Figure 5.18 Conduit penetration through masonry, wall, or floor.

2. Firestop floor penetrations. Sealing of cored or sleeved openings in floor slabs containing pipes, cables, or innerducts requires knowledge of floor thickness, hourly fire rating required, opening size, annular space, sleeves, and materials selection.
3. Firestopping drywall.
   - Requirements for firestopping are variable and are directly dependent upon hourly rating, opening size, annular space, and sleeves.
   - Drywall penetrations are typically required to have firestopping installed symmetrically on both sides of the wall to restore fire rating.

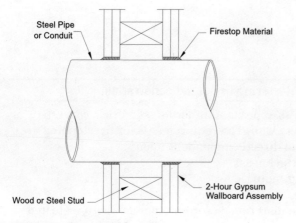

Figure 5.19 Fire seal of drywall.

4   Firestopping all applicable penetrations.
    - Inspect all cable runs for any openings made through fire-rated walls as part of structured cabling system.
    - If sleeves have been used, be sure to firestop the wall penetrations on both sides and each end of the sleeve itself.

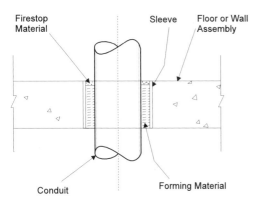

Figure 5.20 Fire seal of wall or floor sleeve.

5   Firestop cable tray. Penetration by a cable tray in a firewall is not normally permitted. Generally the tray is placed against each side of the firewall and a number of appropriately sized sleeves are installed for cables to pass through. If a cabling installer encounters a cable tray penetration directly through firewalls, specifically manufactured seal systems should be obtained. The following diagram demonstrates the components of a qualified seal system.

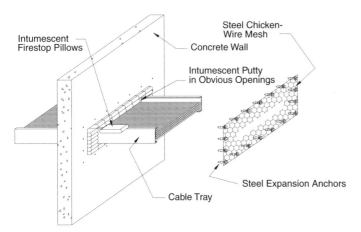

Figure 5.21 Fire-seal cable tray penetration in concrete wall.

Chapter 6

Cable Termination Practices

Pretermination Functions	329
Copper IDC Termination	333
Coaxial Cable Terminations	348
Fiber Termination	351

Overview

Cable termination involves the organizing of cables by destination, forming and dressing cables, and proper labeling, as well as actually creating a connection with a copper or fiber conductor.

Proper cable termination practices are vital for the complete and accurate transfer of both analog and digital information signals. Insulation Displacement Connection (IDC) termination is the recommended method of copper termination recognized by ANSI/TIA/EIA-568-A for UTP cable terminations. This method removes or displaces the conductor's insulation as it is seated in the connection. The specific tools designed for making these connections are required. During termination, the cable is pressed between two edges of a metal clip displacing the insulation and exposing the copper conductor. The copper conductor is held tightly within the metal clip, ensuring a solid connection.

Screw-type terminal faceplates commonly used in voice applications are not recommended by ANSI/TIA/EIA-568-A for UTP terminations and should not be used.

Wirewrap is another method of terminating copper conductors. This procedure wraps the bare conductor around an appropriate terminal post on a terminal block. If done properly, this makes a good electrical connection that is resistant to corrosion and loosening. Copper wirewrap termination is usually practiced in a central office or special application locations. It is not found in structured wiring and will not be covered in this chapter.

Undercarpet flat cable is designed for installation under carpeting. Undercarpet cable makes a transition at walls from round horizontal cables to flat cable in a transition box. The work area termination is the same as for other work area outlets. It should be noted that the length of undercarpet cable should meet the following recommendations for data circuits:

- Avoid high-traffic areas.
- Keep the undercarpet run as short as possible (15 m [50 ft] is the recommended maximum length).
- Maintain at least 152 mm (6 in) clearance from power circuits.
- If power circuits must be crossed, the cross should be at right angles and above the power circuit.

This type of cabling system is mentioned here as information only.

The crimp-style connector for copper and coaxial cable is dependent upon the shape and diameter of the cable. The cable may be round or flat. Care should be taken to ensure that the proper crimp connector for the specific cable is being used.

Optical fiber cables are constructed differently from copper cables. It is not enough to make a sound mechanical connection between the fiber and the connecting hardware, as with a copper connection. The fiber core must be precisely aligned with the fiber core of the connecting cable or within the connecting hardware. This ensures that the maximum transfer of light pulse energy is obtained.

There are several different applications relative to the termination of a cable. This chapter, Cable Termination Practices, will address the following:

- Termination blocks
- Patch panels
- Work area outlets
- Direct connection
- Coaxial cables
- Optical fiber

The following types of cable currently recognized by the ANSI/TIA/EIA-568-A for use in the premises cabling will be covered in this chapter:

- 100 Ω Unshielded twisted-pair copper cable (UTP)
- 100 Ω Screened twisted-pair copper cable (ScTP)
- 150 Ω Shielded twisted-pair copper cable (STP-A)
- Optical fiber cable

Pretermination Functions

Overview

Proper preparation for cable termination not only improves the quality of the job but also decreases the amount of time required for termination. Through proper preparation, the cabling installer can concentrate on doing the job right.

The performance of pretermination functions involves organizing the cable by destination. Cable to be terminated should be placed in close proximity to the point of termination and must also be identified properly to ensure that it is terminated in the correct position.

Forming and dressing the cable involves properly aligning and positioning the cables in a neat and orderly manner for termination. The length of cable needed to reach the termination location must be determined, taking into account enough slack to reterminate if necessary and not placing undue pulling stress on the termination.

Proper cable management results in neat and orderly bundles of cables that are formed into a symmetrical pattern. Besides being aesthetically acceptable, proper cable management provides support and mechanical protection of the pairs.

Cable connection is not complete until all terminations are properly identified and labeled.

Steps—Pretermination

Step	Pretermination Functions
1	Organize cable by destination. ■ Know the wiring scheme. If incompatible parts are used for terminating, drastic results occur. With copper IDC terminations, there are three predominant wiring schemes: T568A, T568B, and Universal Service Order Code (USOC). Only the T568A and T568B wiring schemes are compliant with the ANSI/TIA/EIA-568-A standard. Each of the wiring schemes is shown in figure 6.1.

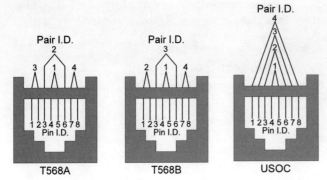

Figure 6.1 Wiring schemes.

- For work area terminations, check that all cables are available and properly labeled at the wall outlet locations. In equipment rooms, ensure that blocks or panels are installed in accordance with the designer's layout.
- Verify that the right products are on hand for the application. Modular furniture requires a different variety of outlets than drywall offices, so care must be taken to ensure that the proper product type and manufacturer is specified.

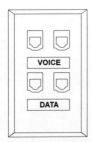

Figure 6.2 Work area outlet.

2. Form and dress cable at the rear of the panel.
   - Prepare the cable for termination by first bringing all cables into a layout which is shaped in a manner that does not allow the cables to be crossed over each other, and is cascaded into a sweeping curve to the destination.
   - On the front side of the rack, dedicate a minimum of one rack space (unit) [44.5 mm (1.75 in)] of cable management for every two rack spaces (unit) [89 mm (3.5 in)] of patch panels.

- Dress the cable by ensuring that all cables are parallel to each other; smooth them with your hand until they form a neat, orderly bundle.
- Cables may be entering the wiring closet from multiple directions, which frequently results in cables of many different lengths. After determining the amount of slack necessary, the cables should be relabeled and cut to uniform lengths.
- Each cable must be carefully remarked, using the same markings, prior to cutting off the excess cable containing the original markings. Attention should be given in this process to ensure that the new labeling is correct.
- Use tie wraps, or, preferably, use hook and loop straps to keep the cables secured. The tie wraps or hook and loop straps should be evenly spaced throughout the dressed length. Care should be taken to tighten tie wraps by hand only.

3. Determine length and slack required.
   - Provide an adequate amount of cable to reach the destination point for termination.
   - When running cables horizontally and the exact location of termination is unknown, leave enough cable to reach the farthest point in the wiring closet plus enough cable to reach the floor from the horizontal cable management. It is important to remember that, through the use of additional cable slack, the exact location of the backboard or rack termination hardware location can be adjusted to allow for additional equipment or hardware. If there is not enough cable slack, then either the location of the termination hardware must be adjusted to fit the horizontal cable, or a new horizontal cable must be repulled. By the relocation of the termination hardware, the organized symmetry of the telecommunications closet can be altered to such a point that future move, add, or change (MAC) activity become extremely difficult.
   - When the exact location of the connecting hardware is unknown, it is recommended that backbone cables running vertically be laid in, with enough slack to reach the floor or ceiling, plus the distance across the backboard. When dealing with backbone cable, remember that the cabling installer generally is dealing with a very heavy, high-pair-count cable that requires a large amount of termination connections. By not allowing an adequate amount of slack in

the backbone cable, situations could arise that could cause a major redesign of the telecommunications closet and the proper design specifications and symmetry cannot be maintained.

4. Use the proper cable management hardware.
  - There are several different types and styles of cable management hardware. All cable management products are designed to properly support the in-place cables and relieve tension, as well as provide support for future cable which may be installed as a result of MACs. Below are examples of cable management hardware.

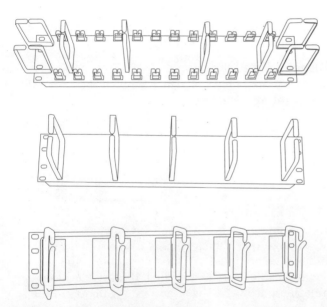

Figure 6.3 Cable management products.

  - Because UTP and ScTP cable are protected from crosstalk and immunity from EMI through the cable's pair twist and lay configuration, care must be taken to maintain the minimum bend radius of the cable. This protects the integrity of the cable being installed. The minimum bend radius for UTP and ScTP cable is four times the cable diameter, while optical fiber cable is ten times the diameter.

Copper IDC Termination

Overview

There are four basic types of IDC termination blocks used in the termination of horizontal and backbone copper cabling.

The most common IDC termination blocks are the 66-type, 110-type, BIX™, and LSA. These comprise the majority of the market; however, other devices are available.

Several manufacturers provide both rack-mountable and wall-mountable IDC termination hardware which can house multiple termination blocks.

To ensure a good connection, care must be taken to closely follow the IDC connecting hardware manufacturer's specifications. Special attention should be given to complying with the proper procedures for:

- Determining the proper method and length of sheath removal.
- Length of pair untwisting permitted. (ANSI/TIA/EIA-568-A recommends a maximum of 13 mm [0.5 in] of untwisted pairs, measured from the last twist to the IDC).

Each type of IDC termination requires a specially designed terminating tool for performing the IDC termination correctly. There are several manufacturers that market termination tools that have the ability to interchange the blades for use on several styles of IDC termination blocks. Caution should be taken to ensure that the brand of tool being used is compatible with the blade. Different blades sometimes look similar, but there are slight differences in design. Improper matching of termination tool handle and termination blade can lead to serious personal injury as well as poor IDC terminations.

The following procedures generally address the proper methods and tools required to terminate each type of IDC termination block.

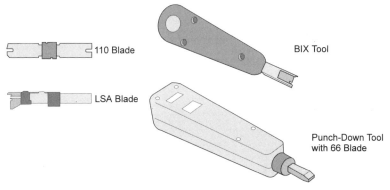

Figure 6.4 66-type, 110-type, BIX, and LSA tools.

334 Chapter Six

Adhere to the manufacturer or customer specifications and guidelines regarding all terminations. Document all termination information properly. This helps to identify the origin, destination, and routing of all cables.

Steps—IDC termination

Step	Copper IDC Termination
1	Determine method and length of sheath removal. Sheath removal can be performed in three ways: • Ringing tool. The ringing tool contains a razor blade which is set to a depth that allows the sheath to be slit deep enough to separate the sheath but not nick the inner pairs. To properly use this tool to the greatest surety, insert approximately 13 mm to 19 mm (0.5 in to 0.75 in) of the cable into the tool. Turn the tool and then remove the severed sheath with electrician snips. • Electrician snips. Using the electrician snips, carefully cut into the sheathing to a depth that exposes the rip cord. Using the rip cord, pull down the sheathing until the proper length of sheath is ready to be removed. Then remove the severed sheath with the electrician snips. • Slitter tool. The slitter tool is carefully inserted between the cable sheath and the pairs. Slide the tool carefully down the sheathing until the proper amount of sheathing has been prepared for removal. Remove the severed sheath using the electrician snips.

Figure 6.5 Sheath removal tools.

2	Using a proper sheath-removal tool, remove the cable sheath in accordance with ANSI/TIA/EIA-568-A standards and the termination equipment manufacturer's specifications.

- Remove only as much of the sheath as is necessary to terminate the cable pairs and ensure that the twist of the pairs is maintained. A common fallacy in the industry is that only 50 to 76 mm (2 to 3 in) of sheath should be removed from the cable. This requirement can vary depending on the block, type, and size of the cable and type of IDC termination hardware manufacturer.

3. Separate, identify, and tie off binder groups.
    - Binder groups are associated with 50 pair and larger pair-count cables.
    - A unique color code identifies each binder group. Cables are grouped in 25-pair increments with each 25-pair group (or subgroup) individually wrapped with a fabric or plastic tape to identify the groupings. See the following color-code chart for clarification.

TABLE 6.1 Binder Color-Code Chart

Pair number	Binder group			
	Tip	Ring	Color	Pair count
1	White	Blue	White-Blue	001–025
2	White	Orange	White-Orange	026–050
3	White	Green	White-Green	051–075
4	White	Brown	White-Brown	076–100
5	White	Slate	White-Slate	101–125
6	Red	Blue	Red-Blue	126–150
7	Red	Orange	Red-Orange	151–175
8	Red	Green	Red-Green	176–200
9	Red	Brown	Red-Brown	201–225
10	Red	Slate	Red-Slate	226–250
11	Black	Blue	Black-Blue	251–275
12	Black	Orange	Black-Orange	276–300
13	Black	Green	Black-Green	301–325
14	Black	Brown	Black-Brown	326–350
15	Black	Slate	Black-Slate	351–375
16	Yellow	Blue	Yellow-Blue	376–400
17	Yellow	Orange	Yellow-Orange	401–425
18	Yellow	Green	Yellow-Green	426–450
19	Yellow	Brown	Yellow-Brown	451–475
20	Yellow	Slate	Yellow-Slate	476–500
21	Violet	Blue	Violet-Blue	501–525
22	Violet	Orange	Violet-Orange	526–550
23	Violet	Green	Violet-Green	551–575
24	Violet	Brown	Violet-Brown	576–600
25	Violet	Slate	No Binder	

- Dependent upon manufacturer, 25-pair binder groups are combined into identifiable master groups.
- Tie off binder groups to keep them identified until ready to terminate.

    Note. One helpful hint is to use a copper pair of the same color as the binder group. Twist it snugly, but not tightly, to both the sheath end and the outside end of the unsheathed cable binder group. This helps to identify binder groups. It also keeps the binder groups together during termination, allowing for easier housekeeping.

4   Fan out and form cable pairs from each binder group.
    - Cable pairs should be uniformly placed so as to be aesthetically pleasing. Also, pairs should not cross or interfere with any other pairs.
    - Wire pairs should be parallel with no tension at the point of connection, and equal tension on all connections.

66-block termination

The 66-type IDC termination block is the choice for connecting voice applications such as PBX, Key Telephone Systems (KTS), and some LANs. Several manufacturers of 66-type termination block designs have updated their termination blocks to handle high-speed data applications and to be compliant with ANSI/TIA/EIA-568-A Category 5 specifications. Caution should be used to ensure that the appropriate 66-type IDC termination block is being installed in new installations.

The 66-type termination block is normally mounted on backboards with an 89-style bracket. There are several different 89-style brackets for use in several applications (i.e., RJ-21X, side-mount and rear-mount 50-pin connector, etc.), but the most common are the 89B- and 89D-style.

Typically, these blocks are mounted on backboards in vertical rows of four blocks each to accommodate up to a 100-pair termination. Several manufacturers provide a preassembled single- or double-sided distribution frame to provide increments up to 2700 pairs per side, for large system installations.

Cable for a 66-type termination block is routed through the 89 bracket to allow the pairs to be fanned out from the rear into the guides in the side of the block. Each five-pair increment is marked with a distinctive groove for ease of identification.

A 66M1-50 block provides the means of terminating two 25-pair or twelve 4-pair cables per block. These blocks have two rows of contacts which are mechanically connected together to provide cross-connection capability. The 66M1-25 blocks have four rows of contacts connected

Cable Termination Practices

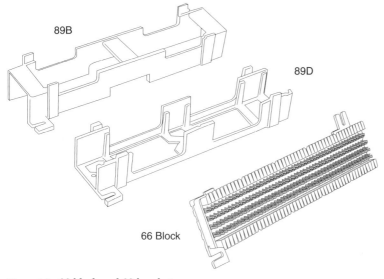

Figure 6.6 66-block and 89 brackets.

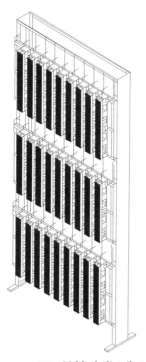

Figure 6.7 66-block distribution frame.

together. Voice applications use bridging clips to make a connection between the left and right set of contacts on a 66M1-50 block. By lifting the bridging clips which opens the circuit, it is easy to test the voice circuit in both directions when troubleshooting.

Note. Bridging clips are not Category 5 compliant. This is for information purposes only.

A fine-tipped, indelible marker is generally used to designate cable-pair identifiers on the fanning strips of 66-type blocks. An additional method of identifying 66 blocks is the use of color-coded hinged covers which can be labeled on the inside of the cover.

110-style hardware

The 110-style IDC termination hardware is used in both voice and data cabling applications. Backbone cabling is commonly terminated on wall- or rack-mounted 110 termination blocks in increments of 50, 100, or 300 pairs, as well as on a 900-pair wall mount.

The majority of patch panels are wired in specific configurations (i.e., T568A, T568B, USOC) and are mainly constructed with 110-style connectors. In addition, work area outlet terminations are manufactured with 110-style hardware.

Cables are routed through the middle pathway of the 110 wiring block from either the top or bottom, and fanned into the wireway from alternate sides. The block wireway is designed such that one row is terminated on the wiring block by punching down from the bottom up. The next row is terminated from the top down. For the specific location of each termination refer to the telecommunications designer's layout.

It should be noted that the 110-style wiring block contains no IDC. The IDC for this type of termination application is in the C-3 (3-pair), C-4 (4-pair), and C-5 (5-pair) connector block that is punched down on top of the 110-wiring block to permit cross-connection. The connecting block pair count is determined by the application (i.e., 4-pair horizontal cabling is terminated on a C-4 (4-pair) connecting block; trunk cables in 25-pair increments are terminated on 5-pair connecting blocks).

The designation strips are then placed in the holder which covers the terminated cables.

BIX hardware

BIX-type termination hardware is similar to the 110 hardware previously described. Unlike 110 hardware which places clips on top of the wiring block, BIX equipment is a one-piece "pass-through" unit which is reversed in its mount after termination of the cable to expose the opposite side to enable cross-connect capability.

Cable Termination Practices   339

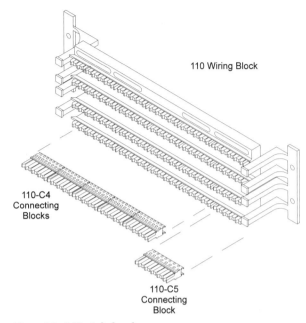

Figure 6.8 110-style hardware.

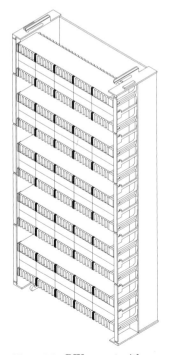

Figure 6.9 BIX mount with connector.

BIX termination block assemblies are available in 50-pair, 250-pair, 300-pair, and 900-pair increments for wall mounting and with floor-frame assemblies for large-size installations.

BIX-type termination hardware is available in both patch panel and work area outlet configurations.

BIX termination block connectors are designed with four slots for inserting small tie wraps to support termination of cables. These connectors can be identified in 2-pair, 4-pair, and 5-pair increments for various applications.

When terminating cables with larger than 25 pairs on BIX termination blocks, the fabric or plastic binder of each of the 25 pairs should be carried to the end of the connector. The designation strips are then placed in the holder that covers the terminated cable.

LSA hardware

LSA-type termination hardware provides silver-plated IDC contacts at a 45-degree angle, with the conductor being held in place by tension in the contacts. This hardware is available in patch panels, work area outlets, and termination blocks as other similar systems. It also provides disconnect modules, connect modules, switching modules, and feed-through modules.

Disconnect modules are normally closed, two-piece contacts which can be disconnected by inserting a disconnect plug into the wire pair. This allows temporary or permanent disconnect of the circuit. A test cord can be inserted into a pair to test circuits both ways when testing is necessary. These modules come in 8-pair or 10-pair increments.

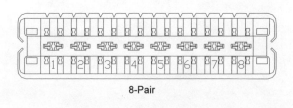

8-Pair

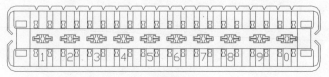

10-Pair

Figure 6.10 LSA 8-pair and 10-pair block.

Connect modules use a one-piece contact, which provides a continuous link between the cable and the cross-connect wiring.

Switching modules consist of a normally open, two-piece contact. Switching modules allow for high-density termination and patch cables.

Feed-through modules consist of a one-piece contact that passes the signal through the module, front to back, and provides a continuous link between feeder and jumper for high-density termination in small areas. These modules are available in 25-pair increments.

For the specific locations of each termination, refer to the telecommunications designer's layout.

Patch panels

Data and voice patch panels are available from various manufacturers and in many different styles and wiring configurations.

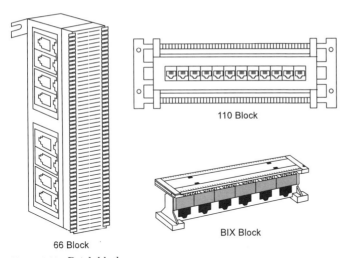

Figure 6.11 Patch blocks.

Patch panels are available which feature 110, BIX, and LSA connections. Common configurations are 24-, 48-, and 96-port.

To properly terminate a 4-pair horizontal cable onto a patch panel, keep sheath removal to the minimum amount required. Category 5 cable pairs must remain twisted to within 13 mm (0.5 in) of the point of termination.

Strain relief of the cables is accomplished by the use of tie wraps or hook and loop straps installed on a cable management bar which is installed at the rear of the patch panel.

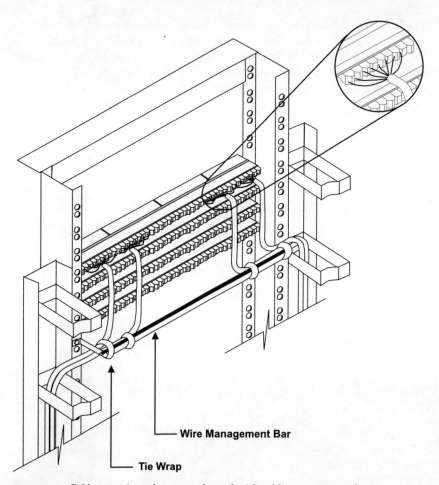

Figure 6.12 Cables terminated on rear of panel with cable management bar

Care should be taken prior to actual termination to verify that the work area outlet wiring scheme (T568A or T568B) matches the patch panel wiring configuration (T568A or T568B), to ensure proper functioning. Identify patch panel termination locations in the space allocated on the patch panel. The manner of labeling patch panels will be provided by the telecommunications designer.

Work area outlets

Many different styles of outlets are provided for work areas. Drywall offices can be terminated in single- or double-gang faceplates with jack inserts ranging from low-density single-port to high-density eight-port devices.

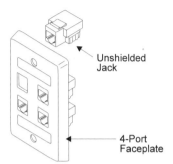

Figure 6.13 4-port faceplate.

Several manufacturers provide specially designed inserts for modular furniture in multiple jack configurations. These inserts are designed to fit in a specific model of furniture.

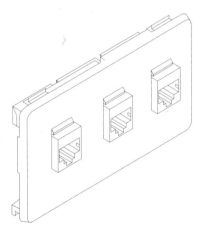

Figure 6.14 Modular-furniture faceplate.

Termination of work area outlets should be completed by removing the minimum required amount of the sheath, according to the manufacturer's specifications. Follow the manufacturer's specifications for proper termination techniques, being careful to ensure that the 13-mm (0.5-in) maximum amount of untwisting of the cable pairs to maintain Category 5 compliance. Carefully coil the remaining slack (minimum of 300 mm [12 in]) into the termination box. Do not kink the cable or exceed the bend radius of four times the cable diameter.

It is always advisable to utilize the same manufacturer's patch panels and work area outlets at a given job site. This will minimize the possibility of component mismatch, especially when cabling for ANSI/TIA/EIA-568-A compliance and meeting the TIA/EIA TSB-67 requirements for basic link performance. Another aspect of component mismatch is the warranty issue for manufacturer compliant installation practices. Additionally, all patch cords should be obtained from one source and pretested, as well as certified Category 5 by the manufacturer.

Direct connection

There are occasions in which a customer request for direct connection of horizontal cable requires that a cabling installer directly field terminate cables with modular plugs. This practice is not recommended for Category 5 applications due to lack of flexibility and excessive downtime if damage occurs to the cabling in the work area.

Field-constructed patch cords

Field-constructed patch cords are not recommended for Category 5 systems. There may be circumstances in which field-constructed patch cords are requested; for example, custom lengths desired by a customer for patching from electronic equipment to a patch panel in order to avoid excess slack.

Step	Field Constructing a Patch Cord
1	Determine the type of cable to be used. ■ Patch cord cable conductors can be either solid or stranded and the cable can be flat or round. The recommended conductor is stranded to allow for maximum flexibility. ■ Cables which do not contain twisted pairs are not Category 5 compliant and must not be used for data applications.
2	Select the proper connector. ■ Modular plugs are available in both flat- and round-cable versions, as well as stranded and solid conductor IDCs. Obtain the correct plug for the type of cable being used.

3   Verify pin-wiring configuration.
- Data cables utilize a straight-through wiring (i.e., Pin 1 to Pin 1; Pin 8 to Pin 8).
- Voice cables are reversed (i.e., Pin 1 to Pin 8; Pin 2 to Pin 7, etc.).
- Certain applications may require unique pin-wiring configurations.
- Consult with the client as to the exact pin-out required for the application.

TABLE 6.2 Data Patch Cord Pin Wiring

Data straight through (T568B configuration shown)		
1	W/O	1
2	O	2
3	W/G	3
4	BL	4
5	W/BL	5
6	G	6
7	W/BR	7
8	BR	8

TABLE 6.3 Voice Patch Cord Pin Wiring

Voice crossover (T568B configuration shown)		
1	W/O	8
2	O	7
3	W/G	6
4	BL	5
5	W/BL	4
6	G	3
7	W/BR	2
8	BR	1

4   Strip jacket to appropriate length.
- Make a clean 90-degree cut end on the cable.
- Remove only enough jacket from the cable to reach the end of the plug and still have the jacket under the cable clamp portion of the modular plug.

5   Use the correct crimp tool.
- Modular plugs are configured in 4-, 6-, and 8-pin combinations. The correct die for crimping all of the wires in one motion is required. Only 8-pin modular plugs are Category 5 compliant.

6   Verify pin-wiring configuration.
- Inspect the connection to make sure all the wires are seated properly and in the correct position.

Screened twisted-pair (ScTP)

Screened twisted-pair cable is provided by several manufacturers for use in areas of high EMI generation and to protect against mechanical damage. It is used for Category 5 compliant installations, and carries the same impedence and electrical characteristics as UTP cable, except it has a mylar screening around all the cable pairs and has a drain wire that is fused to the mylar screening.

A screened twisted-pair cable must be terminated in a screened modular jack. The screened jack is enclosed by a metallic-type, EMI-resistant housing. The dressing block for cable pairs fits within the wall of the connector so that, when terminated, the twisted wires and IDCs are totally enclosed by the metal which forms an EMI shield.

The IDC terminations are made in the same manner as UTP, with the added requirement of terminating the screen shield. ScTP manufacturers provide detailed instructions on the proper termination of the shield. This procedure varies by manufacturer, and the specific method provided for the selected product must be followed for acceptable shield effectiveness over the full 100 MHz bandwidth.

The screen shield is effectively grounded at one end by attaching the drain wire securely to the screen shield of the modular jack, following the modular jack manufacturer's specifications. The other end is terminated to a ScTP patch panel. The IDC terminations are made to the ScTP patch panel in the same manner as UTP, with the added requirement of terminating the screen shield.

> **Note.** ScTP manufacturers provide detailed instructions on the proper termination of the shield. This procedure varies by manufacturer, and the specific method provided for the selected product must be followed for acceptable shield effectiveness over the full 100-MHz bandwidth.

It should be noted that the EMI shielding capability of the ScTP cable is achieved by the internal fusing of the bleed wire to the mylar screening. To ensure that the screening will effectively block EMI, care must be taken to ensure that the patch panel is grounded, following the manufacturer's specifications.

Shielded twisted-pair (STP)

Shielded twisted-pair cable is commonly referred to as IBM-"Type" cable due to the fact that they were specifically designed to be used with the proprietary IBM cabling system (ICS) developed by IBM Corporation for support of IBM computer systems.

- For purposes of clarification, the following types of cable are detailed:
- The Type 1 cable is a 150 W shielded cable comprised of two individually shielded 22 AWG twisted pairs in either a PVC or plenum-rated sheath.

- The Type 1-A cable is a 150 Ω shielded cable comprised of two individually shielded 22 AWG twisted pairs in either a PVC or plenum-rated sheath. Each of the two pairs are shielded by a mylar wrap, and both pairs are then shielded with a braided shield which encompasses the cable. This cable is recognized by ANSI/TIA/EIA-568-A as having a bandwidth of 300 MHz.
- Type 2 cable is the same cable with four nonshielded Category 2 pairs cabled with the two shielded pairs and covered with either a PVC or plenum-rated sheath. Cabling installers should be alert to the fact that the voice pairs in Type 2 cable are Category 2 and not Category 3.
- Type 3 cable is Category 2, low-speed data, twisted-pair cable.
- Type 6 cable is a 150 Ω, two-pair, individually shielded, stranded, 26 AWG construction used for patch cables.
- Type 9 is a 150 Ω shielded cable comprised of two individually shielded 26 AWG pairs.
- There are other types also available, such as Type 5 (optical fiber cable) and Type 8 (undercarpet cable).
- Shielded twisted-pair cable has been enhanced by some manufacturers to support service up to 300 MHz. This enhanced cable is referred to as Type 1-A or STP-A cable.
- Shielded twisted-pair cable is terminated in a hermaphroditic or universal data connector (UDC) as shown.

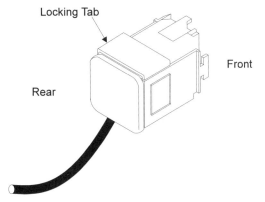

Figure 6.15 Type A connector.

There are two types of IBM data connectors. Type A connectors have many parts, permitting various angles of termination (i.e., 45-degree, 90-degree, or rear exit), plus a detachable locking tab.

> **Note.** These connectors should not be confused with the new Type STP-A connector.

The second connector is the Type B connector, which has only five parts and a built-in locking tab. There is also a new Type B enhanced connector for high-speed cable.

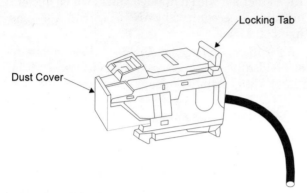

Figure 6.16 Type B connector.

Enhanced STP-A connectors have been designed to support higher bandwidth services through the use of cable pair isolation shielding.

Caution should be exercised to ensure that the proper connector is selected for the cable being installed. If the A connector is used, all grounding should be completed per the manufacturer's specifications.

Every hermaphroditic connector is provided with detailed written and visual instructions for termination procedures. A simplified version of these steps is as follows:

Step	Terminating Hermaphroditic Connectors
1	Strip cable jacket to shield.
2	Slip ring over shield.
3	Mate color-coded wire to color-coded clear plastic dressing block.
4	Snap dressing block into housing; seat firmly with pliers.
5	Snap housing cover into place.

Coaxial Cable Terminations

Overview

Although no longer recognized as an acceptable medium under ANSI/TIA/EIA-568-A for new installations, cabling installers should possess a working knowledge of the various coaxial cables present in the industry and the methods of termination.

A coaxial cable consists of an inner conductor (solid or stranded wire), separated by a dielectric (core) from its outer conductor (single-

or double-braided shield) and covered with either a PVC or plenum-rated outer sheath.

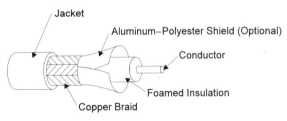

Figure 6.17 Typical coaxial cable construction.

The predominant coaxial cables are RG-6, RG-11, RG-58, RG-59, and RG-62. The table below shows the characteristic impedance and general use of each type of cable.

TABLE 6.4 Coaxial Cable Characteristics

Cable type	Nominal impedance	General usage
RG-58 IEEE 802.3	50 Ω	10BASE-2 Ethernet
RG-6, RG-11, RG-59	75 Ω	CATV, WANG, Video
RG-62	93 Ω	ARCNet, IBM 3XXX

Coaxial cables used in data applications typically have a stranded inner conductor and are terminated using BNC connectors.

The tools required for proper stripping and terminating are a two-step, rotating stripper and a ratcheted crimping tool, which is designed specifically for crimping of the BNC connector.

Figure 6.18 Coaxial cable termination tools.

Although other connector systems (screw-on style) are available, captive-pin connectors which assure positive retention of the center conductor of the coaxial cable are recommended to ensure proper data transfer.

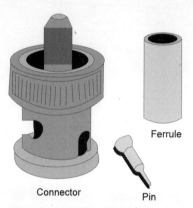

Figure 6.19 Captive pin BNC connector.

Steps—coaxial cable

Step	Coaxial Cable Terminations
1	Determine the proper method and length of sheath removal. ■ Make a straight cut in the termination end of the coaxial cable. ■ Place the connector ferrule over the end of the cable. ■ Adjust the two-step stripping tool to meet the desired cable diameter and stripping requirements. Stripping tool should be adjusted to the dimensions required by the specific conductor being used. Different connectors use different strip lengths. ■ Insert cable into the stripper. ■ Rotate the two-step stripper three to five full turns. ■ Remove the stripper from the cable. ■ Remove the severed sheathing, shielding, and dielectric material. ■ Inspect the cable for stripping quality and ensure that the center conductor and the insulation are not nicked or scored, and that any stray strands of the braided shield are pushed away from the center conductor.
2	Terminate the cable. ■ Seat the connector system pin on the center conductor. ■ Crimp the pin to the center conductor using the small diameter die of the crimping tool.

3   Insert sleeve and connector onto cable.
- Place the connector body onto the cable by aligning it so that its shaft fits over the center conductor pin and between the dielectric and the braid shield.
- Slide the connector ferrule up to cover the exposed braid shield.
- Make sure the crimping tool has the correct die for the coaxial cable being terminated (i.e., RG-6, RG-11, RG-58, RG-59, RG-62, PVC, or plenum-rated sheath).
- Place the crimp tool over the connector ferrule and squeeze the tool until the die is completely closed.
- Inspect the connection for neatness (no exposed braiding strands). The connector has to be tight.

Fiber Termination

Overview

Optical fiber cable has a greater bandwidth than copper cable and can transmit more information through a smaller, lighter cable. It is becoming popular in new telecommunications cabling installations and retrofits. As bandwidths increase, there is an increasing use of optical fiber as horizontal cable ultimately ending at the workstation. The bandwidth of optical fiber is listed in nanometers (nm), which is equal to one billionth of a meter per second.

Optical fiber cables use light generated by a laser or a light-emitting diode (LED) to carry signals. Laser light can be very intense and is invisible to the human eye.

Warning. Never look into a terminated fiber as the light may damage your eyes, even though the light will probably be invisible to your eye. Viewing it directly does not cause pain, but the iris of your eye will not close involuntarily as when viewing such a bright light. Serious damage to the retina of the eye is possible.

Optical fiber cable may also be combined with copper cable in a telecommunications system with the optical fiber carrying information from building to building in a campus installation. The optical fiber cable may be used as backbone, with copper cable used for horizontal cables.

Optical fiber cable is becoming popular for both backbone and horizontal cable because it offers a much lower signal loss than copper cable. It is also immune to EMI and lightning.

In telecommunications, there are two specific types of optical fiber cable used: multimode and singlemode. Each specific type of fiber has its own characteristics.

Multimode optical fiber has an outside diameter of 125 μm. The glass core of the fiber, which carries the optical signal, has a diameter of 62.5 μm. A micron is equal to one millionth of a meter. A protective acrylate coating is typically added to the glass fiber for protection, raising the diameter to 250 μm. ANSI/TIA/EIA-568-A recognizes 62.5/125 μm multimode optical fiber cable for horizontal cabling and backbone applications.

Singlemode optical fiber has an outside diameter of 125 μm. The glass core of the fiber, which carries the optical signal, has a diameter of 8–9 μm. A protective acrylate coating is typically added to the glass fiber for protection, raising the diameter to 250 μm. ANSI/TIA/EIA-568-A recognizes singlemode for backbone applications only.

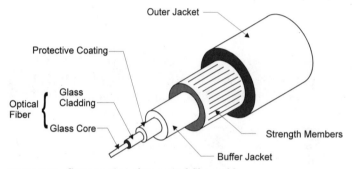

Figure 6.20 Cross section of an optical fiber cable.

An optical fiber, whether singlemode or multimode, has two distinct areas known as the core and the cladding. Although depicted above as being separate, the two areas are a single solid piece of glass. The difference between the two areas is the amount of additional materials, called dopants, which are added to the glass during manufacture to change the index of refraction. The index of refraction changes between the core and cladding enables the optical signal to remain in the fiber core.

Optical fiber cables must be cut (cleaved) precisely prior to termination to yield a square, flat end-face. Proper cleaving ensures maximum transmission of light from the optical fiber and reduces diffusion and reflection of the light at the termination. After termination of the optical fiber by connectorization, the fiber end is then polished. This step helps to ensure good light transmission through the end of the optical fiber and provides the proper alignment of the connection to maintain the integrity of the signal.

Optical fibers are immune to EMI and do not require any shielding, although some optical fiber cables (normally outdoor and aerial cables) may have a metallic armored shield. Optical fiber cables usually do not contain metallic components and, therefore, do not require the same grounding and bonding considerations as copper cable. However, if there is a metallic component as part of the cable, it should be properly grounded and bonded, as with any ground or shield member. Refer to Chapter 1, Section 8 for more information on grounding and bonding.

Specific termination kits are required for optical fiber connectors. Selection of the proper kit is dependent on the cable system, type of cable, specific connector requested by the customer or selected by the telecommunications designer, and the hardware to house the terminated fibers.

ANSI/TIA/EIA-568-A has specified the 568SC connector for new installations. The 568SC connector for a 62.5/125 µm multimode fiber and the adapter is beige in color. The 568SC connector for a singlemode fiber and adapter is blue in color.

When installing fiber to the desktop, the work area outlet may be either a standard four-inch electrical box flush mounted or surface mounted but must be capable of storing at least 1 m (3.3 ft) of two-strand fiber cable. The minimum bend radius of a two-strand fiber cable is 30 mm (1.18 in).

The most common connectors for multimode fiber cable are the ST and SC, which are normally field terminated.

Figure 6.21 Fiber connectors.

Termination procedures for optical connectors in the field vary by manufacturer. Listed below are the various methods:

- Heat-cured termination
- Ultraviolet (UV) light-cured termination
- Crimp termination
- Anaerobic termination

Heat-cured terminations require placing assembled connectors into a curing oven. This is a time-consuming process. Typical curing ovens can hold up to six connectors at a time. The curing cycle for heat-cured connectors can be as long as 20 minutes.

UV-cured terminations are cured under an ultraviolet light; the process takes less than a minute.

Enhancement of optical connectors has resulted in the development of crimp-style connectors, which require no curing process.

Anaerobic termination connectors typically use a two-part chemical application; the first is the epoxy and the second part is the primer or catalyst. When the catalyst makes contact with the epoxy, a hardening of the epoxy occurs. Typically, the hardening takes 10 to 20 seconds, but can take substantially longer.

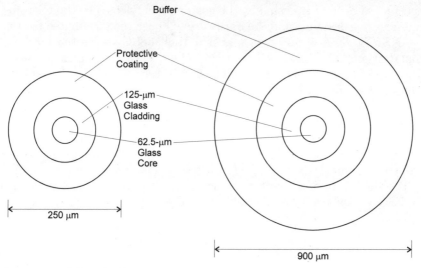

Figure 6.22 Multimode fiber.

The 62.5/125 µm multimode fiber is covered by a buffer coating with an overall diameter of 250 µm or 900 µm. This buffer must be removed prior to connectorization. Removal of the buffer coating is accomplished by use of an appropriate stripper.

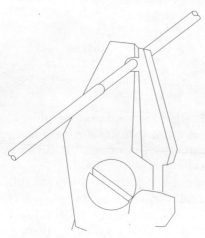

Figure 6.23 Fiber stripping tool.

The following termination procedures are designed to be generic in nature for each of the four methods of termination. Refer to vendor termination instructions, which vary significantly for specific connectors. Termination procedures for both ST- and SC-style connectors are interchangeable in most cases, but the manufacturer specifications have to be followed to ensure a properly aligned low-loss termination.

Steps—Heat-curved termination

Step	Heat-Cured Fiber Termination
Note.	Colors vary by manufacturers. Please check manufacturer's specifications.
1	Slide the boot of the connector over the end of the cable end to be connectorized. There are typically two sizes of connector boots, 3 mm and 900 µm. The 3 mm boot is utilized with pigtails; the 900 µm boot is utilized with 900 µm buffered fiber. The following procedure is for 900 µm fiber connectorization with ST-style connectors.

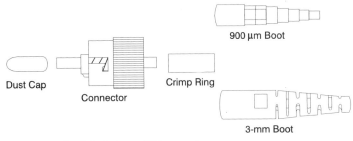

Figure 6.24 Exploded view of ST connector.

2 Carefully mark the length of buffer coating to be removed per the manufacturer specifications. Remove the buffer in 6 to 8

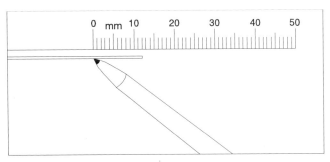

Figure 6.25 Mark the buffer.

mm increments, employing the buffer removal tool. Strip the buffer to the mark to obtain the specified amount of bare fiber. Make sure the buffer coating has no jagged edges and it is cut cleanly. Refer to the instructions packaged with the removal tool for further details.

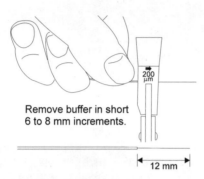

Figure 6.26 Buffer removal.

3. Clean the stripped fiber using an alcohol wipe or bifurcated swab. Fold the wipe over the fiber and squeeze gently on the fiber inside the wipe, while pulling the fiber through the wipe.

    Note. If you are using a bifurcated swab, carefully insert the fiber into the swab, gently squeeze the foam sides of the swab, and pull the fiber straight through the swab.

4. Prepare the epoxy by mixing it in accordance with the manufacturer's instructions.

5. Remove the plugger from the syringe. Load the syringe with the mixed epoxy. Reinstall the plugger into the syringe.

6. Select the connector to be terminated. Remove and discard the small black cap on the rear of the connector. Remove and retain the dust cap from the front of the connector.

7. Insert the epoxy syringe tip into the rear of the connector until it bottoms on the ceramic ferrule. Place a mark, with a black marking pen, on the syringe tip just below the connector. Remove the connector. Place a second black mark on the syringe tip 2 to 3 mm (or whatever distance the manufacturer specifies) from the first mark and toward the

tip of the syringe. These marks enable the technician to repeatedly and accurately fill all connectors.

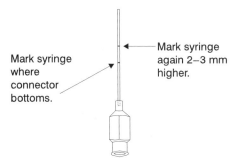

Figure 6.27 Mark syringe.

8 With the syringe pointed up, place the connector back on the syringe tip and, while holding the connector, slowly inject epoxy. The moment that epoxy is visible at the tip of the ferrule, let go of the connector. Continue injecting epoxy and allow the connector to rise on the syringe tip. When the rear of the connector rises past the second black mark, stop injecting epoxy. Remove the connector from the syringe tip.

9 Insert the ferrule into the tube of the load adapter of the curing oven.

10 Place a thin layer of epoxy along the length of the bare fiber. Be sure to coat the junction where the bare fiber enters the buffer. This will keep the epoxy bead intact during fiber insertion into the connector.

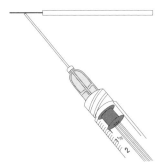

Figure 6.28 Apply epoxy.

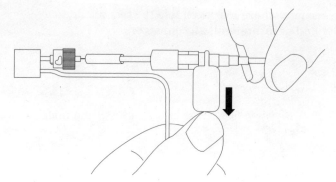

Figure 6.29 Install connector.

11   Slide the fiber into the connector until you feel the buffer seat fully on the rear of the ferrule. Pull the boot back up the fiber cable. Slide the spring clip on the load adapter down, and position the boot back in the jaw of the clip. The load adapter clip now rests on the boot and not directly on the fiber.

12   Remove any excess epoxy at the rear of the connector, being careful not to unseat the ferrule.

13   Place the connector and the load adapter into the oven. Temperature of the oven should be within the 105 to 115°C (220 to 240°F) range. Depending on the type epoxy used, cure time is anywhere from 6 to 20 minutes. Refer to the manufacturer's specifications for details.

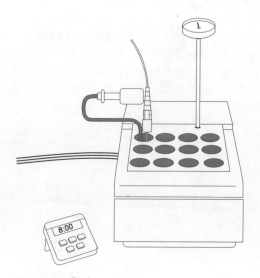

Figure 6.30 Curing oven.

14. Allow the load adapter to cool after the prescribed cure time. The epoxy should change color when cured. Remove the connector from the load adapter. Be very careful not to break the fiber protruding from the tip of the ferrule.

15. Using a precision scribing tool (cleaving tool) included in the termination tool kit, carefully scribe or nick the excess fiber at a point approximately two times the diameter of the fiber from where it exits the bead of epoxy. Pull the scored fiber straight up to complete the separation. Safely dispose of the detached fiber.

Figure 6.31 Scribe fiber.

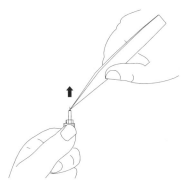

Figure 6.32 Remove excess fiber.

16. Slide the 900 μm boot onto the back of the connector. The connector is now ready to be polished.

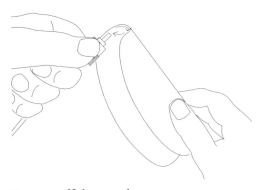

Figure 6.33 Nub removal.

17. Using a 2 μm lapping film formed into a U-shape with the abrasive face inward, remove the fiber nub by making 25 mm (1 in) circles inside the U-shaped disk. Stop when the fiber nub no longer scratches the lapping film.

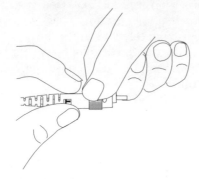

Figure 6.34 Install boot.

18. Select the first connector to be polished and check for excess epoxy on the side of the ferrule. Remove any stray epoxy with a razor blade. Test fit the ferrule in a polishing jig. It should move freely and seat on the jig top. Polishing jigs can be purchased which will polish six connectors at once.

19. Place the silicone rubber pad from the terminating kit on the glass plate provided. Clean the pad with alcohol followed by a dry wipe, and blow with dry air. Place a 3 μm disk and a 0.3 μm disk on the plate.

20. Place a separate polishing jig on each abrasive disk and use each jig only on its corresponding abrasive disk. This prevents contamination of the white adhesive disk.

   Note. If only one polishing jig is used, clean the jig after each step with alcohol.

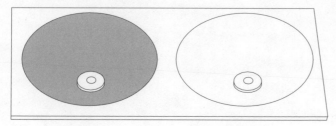

Figure 6.35 Polish jigs on disks.

21. Place a puddle of water 25 mm (1 in) in diameter on each disk and use the polishing jig to uniformly wet the surface.

22. Check the end of the connector. A small amount of epoxy should be present. In cases where no epoxy is visible, go directly to the white abrasive disk.

23. Place the connector into the first jig and onto the 3 μm disk, making one or two initial figure-eight patterns [about 80 mm (3 in) high] using a gentle downward pressure. This step ensures that the fiber nub is flush with the epoxy bead.

24. Once the fiber nub is flush, make single figure eights using medium pressure. Check the epoxy bead after each figure eight to confirm the epoxy bead is gone. Once the epoxy is almost gone, remove and clean the connector with a clean, dry wipe.

25. Place the connector in the second polishing jig and onto the wet 0.3 μm abrasive disk. Make eight to ten figure-eight patterns with only enough pressure to keep the connector steady.

26. Remove the connectors and clean the end face with a clean wipe moistened in isopropyl alcohol (99 percent pure); finish cleaning with a blast of compressed air. Ensure that the air is manufacturer approved as suitable for cleaning fiber terminations (e.g., without Freon™).

27. Insert the connector into the adapter end of a 100X microscope to inspect the end face. Replace any connector that contains a crack. Small pits are acceptable and typically occur when the connector is ground too long during the dark gray abrasive disk polishing step.

28. After passing visual inspection, place the original dust cap over the end of the connector ferrule.

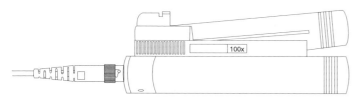

Figure 6.36 100x microscope.

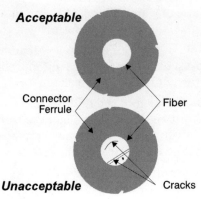
Figure 6.37 Fiber end examples.

Steps—UV-cured termination

Step	UV-Cured Termination
Note.	Colors vary by manufacturers. Please check manufacturer's specifications.
1	Slide the boot of the connector along with the strain relief collar over the end of the cable to be terminated. The following procedure is for a 900 μm fiber.
2	The orientation of the strain relief collar is extremely important. Ensure that the hexagonal end is towards the boot.
3	Carefully mark the length of buffer coating to be removed per the manufacturer specifications. Remove the buffer in 6 to 8 mm increments, employing the buffer removal tool. Strip the buffer to the mark to obtain the specified amount of bare fiber. Make sure the buffer coating has no jagged edges and it is cut cleanly. Refer to the instructions packaged with the removal tool for further details. Clean the stripped fiber with a lint-free tissue soaked in alcohol. Gently squeeze the fiber inside the tissue while pulling the fiber through the tissue.

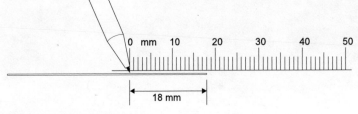
Figure 6.38 Mark the buffer.

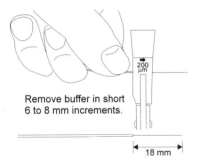
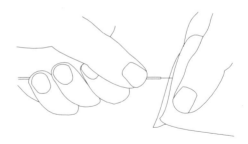

Figure 6.39 Buffer removal.   Figure 6.40 Clean fiber.

4   Clean the stripped fiber using an alcohol wipe or bifurcated swab. Fold the wipe over the fiber and squeeze gently on the fiber inside the wipe while pulling the fiber through the wipe. If using a bifurcated swab, carefully insert the fiber into the swab, gently squeeze the foam sides of the swab, and pull the fiber straight through the swab.

5   Remove and discard the cap on the rear of the UV connector assembly.

6   Insert the syringe tip into the rear of the connector assembly until it bottoms on the ferrule. Place a mark on the syringe tip just below the connector. Remove the connector. Place a second mark on the syringe tip 2 to 3 mm away (or a distance specified by the manufacturer) from the first mark toward the tip of the syringe. These marks will enable the technician to repeatedly and accurately fill all connectors with adhesive.

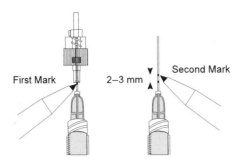

Figure 6.41 Mark syringe.

7   Place the connector back on the syringe tip and inject the UV adhesive slowly until it begins to come out of the tip of the ferrule; let go of the connector. Continue to inject adhesive

allowing the connector to rise up on the syringe tip. When the rear of the connector rises to the second mark, stop injecting adhesive.

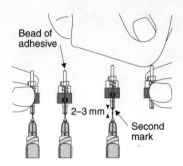

Figure 6.42 Inject adhesive.

8   Remove the connector assembly from the syringe tip. Wipe off the adhesive bead on the end face, employing a lint-free tissue.

9   Hold the fiber vertically, slide the connector onto the fiber and then release it. Gently wiggle the fiber until the connector slides down the fiber and the rear of the ferrule is fully seated on the buffer. When properly seated, approximately 5 mm of bare fiber should be sticking out of the end of the ferrule. The back of the connector assembly should be 3 mm or less from the end of the buffer. A small amount of adhesive should now be present around the fiber.

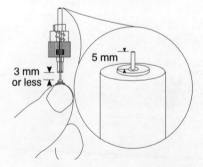

Figure 6.43 Place connector on fiber.

10 Refer to the specific instructions for placing the connector in the UV curing lamp. Some curing lamps have timers and LED

indicators which change color to indicate the curing cycle is complete. The normal cure time under UV lamps is less than one minute.

Figure 6.44 UV curing lamp.

11 Carefully remove the fiber from the curing lamp. Be careful not to break the fiber that is protruding from the ferrule tip.

12 Using a fiber scribe, score the excess fiber where it emerges from the adhesive bead on the ferrule tip. Gently and carefully pull the fiber straight up to remove excess fiber. Safely dispose of the detached fiber.

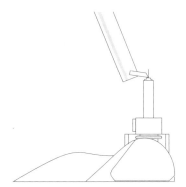

Figure 6.45 Scribe fiber.

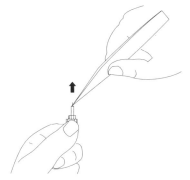

Figure 6.46 Remove excess fiber.

13 Slide the strain-relief collar up against the back of the connector and carefully thread the collar onto the threaded portion of the lead-in tube. Tighten the strain relief until it no longer turns. Do not apply excessive force.

Figure 6.47 Install strain-relief collar.

14 Slide the 900 µm boot onto the back of the ST-compatible connector.

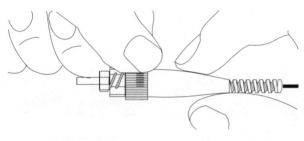

Figure 6.48 Install boot.

15 Clean the surface of the glass plate with alcohol followed by a clean dry tissue. Blow the surface with compressed air.

16 Remove the backing from the 0.3 µm and the 5 µm abrasive disks. Carefully apply the disks to the glass plate, forcing any air bubbles from underneath the lapping film.

17 Place a separate polishing jig on each abrasive disk and use each jig only on its respective disk to prevent contamination of the fine disk.

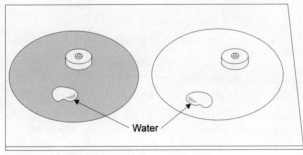

Figure 6.49 Moisten disks.

18 Wet the surface of both disks with water (25 mm in diameter). Use the polishing jig to spread the water evenly on the surface.

19 Take a loose 5 μm disk and form it into a U-shape with the abrasive face inward and, making 25-mm circles, remove the nub of adhesive from the end of the ferrule. Stop when the fiber nub no longer scratches the disk.

20 Test fit the connector ferrule into the polishing jig. It should move freely up and down and be seated on the jig top. Check the condition of the connector's end face. A small amount of the adhesive should be present.

21 Place the connector into the jig on the gray abrasive disk; make one or two figure-eight patterns about 80-mm high. This step is to make sure that the fiber nub is flush with the adhesive bead.

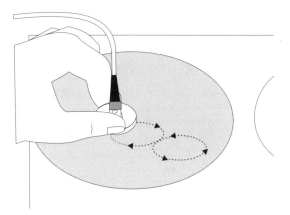

Figure 6.50 Polish fiber.

22 Polish the connector with approximately 20 strokes or until the UV adhesive bead is removed.

23 Gently wipe the end face and ferrule surfaces of the connector with a lint-free tissue soaked in alcohol. Wait at least five seconds for the surfaces to dry.

24 Complete the polishing by placing the connector into the second jig on the 0.3 μm disk and make about 15 figure-eight patterns.

25 Clean the connector again in the same manner as in Step 23 above.

26 Keep the air nozzle approximately 13 mm away from the connector. Ensure that the air is manufacturer approved as suitable for cleaning fiber terminations (e.g., without Freon).

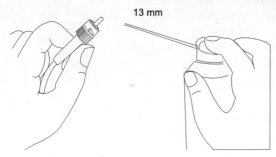

Figure 6.51 Compressed-air cleaning.

27 Secure the connector into the adapter end of the 100X microscope supplied with the tool kit and visually inspect the connector. The surface should have a mirror finish with the edge of the fiber face perfectly circular.

28 If scratches or other imperfections are visible, repolish the connector with the 5 µm lapping film.

29 After passing visual inspection, place the dust cover over the end of the connector ferrule.

Steps—Crimp-style termination

Step	Crimp-Style Fiber Termination
1	Slide the 900 µm boot (small end first) down the fiber until out of the way. This procedure is for termination of either ST or SC crimp-style connectors onto 900 µm fiber.
	Different manufacturers provide distinct tools for crimping connectors to fiber. Read and follow the specific instructions for the particular manufacturer's connector being used.

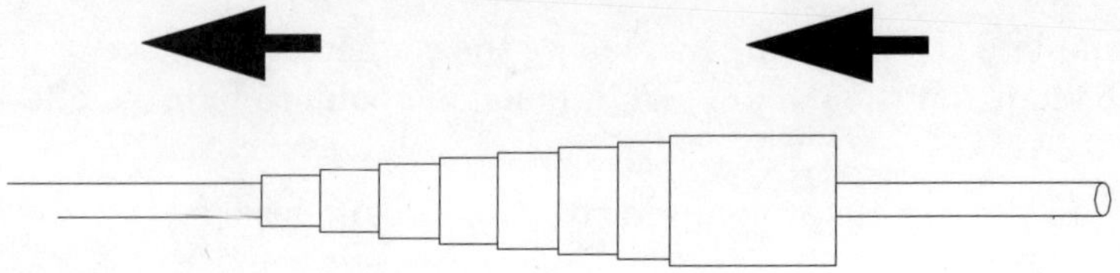

Figure 6.52 Preinstall boot.

Cable Termination Practices 369

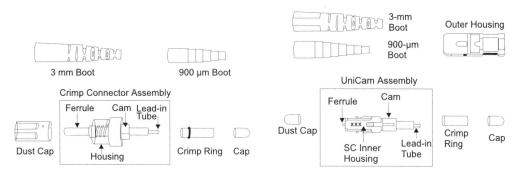

Figure 6.53 ST components.

Figure 6.54 SC components.

2. Carefully mark the length of buffer coating to be removed from the manufacturer's specifications. Remove the buffer in 6- to 8-mm increments, employing the buffer removal tool. Strip the buffer to the mark to obtain the specified amount of bare fiber. Make sure the buffer coating has no jagged edges and is cut cleanly. Refer to the instructions packaged with the removal tool for further details. Mark the buffer an additional 11 mm back from the strip point. Clean the bare fiber with two passes of an alcohol wipe of a lint-free tissue, soaked in isopropyl alcohol (99 percent pure). Do not touch the bare fiber after cleaning and do not remove the 11-mm mark.

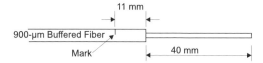

Figure 6.55 Mark the buffer.

3. Cleave the fiber as described in the instructions provided with the cleaving tool. Cleave the fiber to 8 mm or to the specified length. Safely dispose of the fiber scraps.

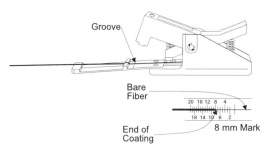

Figure 6.56 Cleave fiber.

4   Insert the crimp connector into the crimping tool following the manufacturer's instructions.

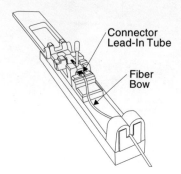

Figure 6.57   Cam tool.

5   Carefully insert the cleaved fiber into the lead-in tube until feeling it firmly stop against the connector's fiber stub. Be sure to guide the fiber in straight and do not bend or angle it.

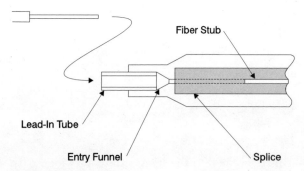

Figure 6.58   Insert fiber.

6   If resistance is felt at the entry tunnel, rotate the fiber back and forth while applying a gentle inward pressure.

7   Carefully, push the 900 µm buffer fiber into the fiber clamp on the tool. Maintaining pressure on the fiber, form a slight bow between the connector and the clamp. This bow is very important, as it helps the fibers make contact in the connector during the next step.

8   Rotate the wrench of the crimp tool past a 90-degree angle to seat the connector.

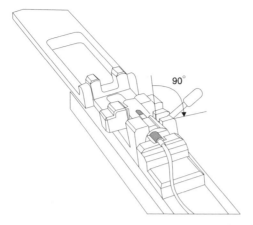

Figure 6.59 Cam fiber.

9. The fiber is now held within the connector by the splice.
10. Carefully flip the crimp handle 180 degrees until it contacts the crimp tube. Push down firmly to crimp. A flat impression can be seen in the crimp tube. The tool cannot overcrimp the connector.

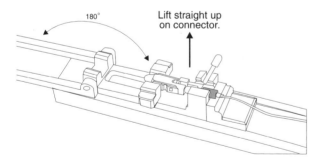

Figure 6.60 Crimp tube.

11. Flip the crimp handle back. Leave the wrench handle down. Remove the connector by lifting it straight up and out of the tool. Do not pull on the fiber.
12. Slide the boot up to the back of the connector until it reaches the "cam."

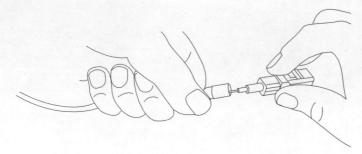

Figure 6.61 Install boot.

13   The outer housing of a multimode connector is beige. To install the connector assembly into the SC outer housing, line up the bevel edges on the inner housing with the key on the outer housing. Using the boot, push the assembly into place. It may be necessary to wiggle the parts to make them snap together.

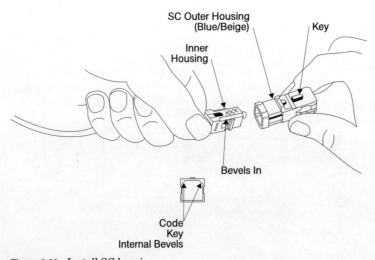

Figure 6.62 Install SC housing.

14   The connector is now ready to use. Leave the front dust cover on until ready to insert the connector into a sleeve.

Steps—Anaerobic-style termination

Step	Anaerobic-Style Fiber Termination
1	Slide the 3 mm boot (small end first) down the fiber until out of the way. Then, slide the crimp collar down the fiber. This procedure is for termination of either ST- or SC-style connectors onto 3 mm jacketed fiber.
2	Following the manufacturer's specifications, measure and mark the jacket the specified distance from the end of the fiber. Using the proper tool, sever the jacket (or sheath) and remove it from the cable.
3	Following the manufacturer's specifications, measure and mark the aramid yarn the specified distance from the end of the fiber.
4	Carefully mark the length of buffer coating to be removed per the manufacturer's specifications. Remove the buffer in 6 to 8 mm increments, employing the buffer removal tool. Strip the buffer to the mark to obtain the specified amount of bare fiber. Make sure the buffer coating has no jagged edges and is cut cleanly. Refer to the instructions packaged with the removal tool for further details. Clean the bare fiber with two passes of an alcohol wipe or a lint-free tissue that has been soaked in isopropyl alcohol (99 percent pure). Do not touch the bare fiber after cleaning.
5	Insert the syringe tip into the rear of the connector assembly until it bottoms on the ferrule. Slowly squeeze the adhesive into the connector until the adhesive appears on the tip of the ferrule. Withdraw the syringe slowly while continuing to lightly squeeze adhesive into the barrel of the connector.
6	Set aside the connector and paint the bare fiber with the primer (catalyst). Ensure that the bare fiber, as well as the exposed buffer coating, is primed for insertion.
	Caution. Do not simply dip the fiber into the bottle of primer, as a slight bump of the bare fiber could cause it to break off and fall into the primer, creating a safety hazard.
7	Carefully and steadily, insert the fiber into the connector until the connector reaches the buffer tube.
	Note. Depending on the manufacturer and type of adhesive and primer used, there may be only 15 seconds for the

insertion to take place before the catalytic action hardens the chemicals and causes a solid bond.

8. Very carefully slide the crimp ring up the buffer tube and then fan the aramid yarn around the connector base. Slide the crimp connector up onto the base, and with the specified crimping tool, crimp the crimp ring onto the body of the connector. Depending on the manufacturer, a second crimp may be required around the jacket and aramid yarn.

9. Slide the boot up the jacket to the back of the connector and over the crimp ring.

10. Using a fiber scribe, score the excess fiber where it emerges from the adhesive bead on the ferrule tip. Gently and carefully pull the fiber straight up to remove excess fiber. Safely dispose of the detached fiber.

11. Clean the surface of the silicone pad with alcohol followed by a clean, dry tissue. Blow the surface with compressed air.

12. Using a 5-mm lapping film formed into a U-shape with the abrasive face inward, remove the fiber nub by making 25 mm circles inside the U-shaped disk. Stop when the fiber nub no longer scratches the lapping film.

13. Place the connector into the polishing jig on the 3 mm abrasive disk; make figure-eight patterns about 80 mm in diameter. This step is to ensure that the adhesive bead has been removed. Clean the end of the ferrule with alcohol and inspect the connection with a microscope to ensure no shattering of the fiber's end.

14. Place the connector into the polishing jig on the 0.3 μm abrasive disk and make figure-eight patterns about 80 mm in diameter. This step is to ensure that the end of the fiber is polished to eliminate power loss due to irregular polishing. Clean the end of the ferrule with alcohol and inspect the connection with a microscope to ensure that connection is clear and without blemishes. The surface should have a mirror finish with the edge of the fiber face perfectly circular.

15. If scratches or other imperfections are visible, repolish the connector with the 0.3 μm lapping film.

16. After passing visual inspection, place the dust cover over the end of the connector ferrule.

Chapter 7

Splicing Cable

Copper Cable   375
Optical Fiber Cable   389

Copper Cable

Overview

Intrabuilding splices are generally constructed in equipment rooms, telecommunications closets, main terminals, entrance facilities, and cable trays. Cable splice locations and other splicing details are usually specified by the telecommunications distribution designer in work order prints. These work order prints must be strictly followed because the splicing work could affect future cabling system plans. Where questions arise concerning cable pair counts or specific splicing methods, seek direction from the immediate supervisor. For the purpose of this section, the cable splicing techniques described are for in-building applications using single sheath cable (yet to be activated), and fire-retardant 25-pair connector modules in a two-bank, in-line configuration.

Intrabuilding copper cable splicing is allowed only in backbone cable—never for horizontal cable (horizontal cable extends between the telecommunications closet and the work area). Backbone cables usually have expanded polyethylene-polyvinyl chloride (PVC) insulation, and will normally be Communications Riser Rated (CMR).

CMR cable consists of 24 AWG copper conductors enveloped by a core wrap and an overall shield, and are bonded to a fire-retardant sheath. A fire-retardant sheath is required by the *National Electrical Code*®

(NEC) to mitigate the spread of fire from floor to floor. The insulated cable, commonly known as plastic insulated conductor (PIC) cable, is designed for ease of cable pair identification. PIC cable is bundled in 25-pair color-coded binder groups with color-coded pairs in each binder (see color-code chart in Table 7.1 for cable sizes up to 600 pairs). Intrabuilding PIC backbone cables commonly range in size from 25 to 1800 pair. The first binder group of the cable is normally located near the center of the cable and successive groups are layered toward the cable sheath.

TABLE 7.1 Telecommunications Cable Color Code

Pair Number	Binder group			
	Tip	Ring	Color	Pair Count
1	White	Blue	White-Blue	001–025
2	White	Orange	White-Orange	026–050
3	White	Green	White-Green	051–075
4	White	Brown	White-Brown	076–100
5	White	Slate	White-Slate	101–125
6	Red	Blue	Red-Blue	126–150
7	Red	Orange	Red-Orange	151–175
8	Red	Green	Red-Green	176–200
9	Red	Brown	Red-Brown	201–225
10	Red	Slate	Red-Slate	226–250
11	Black	Blue	Black-Blue	251–275
12	Black	Orange	Black-Orange	276–300
13	Black	Green	Black-Green	301–325
14	Black	Brown	Black-Brown	326–350
15	Black	Slate	Black-Slate	351–375
16	Yellow	Blue	Yellow-Blue	376–400
17	Yellow	Orange	Yellow-Orange	401–425
18	Yellow	Green	Yellow-Green	426–450
19	Yellow	Brown	Yellow-Brown	451–475
20	Yellow	Slate	Yellow-Slate	476–500
21	Violet	Blue	Violet-Blue	501–525
22	Violet	Orange	Violet-Orange	526–550
23	Violet	Green	Violet-Green	551–575
24	Violet	Brown	Violet-Brown	576–600
25	Violet	Slate	25th binder is not used.	

Note. Where the NEC applies, only "Listed" cable can be used within a building structure, except for an entrance cable that is within 15 m (50 ft) from its point of entrance. The only exception to this would be to extend unlisted or outside cable in rigid or intermediate metallic conduit. This would allow 15 m (50 ft) of exposed cable where the cable emerges from the rigid metal conduit or intermediate metal conduit. Listed copper cables are identified by markings on the outer sheath such as CM (general use), CMR (riser rated), and CMP (plenum rated). CM is the abbreviation of communications media.

Note. The cable binders for cable up to 600 pairs have a white overall binder on each grouping. Cable extending above 600 pairs will begin with a red overall binder on each grouping and continue the color sequence assigned to the tip conductor.

Planning is critical to the splicing operation. Several items must be considered, such as cable placement, support structure for the splice, and selection of the closure. Generally, the planning is done by the telecommunications designer who constructs the prints and ensures that the parts are ordered. The splicing operation can be done with one or two widely used types of equipment: MS2 by 3M or type 710 by Lucent Technologies. Both are IDC or insulation displacement contacts.

When the parts are on the job site, the cabling installer must then set up the splicing operation properly to accommodate an in-line, butt, or branch splice in a horizontal or a vertical orientation. The cabling installer must position the cable, rig the support structure, and ensure that the proper materials are used—such as the right splicing modules and closure. For example, too many cable conductors spliced in a closure without adequate splice banks may stress the wires. Reentry and churning (repetitive activity) conductors without proper planning can lead to deterioration of the cables at their weakest point—the splice.

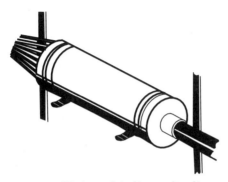

Figure 7.1 Horizontal, in-line vault splice supported in a rack.

Once the cables are set up, they must be prepared for opening of the sheath (outer cable covering). A decision needs to be made as to what method of splicing will be used and the number of banks within the splice. These decisions will provide the means of selecting the properly sized closure and the opening of the sheath to the length recommended in manufacturer instructions. Once the sheaths are open, the cable shields bonded together, the closure endplates installed and the binder groups marked, the chosen splicing method can then be used to join the cable pairs.

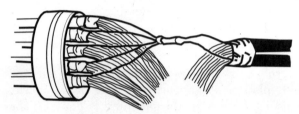

Figure 7.2 Splice opening with sheath bonds and endplate installed

Two common modular (groups of pairs) splicing techniques are:

- In-line.
- Foldback.

With the in-line splicing method, wire is placed in a straight-across arrangement and provides for little wire slack. The in-line method is not designed to be rearranged and should receive minimum handling. The foldback splicing method allows the conductors to be folded into the splice which in turn provides for maintenance, rearrangement, and transfer of the conductors. The foldback method typically requires more cable be stored within the splice and could increase the size of the needed splice closure.

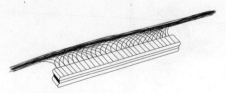

Figure 7.3 In-line splice

Figure 7.4 Foldback splice

Most copper cable splicing is performed with modular connectors. Modular connectors are available in several varieties—in-line, branch, and half-tap. They are for outside plant or intrabuilding use and, depending upon the manufacturer, accommodate 19 to 28 AWG wire. In addition, these connectors are available in several pair sizes, from 1 pair to 25 pair and can be placed in 1-, 2-, 3-, or 4-bank configurations within the splice. Testing the cable and the splice can be done as it is being constructed or afterwards, in accordance with the supervisor's instructions. As the splice is constructed, each of the modules is labeled with the cable number and pair count.

Notes

1. Only use fire-retardant modular connectors for intra-building splices.
2. In most cases, backbone cable will be 24 AWG.

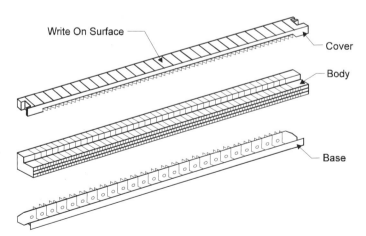

Figure 7.5 Modular connector

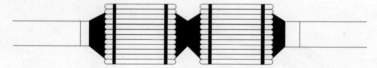

Figure 7.6 2-Bank splice

After the conductors are spliced, they must be prepared using a polyethylene wrap before the splice closure is installed. The splice wrapping should be snug, yet not overtightened, so that modules and cable pairs do not become lodged in the splice closure seams. Once the splice is wrapped, a splice closure must be installed over all intrabuilding splices. An intrabuilding splice closure is a strong, lightweight, fire-retardant covering that protects nonpressurized splices. The closure shields the splice against humidity, moisture, and may even resist temporary immersion in water. When the closure is installed, it must be properly supported, grounded, and tested for air leaks according to manufacturer's recommendations. Labels must also be affixed to all cables entering the splice, indicating cable number and pair counts. Care must also be taken to clearly designate the "In" and "Out" for the spliced cables.

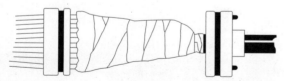

Figure 7.7 Wrapped splice with endcaps

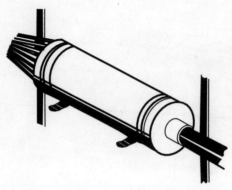

Figure 7.8 Completed splice in cable rack

Safety

Follow the splicing rig and closure manufacturer's guidelines when performing splicing operations. In addition, the items listed below should be enforced:

- Follow all OSHA guidelines.
- Wear safety glasses.
- Wear a hardhat.
- Check cable sheath for presence of unsafe voltages.
- Ensure that there is adequate lighting.
- Wear safety gloves when opening cable sheath.
- Keep cable shields bonded between the splice at all times.

Steps—Copper splicing

Step	Copper Splicing
1	The splice hardware will have been selected by the designer or supervisor to match the tools on hand prior to starting the job. However, it never hurts to ensure the right materials and tools are available for the job. Ensure that the appropriate modular connectors have been selected for splicing, making sure that the tools accommodate the modular connectors to be used. For an intrabuilding splice, ensure that the modules are of the fire-retardant type, and that the gauge of the wire within the cable corresponds to the range of sizes accepted by the connector module. Also, ensure that the appropriate fire-retardant closure has been selected, including endcaps. To calculate splice and closure size, consider the following factors: - Cable pair count - Wire gauge - Connector type - Number of banks of connectors - Splicing method
2	Determine the scope of setting up the splice (also known as "sizing-up" the job). Consider the space needs of the splice closure, the working space, and the cable pathway leading to the splice. Do not bend the cable such that it twists or kinks. As a rule, the bending radius of the cable should not exceed 10 times the cable diameter. Follow the cable manufacturer's guidelines when bending cables.

The bending radius is the essential factor in determining the size of the splice box or maintenance hole into which the splice will ultimately be placed. The following guidelines may be useful in considering space for the splice.
- Overhead splice
  - 686 mm (27 in) between closure and nearest overhead obstruction (813 mm [32 in] in hardhat areas).
  - 635 mm (25 in) between closures where positioning does not permit closure to be lowered.
- Horizontal splice
  635 mm (25 in) above the floor. (There also can be a temporary position used only for splicing.)
- Vertical splice
  Limited by space required to position cable, closure, and supporting structure.

3. Check the work area to ensure that the cabling installer is working safely. Verify that there is adequate work area and that dangerous power conditions do not exist. The cable shield should be checked for foreign voltages since neither the other end of the cable nor the pathway it is taking can be seen. Foreign voltages can be transmitted by accidental contact with a power cable or electromagnetic induction.

4. Safely set up the splicing area using ladders or scaffolding, where necessary. Use warning cones and tape if working in an area that is accessible to the public. Set up portable lighting to clearly see the color-coded cable pairs and binder groups to ensure splice accuracy. Have tarpaulins available to protect both the splice in harsh environments and the splice area from debris while performing the work operation. It will be useful to have a garbage bag and a scrap wire bag handy while constructing the splice.

5. Install a support structure (e.g., trapeze, ladder rack, cable tray, or cable rack) for the splice, if necessary. Move the cables to the support structure to position them for splicing. Remember not to twist or kink the cable. Maintain the proper bending radius of the cable and allow room for the appropriate splice closure.

6. Place a mark on the cables at the center location of the splice. Next, mark the cables at the proposed location of the splice sheath opening in accord with the manufacturer's instructions for the selected closure. Scuff the sheath of the cables, using a carding brush or similar tool, for approximately 229 mm (9 in)

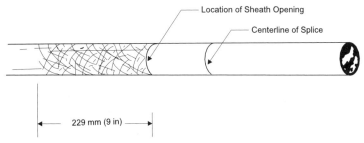

Figure 7.9 Scuffed sheath with markings on the sheath

away from the marking of the sheath opening toward the area where the endcaps will be placed.

7. At this time, it may be easier to start installing the closure endcaps. Follow the manufacturer's instructions for installation. (These steps continue as if the endcaps will be installed after the conductors are spliced.)

8. Ring (circular cut) one of the cables with a cable knife where the splice opening will begin.

   Note. You may feel the cable shield under the sheath as you ring the cable.

   Warning. Do not put more pressure on your knife than necessary to extend the knife blade to the metallic shield.

At a 90-degree angle from this circular cut, use the cable knife and cut into the sheath again—running the knife to the end of the cable.

Remove the outer sheath of the cable and discard it. Next, nick the metallic shield at the sheath ring to remove the shield. By putting enough pressure on the shield, it will sever as the shield is removed. Care must be taken, since the shield may be very sharp. Be careful that the core wrap and binder tapes do not unravel from the cable end such that the cable pairs of different binder groups can become mixed.

Secure the core wrapper at the end of the cable with vinyl tape. Place three wraps of vinyl tape (adhesive side out) around the core wrapper and adjacent to the cable sheath. Slide half of the tape's width under the shield. Be sure that any exposed edges of the shield are trimmed off or bent up and away from the cable core.

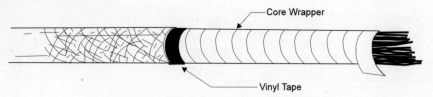

Figure 7.10 Sheath opening with core wrapper in place and vinyl tape protection from shield

9. The hardware can be found within the packaging of the closure for connecting the cable shields. If it is not included with the package, obtain the proper materials to bond the cable sheaths. In most cases a bonding braid (equivalent to a minimum 6 AWG) and an insulating sleeve are provided along with the shield connector. Slide a length of insulating sleeve over the bonding braid, which will connect the shields of the cables to be spliced. In accordance with the manufacturer's instructions, install a shield connector on the back side of each of the cables from where you are working. This may require cutting a small piece of the sheath and shield with tabbing shears.

   Attach the bonding braid to the connector stud of each cable. The braid should exit the side of the connector. An alignment hole can be punched through the braid with a blunt object, such as a pencil, approximately 13 mm (0.5 in) from the end of the braid. Use vinyl tape to cover over the shield connectors and extending 19 mm (0.75 in) out from the sheath opening (onto the inner sheath or core wrap).

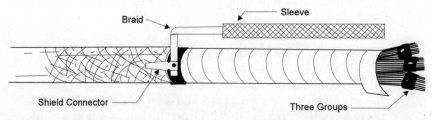

Figure 7.11 Sheath opening with shield connector, braid, and insulating sleeve

10. Remove the core wrap to the taped edge of the cable butt and secure the ends of each binder group with vinyl tape to prevent loss of conductor pair identification.

11. Identify each 25-pair binder group within the cables by using the appropriate color-coded cable ties or scrap wire. Scrap-wire pairs from junk cable can be wrapped around a corresponding

binder group (two times), the ends of the wire twisted and cut to provide identification. Place the cable tie or twisted wires approximately 25 mm (1 in) from the butt of the cable. When placing the cable ties or twisted wires, ensure that they are loose enough so that they will turn freely.

12. Align the cable binder groups from both cables. The binder groups can be aligned by grasping the core of the cable and turning it within the cable sheath.

13. You are now ready to align the cables and secure them so that the splice opening does not spread or collapse during the splice operation. First, ensure that the cables are aligned to the splice location and that the splice opening meets the instructions of the splicing rig and the splice closure. Secure the cables in place by tying them to the cable rack.

14. Determine whether the splice will be constructed using the in-line or foldback method. With the in-line splicing method, wire is placed in a straight-across arrangement. The in-line method is not designed to be rearranged and should receive minimum handling. The foldback splicing method allows the conductors to be folded into the splice which provides for maintenance, rearrangement or transfer of the conductors. As mentioned in the Overview paragraphs, the in-line, 2-bank splicing method will be described.

15. Attach the support tube and related components of the splicing rig to the cable sheaths. Follow the manufacturer's recommendations for setting up the splicing rig. Usually, this will involve strapping a support tube to the cables, attaching a traverse bar with a clamp, and adjusting the splicing head to the position of the cable with the head clamp.

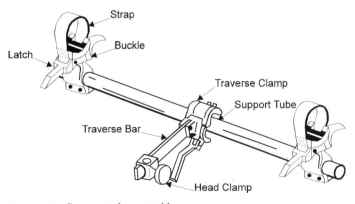

Figure 7.12 Support-tube assembly

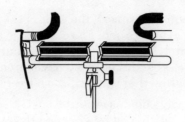

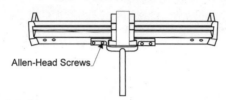

Figure 7.13 Double splicing-head setup

16   Select the back, bottom, 25-pair group from the feeding cable. The position of the splicing head is very important to the operation so that the wires will fit snugly into splicing head and connector module.

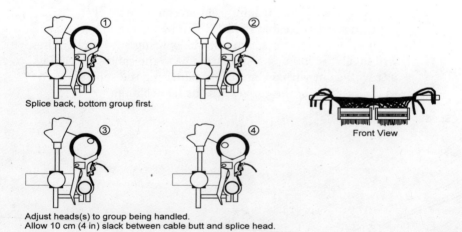

Figure 7.14 Position of the splicing head to the cable groups

17   Place a connector module base into the splicing head. Lay the pairs of the feeding cable into the connector module base following the color code.

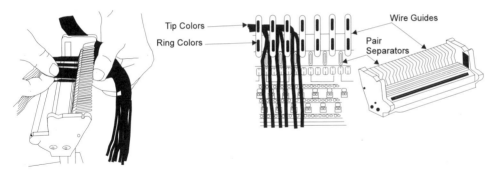

Figure 7.15 Cable pairs being placed into splice head

18. After the 25 pairs have been placed into the connector module base, visually check to see that each wire (i.e., ring, tip) is positioned properly.

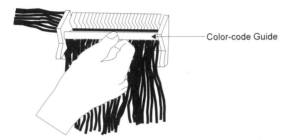

Figure 7.16 Module being checked for correct color code

19. Place the connector body over the connector base which holds the arranged cable pairs. Follow the same process for the group of wires from the away cable side of the splice. Place the connector cover over the connector body. Position the crimp head onto the splice head and crimp the connector module. The binder group pairs may be electrically tested from the splice point if the equipment is available, and the cable ends

Figure 7.17 Crimper being placed on splicing head

away from the splice have been cleared of potential trouble (e.g., shorts and crosses).

20   Continue splicing the cable groups according to the designer's work-order prints. The binder groups can be tied with a color-coded wire tie, scrap wire, or cotton sleeving with a clove hitch. Mark each module cover with the cable number and pair numbers using an indelible marking pen.

21   Tie down the splice bundle in the center—working any slack toward both cable butts.

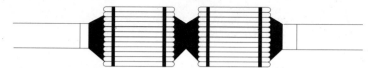

Figure 7.18   Spliced cable after being tied down

22   Cover the splice with a polyethylene wrap. The polyethylene wrap material is usually on a 100 mm (4 in) wide roll. Wrap the splice so that the connector modules and cable pairs do not become lodged in the closure.

Figure 7.19   A splice wrapped with polyethylene and endcaps

23   Install the closure according to the manufacturer's instructions. Part of the closure may have been installed earlier in the splicing operation. A grounding lug will usually be provided in one of the endplates and must be connected to the cable sheath bonding braid. The grounding lug on the exterior of the closure must be bonded to the location indicated by the work order prints or by the distribution designer. Ensure that the cable and splice closure are properly supported and test the closure for air leakage according to the manufacturer's instructions.

24   Label the cables entering the splice with the cable number and pair numbers. Labels should be visible and not be easily removed.

25 Remove the ladders, scaffolding, etc. and clean up the work area.
26 Cable testing from the cable ends can now take place if the cable has not previously been tested from the splice point.

Optical Fiber Cable

Overview

In customer premises applications, a telecommunications distribution designer avoids the requirement of fiber-to-fiber field splicing by installing a continuous length of cable. This is normally the most economical and convenient solution; however, splices cannot always be avoided because of the cable plant layout, length, raceway congestion, requirements for a transition splice between nonlisted OSP (outside plant) cables and listed cable at the building entrance point and, lastly, because of unplanned requirements such as damaged cable during the installation or in cable dig-up.

The figure below shows examples of splices required due to cable routing.

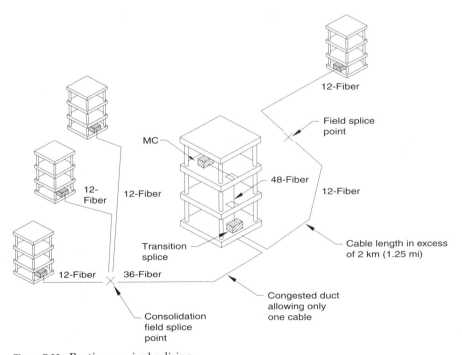

Figure 7.20 Routing required splicing

There are two major categories of field-splicing methods for optical fibers—fusion and mechanical. Both single-fiber and mass-fiber (typically 12 fibers) splicing methods are available. Both of these methods are field proven and have excellent long-term reliability when completed according to manufacturer's instructions.

For outside plant splice locations, the splices and stripped cables are usually protected and secured by a splice closure. For splicing inside a building, a splice enclosure is often used which is secured to a rack or wall. In both cases, the splice closure or enclosure contains the fiber splices in splice trays or organizers, typically in groups of 6, 12, 24, or more fibers per splice tray or organizer.

Splicing can occur between two optical fiber cables—loose-tube cables containing 250 µm coated fiber and tight-buffer cables containing 900 µm buffered fibers. Both mechanical or fusion splice methods can perform 250-µm-to-250-µm splicing, 250-µm-to-900-µm splicing, or 900-µm-to-900-µm splicing. Typically, multimode fibers are 50/125 µm to 62.5/125 µm while singlemode fibers are 8–9/125 µm. Mechanical or fusion splicing can accommodate both multimode and singlemode fiber.

Note. While it may be possible to physically splice multimode to singlemode fibers, the splice loss will be unacceptable. The fiber type should typically be the same, i.e., 62.5 µm to 62.5 µm or 8–9 µm to 8–9 µm.

Planning is critical to the splicing work operation. Several items must be considered such as cable placement, mounting location for the splice closure, or enclosure and selection of the splice hardware and splice method. Generally, the planning is done by the designer who constructs the prints and ensures that the parts are ordered. When the parts are on the job site, the cabling installer must then set up the splicing operation properly to accommodate either an in-line, butt, or branch splice. This is the point at which the cabling installer must position the cable, mount the splice hardware, and ensure that the proper materials are used. The size and type of splice hardware is determined by the:

- Environment in which the hardware will be placed.
- Number of fibers and cables to be spliced.
- Splice trays designed to hold the particular splicing method chosen.

Based on the splice hardware used and the type of cable being spliced, the cabling installer must determine the proper amount of each cable element to be stripped (outer jacket, strength members, buffer tube, or subunit jacket) before splicing can begin. This is a critical element, because the cables must be properly secured to the closure or enclosure and the buffer tubes or subunit elements must be correctly routed from the splice hardware entrance point to the splice trays. Leaving too much buffer tube or too many subunits may cause difficulty in storage of these elements after splicing is complete; leaving too little may cause difficulty in providing an acceptable work area for completing the splice.

Safety

Follow the hardware and manufacturers' guidelines when performing splicing operations. The items listed below should also be observed:

- Follow all Occupational, Safety, and Health Administration (OSHA) guidelines.
- Wear a hardhat when in a construction area.
- Wear safety gloves when opening cable sheath.
- Ensure that there is adequate lighting.
- Properly discard small pieces of fiber to ensure they do not accidentally penetrate the skin.
- Good housekeeping.

Steps—Length determination

Step	Length Determination
1	Determine the proper cable strip length. Strip length is determined by the type of hardware in which the optical fiber cable and splices will be housed. Consult the instructions for the specific optical fiber hardware.
2	Determine the length of outside jacket to remove. This is typically governed by Steps 3 through 5. While the length varies, depending on the splice hardware, the typical length is from 1–3 m (3–10 ft).
3	Determine the length of strength member (aramid yarn) to remove. The strength members should be secured to splice hardware, using securing hardware provided, close to the area where the jacket is removed.
4	Determine the length of subunit needed. Typically, a length of cable subunit remains which protects the fibers as they are routed, from the point where the jacket is removed to the splice trays. This length varies greatly, depending upon the splice hardware which is used. Consult the hardware instruction sheet. Subunits are associated with OSP cables and high fiber-count indoor cables. For low fiber-count indoor cables, subunits

typically do not exist and, thus, the length applies to buffered fibers. Additionally, some manufacturers recommend the use of a specific tubing to route the fibers to the splice trays; once again, the entire length of subunit jacket is removed and replaced with this tubing.

5   Determine the length of coated or buffered fiber. This length is determined by the fiber length which will be stored inside the splice tray.

Steps—Stripping

Step	Stripping the Cable
1	Mark the cable jacket where the jacket is to be removed by notching it with your cable knife.
2	Most cable jackets can be removed with a standard cable jacket removal tool.
3	Either remove the length of cable jacket in short sections of jacket at a time, using a series of ring cuts, or make longer sections which will typically require splitting the cable jacket lengthwise. Only cut the outside jacket. A cable knife or hook-type blade can be used for this operation.
4	Cut the strength members to the correct length for securing to the splice hardware, using high quality snips.
5	Usually, the aramid yarn is secured in the splice hardware before continuing to strip the cable.
6	Remove a required length of subunit (multifiber units for high fiber-count indoor cables) jacket. For multifiber units, the thin jacket is removed by either using the coax ring cutter or by using a cable knife.
7	Secure the subunit to the splice tray and dress the fibers into the tray. The subunit should be secured to the splice tray per the manufacturer's instruction. Commonly used methods are either small cable ties or a metal tab in the splice tray. No matter what device is used, ensure the subunit is not deformed or crushed.
	It is at this stage that the cabling designer's splice plan must be fully understood and followed, (i.e., which fibers from which cable are to be spliced together). The proper subunits must be routed to the correct corresponding trays, (i.e., in a through splice, fibers 1–12 from cable 1 are routed to the same splice tray as fibers 1–12 from cable 2).

Step	
8	Continue securing the subunits to the splice trays and dressing the fibers in the tray, until the entire cable is stripped and loaded into the splice trays.

Steps—Single-fiber mechanical

Step	Single-Fiber Mechanical Splicing
1	A number of mechanical splices are available on the market which have their own special tools and procedures. Familiarize yourself with the manufacturer's instructions prior to beginning the splicing operation. Most products follow the same basic procedures and are explained below.

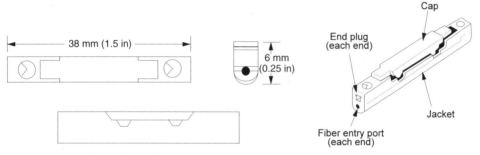

Figure 7.21 Sample mechanical splice

2	Most manufacturers produce a special assembly tool which may or may not be required; however, this tool typically simplifies the splicing operation and helps ensure a successful splice.

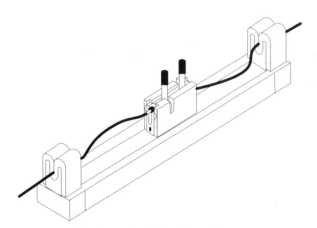

Figure 7.22 Sample mechanical splice tool

3. Ensure that the mechanical splice is in the open position, so that the cleaved fibers can be inserted into the splice. This may be accomplished by using a key, rotating a cam, or ensuring the cap is in the proper position. If using a splice assembly tool, load the splice into the tool at this time.

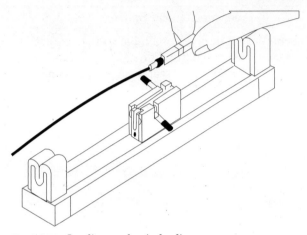

Figure 7.23 Loading mechanical splice

4. Wipe the filling compound from the fibers with a tissue soaked in isopropyl alcohol (99 percent pure).

5. Strip the coating or buffer from the fiber strand to the required length, using the cable manufacturer's recommended fiber stripper. For 250 μm coated fibers the most commonly used tool is the Miller tool. For 900 μm buffered fiber, either

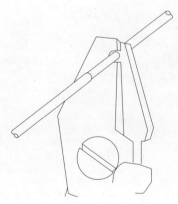

Figure 7.24 Fiber-stripper tool

the Miller tool or a No-Nik® tool may be required. The proper strip length is determined by the requirements of the mechanical splice and the cleaver being used; however, 25 to 50 mm (1 to 2 in) is typically all that is necessary.

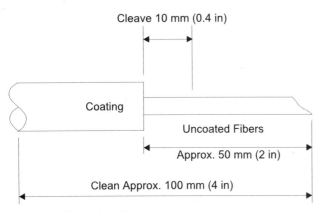

Figure 7.25 Cleave length

6. Gently wipe the fiber with a tissue soaked in isopropyl alcohol (99 percent pure). This removes any fragments of dirt remaining on the fiber. This is an important step because the fibers are aligned within the splice by the outside dimension of the stripped fiber.

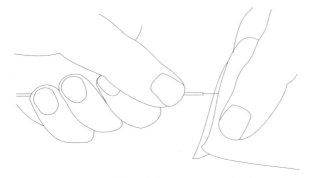

Figure 7.26 Isopropyl alcohol (99 percent pure) wipe

7. Cleave the fiber to the prescribed length for the splice with the cleaving tool. A number of different cleaving tools could be

used for this operation. Follow the manufacturer's instructions. It is important that tools remain clean and in proper working order. Observe the manufacturer's length recommendations closely, because this ensures that the fibers butt together correctly in the splice.

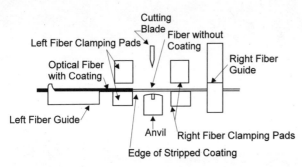

Figure 7.27 Cleaving fiber

8   Discard any broken fiber scraps by depositing them into an approved container.

9   If using the splice assembly tool, push the fiber down into the fiber retention or foam clamps.

10 Holding the fiber by the coating next to the bare fiber, slide the fiber into the mechanical splice until it stops.

11 Repeat Steps 4 through 9 for the second fiber that is to be spliced.

12 Holding the second fiber by the coating next to the bare fiber, slide the fiber into the mechanical splice until it butts against the first fiber inserted.

13 It is important to ensure that the fibers butt against each other in the center of the splice. Various manufacturers have different procedures to help ensure that this occurs, especially when using an assembly tool. Follow their instructions. Typically, this is accomplished by observing the proper cleave length stated in Step 7.

14 Close the splice. This is accomplished by rotating the keys or cam or pushing down on the cap, depending on the product being used.

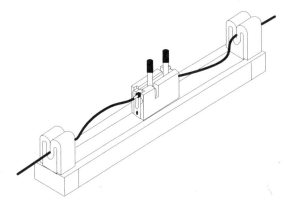

Figure 7.28 Close mechanical splice

15. If using an assembly tool, remove the fibers from the retention or foam pads. Then remove the splice by lifting the splice out of the tool.

16. Secure the mechanical splice in the splice tray by following the manufacturer's instructions. Ensure the splice is secured and that proper bend radius on the fiber is maintained.

17. Repeat Steps 3 through 16 until all fibers have been spliced.

18. Using an ODTR, check each splice for acceptable loss. BICSI and TIA stipulate the splice loss shall be 0.3 dB or less. Note that loss measurement values using an OTDR can be directional in nature. If extremely accurate loss readings are needed, the average from both directions will be needed; however, splices in premises applications are not that critical and one-direction measurements should be adequate.

19. If the splice loss is above 0.3 dB, go back and remake the splice. Common problems that could occur, are:
    - The fiber ends are separated and can be corrected by simply repositioning the fibers.
    - The fiber in the trays may be too tight, causing loss that can be corrected by relaxing the bend.
    - The cleave is not of acceptable quality, requiring repeating Steps 3–16.

20. Shut the splice closure or enclosure per the manufacturer's instructions.

21. Secure the splice closure or enclosure and any cable slack to the mounting location.

Steps—Single-fiber automatic fusion splicers

Step	Single-Fiber Automatic Fusion Splicers
1	A number of fusion splicers are available on the market that provide special tools and procedures. Familiarize yourself with the manufacturer's instructions prior to beginning the splicing operation. While most fusion splicers have some degree of automation, this section on single-fiber automatic fusion splicers is dedicated to splicers which automatically position the fibers in the XYZ-axis and also provide automatic splice loss measurements. There are basically two distinct types of automatic fusion splicers: the local injection detection (LID) method and the profile alignment system (PAS) method. While these two distinct methods exist, they both follow the same basic procedures and steps. These are explained below.
2	Most automatic fusion splicers are menu driven and allow for various degrees of automation; however, these procedures are based on full automation. Most automatic fusion splicers furnish their own fiber cleaver design; however, an alternate cleaver can be used if care is taken in following the prescribed cleave length.

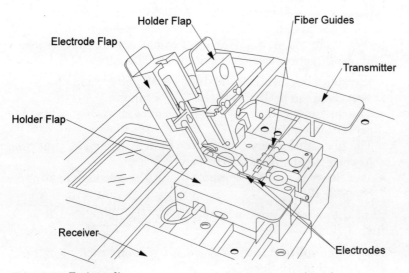

Figure 7.29 Fusion splicer

3. Open the fusion splicer and connect it to an approved power source. Some splicers are equipped with either an internal or external battery supply normally capable of eight hours of operation.

4. Turn on the fusion splicer per the manufacturer's instructions. Most units will go through some sort of internal diagnostics. Depending on the machine and the power supply, some machines have an automatic cutoff function to conserve battery power.

5. Most machines will have some sort of parameter-setting capability. Specifically, choose whether the splicer is to be set up for multimode or singlemode operation. While a number of parameters are available for changing, it is suggested that the cabling installer initially try splicing with the manufacturer's default settings.

6. Select fully automatic, semiautomatic, or manual operation. It is suggested that fully automatic should be used unless some special circumstance exists.

7. Some fusion splicers may or may not have different sized V-grooves used for alignment based on 250 μm coated or 900 μm buffered fibers. Ensure that the proper V-groove is used, based on the type of fibers being spliced.

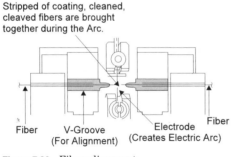

Figure 7.30 Fiber alignment

8. Wipe the filling compound from the fibers with a lint-free tissue soaked in isopropyl alcohol (99 percent pure).

9. If using heat-shrink tubing for mechanical protection of the completed splice, slide the tubing over one of the two fibers to be spliced.

10. Strip the coating or buffer from the fiber to the required length, using the cable manufacturer's recommended fiber stripper. For 250 µm coated fibers, the most commonly used tool is the Miller tool. For 900 µm buffered fiber either the Miller tool or a No-Nik® tool may be required. The proper strip length is determined by the fusion splicer and the cleaver being used; however, 25 to 50 mm (1 to 2 in) is typically all that is necessary.

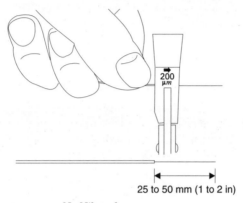

Figure 7.31 No-Nik tool

11. Gently wipe the fiber with a lint-free tissue soaked in isopropyl alcohol (99 percent pure). This removes any fragments of dirt remaining on the fiber. This is an important step because the fibers are aligned within the V-grooves of the fusion splicer in relation to the outside dimension of the stripped fiber.

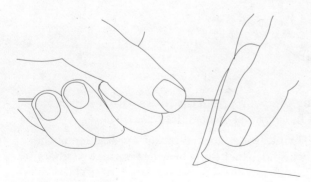

Figure 7.32 Wipe fiber

12 Cleave the fiber to the prescribed length for the splice, using the cleaving tool. A number of different cleaving tools could be used for this operation. Follow the manufacturer's instructions. It is important that tools remain clean and in proper working order. Observe the manufacturer's length recommendations closely, because this ensures that the fibers have enough travel in the fusion splicer to select the proper Z-gap and to butt against each other during splicing.

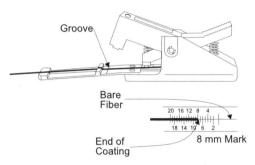

Figure 7.33 Cleave fiber

13 Discard the broken fiber stub properly.

14 Open the flaps over the electrode and the fiber-holding V-grooves.

15 Grasp the fiber by the coating next to the bare fiber; place the fiber into the fusion splicer. Some units may mark the V-grooves where the end of the coating or buffer should be placed. Typically, the end of the fiber is placed so that it stops between, but not past, the upper and bottom electrodes.

16 Close the fiber-holding V-groove flap but not the electrode flap.

17 Repeat Steps 7 through 16 for the second fiber to be spliced.

   Note. If using heat-shrink tubing for protection, only one tube is needed per fiber splice, (i.e., it is not needed on the second fiber). If the first fiber was inserted into the left-hand side of the splicer, the second fiber is inserted into the right-hand side or vice versa.

18 Close the flap over the electrodes.

19 Push the button that causes the machine to fusion splice the fibers. In this mode of operation, the fusion splicer will clean the fiber endfaces (prefuse), determine the end-face quality, automatically align the fibers in the X- and Y-axis, set the

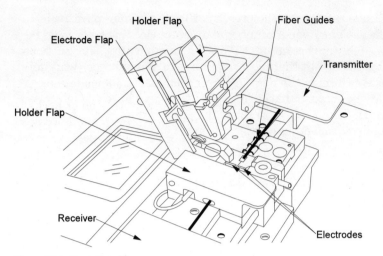

Figure 7.34 Inserting fiber

proper Z-gap between the fibers, and then perform the fusion operation at the time they are butted. Lastly, the fusion splicer will provide a splice loss based on either local injection detection or profile alignment calculations.

20. It is possible that the fusion splicer may not be able to complete the splicing operation because of a number of problems. Often the unit will detect the problem and provide the corrective action to be taken. Listed below are common problems and solutions:
    - End-face quality unsatisfactory. Inspect the end-faces of the two fibers to determine which one should be recleaved. Inspect the fibers using both the X- and Y-axis views.
    - Cannot align the fibers. Inspect the alignment of the two fibers in both the X- and Y-axis. They should be closely aligned in the center of the viewer. If not, one or both of the V-grooves may be dirty, the fiber may be dirty, or the correct strip and cleave length may not have been obtained.
    - Cannot find the proper Z-gap. Inspect the positioning of the two fibers in the viewer to ensure they are close to the center of the viewer, allowing enough travel for them to be butted. One or both fibers may not have been inserted to the correct location in the V-grooves.
    - Refer to the manufacturer's instructions for other error messages and corrective actions.

21. Visually check the quality of the splice. If a good splice has been made, the cladding (outside of the fiber) should have a

smooth surface, (i.e., no protrusions or neckdown). There should be no visible faults or shadows inside the fiber image.

Note. The typical bright lines in the center of the fiber are the result of the lens effect.

22 The most common occurrence is either a protrusion or neckdown or, possibly the fibers are not joined. You should break and redo the splice if this occurs. If this problem persists, refer to operating instructions which indicate whether the autofeed is too high or low, the fusion time too long, or the current too high. Refer to the manufacturer's instructions on how to change these settings.

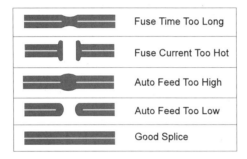

Figure 7.35 Common occurrences

23 Check the value of the splice loss. According to TIA and BICSI, the splice loss should be equal to or less than 0.3 dB. Specific customers may have established their own splice loss values. Consult the work specifications provided by the designer.

24 If the splice loss is above 0.3 dB, go back and correct the splice. Try to remake the splice before changing the manufacturer's default parameters. If high losses continue, additional common problems can occur.
    - The settings are for multimode fiber and singlemode fibers are being spliced or vice versa.
    - During the splicing operation, the default settings are changed. Check the fusion splicer settings to ensure that they are the same as those stated in the operating instructions.
    - If the temperature or humidity is extremely high or low, if may be necessary to either shorten or lengthen the fusion time or raise or decrease the fusion current. Normally, only

adjust the fusion time and fusion current to obtain a satisfactory splice in any environment.
- The electrodes are dirty or worn. Refer to the manufacturer's instructions on cleaning or replacing the electrodes.

25. Record the splice loss value. Some fusion splicers have the ability to store the splice loss value. Follow the manufacturer's instructions.

26. Open the flaps over the electrodes and the fiber-holding V-grooves; remove the completed splice from the fusion splicer.

27. If using heat-shrink tubing protection, slide the device so that it is centered over the splice. Place the tubing into either a heater or use a heat gun to shrink the tubing down onto the splice. Follow the manufacturer's instructions, but ensure that the completed covering does not cause bending of the fibers or damage to the fiber coating.

28. Secure the fusion splice in the splice tray by following the splice tray manufacturer's instructions. Ensure that the proper fiber bend radius is maintained. If using a tube of silicone rubber to protect the fusion splice, complete all splices in that splice tray. Then apply the silicone over the splice organizer and splices—typically, twelve splices per organizer.

29. Repeat Steps 8 through 28 until all fibers have been spliced.

30. Shut the splice closure or enclosure, per the manufacturers' instructions.

31. Place the splice closure or enclosure (and any cable slack) to its mounting location.

Steps—Single-fiber semiautomatic fusion splicers

Step	Single-Fiber Semiautomatic Fusion Splicers
1	A number of fusion splicers are available which have their own special tools and procedures. Familiarize yourself with the manufacturer's instructions prior to beginning the splicing operation. While most fusion splicers have varying degrees of automation, this section on single-fiber semiautomatic fusion splicers is dedicated to splicers which automatically feed (abuts and aligns) during the fusion operation. However, the operator may have to prefuse, align, and set the Z-gap of the fibers. Machines follow the same basic procedures and steps as explained next.

2. Most semiautomatic fusion splicers are menu driven and allow for various degrees of automation; however, these procedures are based on manual operation, except for autofeed. Most fusion splicers furnish their own fiber-cleaver designs, but an alternate cleaver can be used if care is taken in following the prescribed cleave length.

3. Open the fusion splicer and connect the fusion splicer to an approved power source. Some splicers are equipped with either an internal or external battery supply normally capable of eight hours of operation.

4. Turn on the fusion splicer per the manufacturer's instructions. Most units will go through some sort of internal diagnostics. Depending on the machine and the power supply, some machines have an automatic cutoff function to conserve battery power.

5. Almost all machines will have some sort of parameter-setting capability. Specifically, choose whether the splicer is to be set up for multimode or singlemode operation. While a number of parameters are available for changing, it is suggested to initially try splicing with the manufacturer's default settings.

6. Select semiautomatic or manual operation. Use semiautomatic unless some special circumstance exists. However, these procedures will follow manual operation and can be used depending on the degree of automation that your fusion splicer provides.

7. Some fusion splicers may or may not have different sized V-grooves used for alignment based on 250 μm coated or 900 μm buffered fibers. Ensure that the proper V-groove is used, based on the type fibers being spliced.

8. Wipe the filling compound from the fibers with a lint-free tissue soaked in isopropyl alcohol (99 percent pure).

9. If using heat-shrink tubing for mechanical protection of the completed splice, slide the tubing over one of the two fibers to be spliced.

10. Strip the coating or buffer from the fiber to the required length, using the cable manufacturer's recommended fiber stripper. For 250 μm coated fibers, the most commonly used tool is the Miller tool. For 900 μm buffered fiber, either the Miller tool or a No-Nik® tool may be required. The proper strip length is determined by the fusion splicer and the cleaver being used; however, 25 to 50 mm (1 to 2 in) is all that is required.

11. Gently wipe the fiber with a lint-free tissue soaked in isopropyl alcohol (99 percent pure) removing any fragments of dirt remaining on the fiber. This is an important step, because the fibers are aligned within the V-grooves of the fusion splicer in relation to the outside dimension of the stripped fiber.

12. Cleave the fiber to the prescribed length for the splice using the cleaving tool. A number of different cleaving tools could be used for this operation. Follow the manufacturer's instructions. It is important that tools remain clean and in proper working order. Observe the manufacturer's length recommendations closely, because this ensures that the fibers have enough travel in the fusion splicer to select the proper Z-gap and to butt up against each other during splicing.

13. Discard the broken fiber stub properly.

14. Open the flaps over the electrode and the fiber-holding V-grooves.

15. Grasp the fiber by the coating next to the bare fiber; place the fiber into the fusion splicer. Some units may mark the V-grooves where the end of the coating or buffer should be pulled. Typically, the end of the fiber is placed so that it stops between, but not past, the upper and bottom electrodes.

16. Close the fiber-holding V-groove flap but not the electrode flap.

17. Repeat Steps 7 through 16 for the second fiber to be spliced.

18. Close the flap over the electrodes.

19. Using the Z-gap controls, bring the fibers to the proper Z-gap based on the manufacturer's instructions to perform the prefusion operation.

20. Inspect the endface of each fiber to ensure a clean right-angle cut.

21. If the fusion splicer has X- and Y-axis controls, visually align the fibers in both axis. If the fusion splice has fixed V-grooves, the fibers in the X- and Y-axis cannot be aligned by simply relying on the V-grooves to perform the alignment. If the fibers are not aligned, one or both of the V-grooves may be dirty, the fiber may be dirty, or the correct strip and cleave length may not have been obtained.

22. Set the Z-gap to the proper length based on the manufacturer's instructions. If the fibers cannot be moved to the proper gap, then one or both fibers may not have been inserted to the correct location in the V-grooves.

23 Press the fusion button.

24 Visually check the quality of the splice. If a good splice has been made the cladding (outside of the fiber) should have a smooth surface, i.e., no protrusion or neckdown. There should be no visible faults or shadows inside the fiber image.

> **Note.** The typical bright lines in the center of the fiber are the result of the lens effect.

25 The most common occurrence is either a protrusion or neckdown or, possibly the fibers are not joined. Break and redo the splice if this occurs. If this problem persists, refer to the operating instructions which indicate whether the autofeed may be too high or low, the fusion time too long, or the current too high. Refer to the manufacturer's instructions on how to change these settings.

26 Depending upon the specific semiautomatic fusion splicer, it may or may not be capable of providing automatic splice loss readings. If this feature is available, check the reading of the splice loss using the values provided. If not, use an OTDR. According to TIA and BICSI, the splice loss should be equal to or less than 0.3 dB. Specific customers may have established their own splice loss values. Consult the work specifications provided by the designer.

27 If the splice loss is above 0.3 dB, go back and correct the splice. You should attempt to remake the splice before changing the manufacturer's default parameters. If high losses continue, common problems that can occur in addition to the ones stated above are:
    - The settings are for multimode fiber and singlemode fibers are being spliced or vice versa.
    - During the splicing operation, the default settings were changed. Check the fusion splicer settings to ensure that they are the same as those stated in the operating instructions.
    - If the temperature or humidity is extremely high or low, it may be necessary to shorten or lengthen the fusion time or raise or decrease the fusing current. Normally, only adjust the fusion time and fusion current to obtain a satisfactory splice in any environment.
    - The electrodes are dirty or worn. Refer to the manufacturer's instructions on cleaning or replacing the electrodes.

28 Record the splice loss value. Some fusion splicers have the ability to store the splice loss value. Follow the manufacturer's instructions.

29. Open the flaps over the electrodes and the fiber-holding V-grooves. Remove the completed splice from the fusion splicer.

30. If using heat-shrink tubing protectors, slide the device so that it is centered over the splice. Place the tubing into either a heater or use a heat gun to shrink the tubing down onto the splice. Follow the manufacturer's instructions but ensure that the completed covering does not cause bending of the fibers or damage to the fiber coating.

31. Secure the fusion splice in the splice tray by following the splice tray manufacturer's instructions. Ensure that the proper bend radius is maintained. If using a tube of silicone rubber to protect the fusion splice, complete all splices in that splice tray, apply the silicone over the splice organizer and splices—typically, twelve splices per organizer.

32. Repeat Steps 8 through 31 in this procedure until all fibers have been spliced.

33. Shut the splice closure or enclosure, per the manufacturer's instructions.

34. Place the splice closure or enclosure (and any cable slack) to its mounting location.

Summary

In today's infrastructure design, far less splicing is required but is called for in some situations.

When splicing is required, the cabling installer should consider investing in the necessary splicing equipment and in-depth training from a manufacturer or distributor providing the splicing equipment. This is true for copper and fiber.

Chapter 8

Testing Cable

Copper Cable	409
Horizontal Cabling—Unshielded Twisted-Pair (UTP), Screened Twisted-Pair (SCTP), and STP-A Cables	414
Backbone Cable	426
Coaxial Cable—Data	428
50-Ohm Coaxial Cable	429
75-Ohm Coaxial Cable	431
Optical Fiber Cable	432
Light Source and Power Meter Testing—Channel or Link	435
Light Source and Power Meter Testing—Patch Cables	449
Optical Time Domain Reflectometer	450

Copper Cable

Overview

Characterizing cabling provides useful data for the support of future equipment application. The verification of the transmission performance of cable plant through field measurements results in a more professional hand-off from the cabling installation contractor to the customer. This should improve the overall effectiveness of providing premises networking solutions, as well as a baseline for future troubleshooting, resulting in higher customer satisfaction.

This chapter addresses the field testing of unshielded twisted-pair (UTP), screened twisted-pair (ScTP), coaxial cables, optical fiber cables, and shielded twisted-pair 150 Ω (STP-A) cable. This media is defined in ANSI/TIA/EIA-568-A, *Commercial Building Telecommunications Cabling Standard*.

Introduction

Testing is an organized, systematic method utilized to verify that the installation has been completed in accordance with all of the terms and conditions of the contract and industry standards. This method may be divided into three distinct phases. They are

- Visual inspection.
- Testing with the various field testers required for each type of cable.
- Documentation.

The visual verification (phase 1) should include all pathways and spaces (where possible), telecommunications closets, and equipment rooms. Items to be inspected include, but are not limited to, the following:

- Infrastructure
- Grounding and bonding
- Cable placement
- Cable termination
- Equipment and patch cords
- The proper labeling of all components

Once the visual inspection has been completed and all discrepancies corrected, it is now possible to move to phase 2, testing with the various electronic field test equipment. The following is a list of the types of cabling field test equipment:

- Volt-Ohm-Milliampere multimeter
- Induction amplifier/tone generator
- Wire map testers
- Cable-end locator kit
- Certification field tester
- Time domain reflectometer (TDR)
- Optical fiber flashlight
- Infrared conversion card
- Low intensity laser
- Strand identifer
- Optical light source and power meter
- Optical time-domain reflectometer (OTDR)

- Telephone test set (butt set)
- Appropriate test adapters, leads, and cables

During this phase, it is important to remember that the field test equipment selected must be appropriate for the type of cable to be tested and provide the required test documentation.

The third phase, documentation, is probably the most prone to neglect. Documentation should be in the form requested by the customer or, if not specified, then in accordance with ANSI/TIA/EIA-606, *Administrative Standard for the Telecommunications Infrastructure of Commercial Buildings.*

Field testers

Each field tester is designed to perform a specific function or a range of functions required for the testing and certification of a specific cable type. They vary greatly in capability and price. It is necessary to select the most cost-effective unit available to perform the specific tests required.

Volt-ohm-milliampere multimeter. The multimeter is probably the most basic and also the most widely used field tester available. Both analog and a more accurate digital type are available. It can be used to measure voltage, current, and resistance in copper wires. With the use of a shorting device on one end of the pair, continuity can be tested. When testing coaxial cable, it is possible to calculate the length of the cable by determining the actual resistance of the loop. Since most coaxial cable LAN applications require only continuity and length verification, the multimeter is an ideal choice for this type of testing. The analog meter has the advantage of being able to see a signal that is fluctuating, where a digital meter would never lock in.

Induction amplifier/tone generator. Also known as a toner or cable tracer, this field test equipment provides the ability to identify a specific pair by generating a tone on one end of the pair, with an inductive amplifier identifying it at the opposite end. Some induction amplifiers also provide the ability to trace the pair throughout its entire length by this same method. This field test equipment is primarily used for cable identification and troubleshooting. Most units are now a combination of tone generator and continuity tester, commonly known as a wand and toner set.

Wire map testers. Wire map testers, also known as pair scanners, are low-cost cable testers that usually test for opens, shorts, crossed pairs, and miswires such as reversed pairs in either a 4-pair or 25-pair cable.

Some testers in this category will also test for split pairs. These units are good for quick, basic tests, but lack the sophisticated diagnostic capabilities of more expensive testers. Most testers in this category are designed exclusively for UTP/ScTP.

Cable-end locator kit. Sometimes called an office locator kit, this is a set of numbered 8-pin modular plugs, which can be identified by the cable test equipment. The standard practice is to insert the plugs into outlets in the work area, then search with the tester until it finds the plug at the opposite end of the cable.

Certification field testers. Certification field testers are used to verify that a cabling system meets the transmission performance requirements as specified in TIA/EIA TSB-67. All of these units will test a cabling system up to at least 100 MHz, and in the autotest mode, include length, attenuation, wire map, and near-end crosstalk (NEXT) tests. When the field tester is operated in autotest mode, it compares the actual measured values with required values for Category 3, 4, or 5 and displays pass or fail for the entire battery of tests. The tester will also display pass or fail and the actual tested values for each parameter. These testers are capable of other measurements, including impedance, capacitance, resistance, delay, delay skew, equal level far-end crosstalk (ELFEXT), and attenuation-to-crosstalk ratio (ACR) calculations. In addition to the TIA/EIA TSB-67 standard, they also include the classes of ISO/IEC and the pass/fail criteria in their database. Each certification field tester is configured to test continuity and length of coax. Several have the ability, with an add-on module, to perform the power meter test for attenuation of optical cable fiber. Certification testers can store test data and export it to a database or output it to a printer. Most field testers also have the ability to be controlled by a PC and download and store directly to the PC hard drive or floppy disk.

Certification test sets. Certification test units will test a UTP and ScTP cabling system to at least 100 MHz and will measure and record the following parameters:

- Wire map
- Length
- Attenuation
- NEXT

Additional parameters which may be measured or calculated and recorded include

- Return loss
- ELFEXT
- ACR
- Propagation delay
- Delay skew
- Power sum NEXT
- Power sum ACR
- Power sum ELFEXT

These additional parameters are expected to be included as part of TIA/EIA TSB-95 and ANSI/TIA/EIA-568-A-5, "Enhanced Category 5E."

The autotest feature compares the actual measured values with required values for Category 3, 4, or 5 performance and displays pass or fail for the entire battery of tests. The certification test equipment will also display pass or fail and the actual measured values for each test, individually. These units are capable of other tests, including impedance, capacitance, and loop resistance. Certification testers can store data and export it to a database or output it to a printer.

Time domain reflectometer (TDR). The TDR locates and tests all cable defects, splices, and connectors and gives loss values for each occurrence. In addition, the TDR is used to measure the electrical length of a cable. Developed originally for use on coaxial networks, it is also an excellent troubleshooting tool for use with UTP, ScTP, and STP-A. The measurement of the cable is accomplished by injecting a high rise-time pulse into the cable and then looking for the reflections caused by impedance mismatches to be returned. Mismatches, caused by kinks, splices, etc., are displayed as well as a large mismatch at the end of the cable. The reflections are displayed either on a screen or in the form of a printout.

Optical fiber flashlight. This is also called an optical fiber LED or visible light source. It is used to test and troubleshoot continuity in optical fiber strands by determining whether there is a fiber break in the cable strand. It generates a safe light that can be checked with the naked eye.

> **Warning.** Never look at optical fiber strands, unless personally confirmed that it is disconnected from potential laser light sources. Those sources can cause eye damage.

Infrared conversion card. An infrared conversion card allows the cabling installer to visually detect an infrared signal (fiber optic) when that signal is directed at the card's phosphorus material.

Low intensity laser. Also known as the hot red light, operates in the visible light range. It is used to identify individual fibers within a cable by passing a red light down the fiber. When used as a troubleshooting tool, the fiber strand will glow red at the point of break.

Strand identifier. This clamp-on unit inserts a microbend into the optical fiber cable and thereby is able to detect the light escaping from the fiber. It is able to detect the presence of light as well as the transmit and receive direction on singlemode and multimode cable.

Optical light source and power meter. An optical light source and power meter are required to perform loss (attenuation) tests on optical cable. Some hybrid units also include the ability to measure the length of the fiber.

Optical time-domain reflectometer (OTDR). The OTDR locates and tests all cable defects, splices, and connectors and gives loss values for each occurrence. These sets vary in complexity from the complete standalone unit to a mini OTDR, which may require the use of a laptop computer.

Telephone test set (butt set). A telephone test set is used to test voice circuits and the following functions:

- Simulate the user's telephone equipment
- Identify circuits
- Circuit diagnostics and troubleshooting

Test adapters, leads, and cables. The cabling installer must have proper adapters available to connect the test equipment to the cable under test.

Horizontal Cabling—Unshielded Twisted-Pair (UTP), Screened Twisted-Pair (SCTP), and STP-A Cables

Introduction

Horizontal cabling as specified in ANSI/TIA/EIA-568-A is composed of UTP and ScTP Categories 3, 4, and 5, STP-A, and multimode fiber. This section will cover the testing of the copper media used in horizontal cabling. Test procedures are those specified in TIA/EIA TSB-67.

Continuity testing

This is the most basic test to establish proper cabling installation. It is also referred to as a wire map test when utilizing handheld testers. At a minimum, it searches for

- Open circuits, which indicates incomplete terminations, faulty connectors, or broken cables.

  Note. At least one pair must be available for the field tester to recognize the remote.

- Short circuits, which indicates faulty connectors or crushed cables.
- Improper termination.
- In the case of ScTP, the drain wire continuity.

A multimeter can be used to measure dc loop resistance; however, this is impractical because of the time involved in checking combinations. Wire map field testers, also known as pair scanners, are more practical because they quickly identify open or short circuits and verify wiring positions for multiple pairs. Pair scanners do not test cabling performance but provide the functions of a continuity tester. Some units additionally indicate cable length.

Step	Utilizing a Pair Scanner
1	Disconnect application equipment.
2	Attach pair scanner to one end of cabling. (This can also include patch cords.) Attach the pair scanner's remote device to the opposite end.
3	If the pair scanner is unable to locate the distant end, locate the break in the cable utilizing a TDR or the TDR function of a certification field tester.
4	Diagnose and fix faulty cable runs.
5	Maintain a record of the test results for the customer's records.

Cable testers

Attenuation and NEXT test units have been available for several years, but were not based on industry standards. In October 1995, TIA/EIA TSB-67 established requirements for such measurements.

Certification field testers are much more complicated than pair scanners and require training and experience to ensure proper results. Requirement limits are designated for three TIA performance categories—Category 3, Category 4, Category 5, and two ISO/IEC cabling classifications—Class C and Class D.

Setup. Basic procedure:

- Select test standard and cable type.
  - Unshielded
  - Screened
  - Shielded
  - Coaxial
- Select average cable temperature (if available).
- Select conduit setting (if available).
- Select category or classification when applicable.
  - Category 3
  - Category 4
  - Category 5
  - Class C
  - Class D
- Select test configuration when applicable.
  - Basic link
  - Channel

Calibration. Different field test instruments involve slightly different calibration procedures. It is important that the stated procedures are followed, since proper calibration of field testers is required to ensure accurate measurements. Although the recommended time interval between field calibrations varies by manufacturer, a good rule of thumb

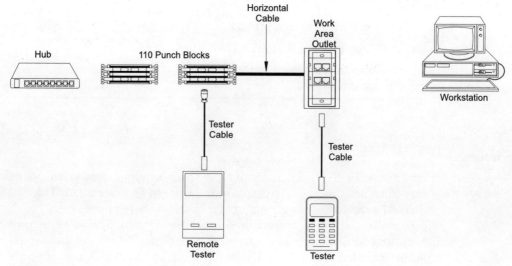

Figure 8.1 Basic link test configuration.

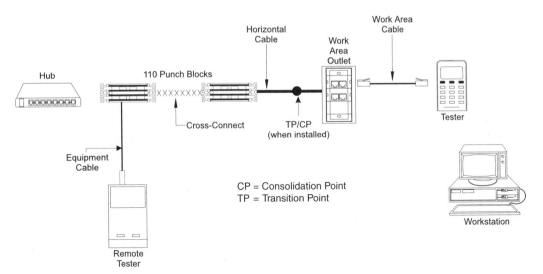

Figure 8.2 Channel test configuration.

is to perform the field calibration at least once a day or before beginning a new series of tests. A short length of tester equipment cord (supplied with field tester) or adapters are used to connect the main unit to the remote unit; the calibration sequence is activated from the main unit.

NVP calibration. It is necessary that the correct cable type be selected and the Nominal Velocity of Propagation (NVP) be determined using the field tester calibration procedure. If cable information is not available, TIA/EIA TSB-67 outlines a simple calibration procedure for establishing the NVP, using a known length of cable. This will permit accurate length measurements. Figure 8.3 illustrates the NVP calibration setup.

To calibrate

- Reel off a minimum of 15 m (50 ft) of cable and attach category-specific modular jacks to both ends. (Longer lengths will produce a more precise NVP.)
- Verify the length of this cable utilizing a tape measure or the cable sheath markings.
- Attach this connectorized cable to the main unit and to the remote unit of the tester.
- Select NVP calibration from the menus and enter the length of the cable.

- Make sure that the field tester is set to the appropriate units (metric or United States customary units) before calibration.

Note. If moving on to another installation, the cabling installer may be testing a different cable, which requires reestablishing NVP (see TIA/EIA TSB-67, Annex D, Paragraph D.4 for additional information).

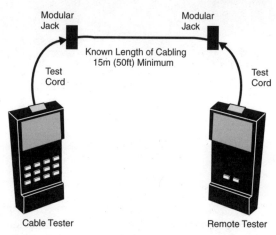

Figure 8.3 Calibration of NVP

Sanity checks. Any tester should undergo a regular performance evaluation or some form of sanity check. The cabling installer should develop some expertise to assess whether a tester is providing useful results. One simple method is to construct a sample link to be used as a reference. Make sure that it is physically protected and not subject to movement. Expect some variation with temperature, humidity, and time, but this is a valid check for malfunctioning units.

A link such as the one required for NVP calibration may be used as a reference link and will serve as a simple test to verify the consistency of a field tester. If this same link is used periodically to examine tester repeatability, one will get a useful indication whether the field tester is functioning properly.

Testing. Most measurements are carried out using the autotest function, which automatically completes a test series.

TIA/EIA TSB-67 and ANSI/TIA/EIA-568-A and addenda requirements

The following lists some of the TIA/EIA TSB-67, ANSI/TIA/EIA-568-A, and addenda tests and requirements:

- Continuity and wire map testing
  - Cable pairs must be properly connectorized to meet requirements.
- Length
  - No more than 94 m (308 ft) for a basic link (installed horizontal cabling, measured connector to connector). This length includes the two test cords provided that are each approximately 2 m (6.5 ft) long.
  - No more than 100 m (328 ft) for a channel (installed horizontal cabling, cross-connect, and equipment cords).
  - Length measurements are of limited accuracy because they are computed (estimated) from delay measurements, which may utilize an NVP value of limited accuracy.
- Attenuation
  - Evaluated across a frequency range.
  - Compared against a category or classification limit through the frequency range.
  - The category or classification limit determines pass/fail of each pair.

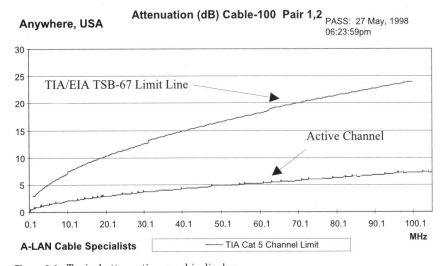

Figure 8.4 Typical attenuation graphic display.

- NEXT
  - Evaluated across a frequency range.
  - Compared against a category or classification limit through a frequency range.
  - NEXT measurements must be made at both ends of the cabling. Some testers will measure NEXT at both ends during the same autotest.

- The category or classification limit determines pass/fail of the cabling.
- Delay
  - Evaluated at all frequencies from 1 MHz to the highest frequency for a given category.
  - Compared against a category limit for each frequency.
  - The category limit determines pass/fail of the cabling.

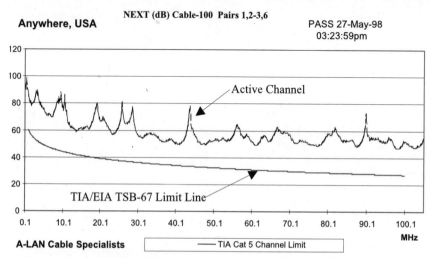

Figure 8.5 Typical NEXT graphic display.

- Delay skew
  - Evaluated from 1 MHz to the highest point and frequency of each category.
  - Compared against a category or classification limit for frequency.
  - The category or classification limit determines pass/fail of the cabling.

Additional tests

Although the following tests are not specified for field testing in ANSI/TIA/EIA-568-A and TIA/EIA TSB-67, they may be helpful in characterizing a cable segment as well as in troubleshooting.

- dc loop resistance
  - A measurement of the dc loop resistance of each cable pair.
  - Nominal dc loop resistance values are:

Twisted pair	Ohms/100 m (328 ft)	Ohms/305 m (1000 ft)
24 AWG	18.8 Ω	57.2 Ω
22 AWG	11.8 Ω	36.0 Ω

- Cable impedance
  - Impedance is a measure of the total opposition to current flow within an electrical circuit. It should remain constant regardless of the amplitude and frequency of the signal, and independent of the cable's length.
  - The defined characteristic impedance for UTP cable is 100 Ω ± 15 percent. When looking at the graph produced by a TDR, a graph of the absolute impedance is shown. This is used to determine length and distance to impedance discontinuities.

  Note. The accepted parameter for determining impedance variations discontinuities in installed cabling is return loss.

- Capacitance
  - Defined as an electrical component's tendency to store energy.
  - Cables are rated nominally in picofarads (pf).
  - The mutual capacitance of UTP is rated at 15–25 pf/ft.
  - The mutual capacitance by cable type as specified is ANSI/TIA/EIA-568-A and shall not exceed:

Cable category	Maximum mutual capacitance
3	20 pf/ft
4	17 pf/ft
5	17 pf/ft

- Attenuation to Crosstalk Ratio (ACR)
  - ACR is a mathematical formula that calculates the ratio of attenuation to near-end crosstalk for each combination of cable pairs.
  - Information provided:
    - Pass/Fail (ISO/IEC 11801 only)
    - Worst-case ACR
    - Worst-case frequency
    - Margin

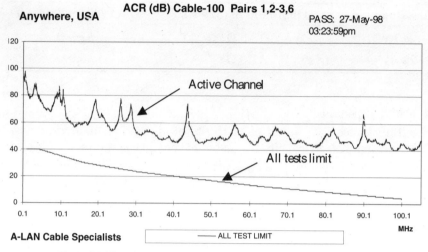

Figure 8.6 Typical ACR graphic display

- Return Loss
  - Measures the difference between the test signal's amplitude and the amplitude of signal reflections returned by the cable.
  - The return loss test indicates how well the cable's characteristic impedance matches its rated impedance.

Tester performance

Individual parameter failures can be one of two types: "FAIL" or "*FAIL." The "*" indicates that the failure is subject to some degree of uncertainty because the measured value is so close to the specified limit that the field tester is not sufficiently accurate to resolve the measurement. It is a good idea to investigate any "FAIL" or "*FAIL" parameters to understand the underlying cause. Note that any "*FAIL" of a single parameter will result in an overall "FAIL" for the link.

"*PASS" indicates that the result is better than the specified limit, but questionable due to the accuracy of the instrument. A "*PASS" for any parameter will be treated as a "PASS" and, if all other parameters are acceptable, the result is an overall "PASS" for the link under test. The "*FAIL" and "*PASS" areas of uncertain measurements are often referred to as the gray zone and are illustrated in the following figure.

Interface adapters. Sometimes an interface adapter is necessary to attach field testers to the channel/link under test. These adapters are an extra element in the link configuration and may affect the accuracy

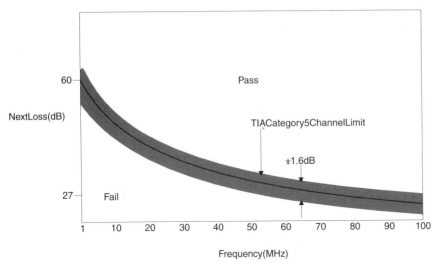

Figure 8.7 Uncertainty zone for certification field testers

of a measurement. In general, unknown adapters should be avoided. Test instrument vendors should provide the adapters or adapter cords. If adapters must be used, ensure they are rated for at least the same level as the test being performed (i.e., do not use a Category 3 adapter to test a Category 5 cable).

Durability. Durability is an important issue in keeping track of the life cycle of test interfaces. Any jack or plug has a limit on the number of reliable connect/disconnect cycles. Interfaces should be dated to give the user an indication of age, so reliable limits are not exceeded. Additional interfaces should be ordered and available for replacement at suitable intervals. If measurements are not repeatable, or if the failure rate of links appears to increase, it may also be time to replace the plug/jack interfaces.

Downloads. All certification field testers are available with download software that will allow the operator to either download directly to a printer or to the floppy or hard drive of a computer. The connection is made through the communications port of the computer. As the software from each manufacturer varies, refer to the field-tester manual for detailed instructions on the download process. Some field testers also have the ability for direct computer control of the field tester during the autotest phase. One advantage of this arrangement is that the computer keyboard may input the test identification data rather than from the field tester itself. Note the following about data storage: Most field testers give one the option of either saving or not saving the individual test traces along with the basic pass/fail test data. If saving the traces, the number of tests that may be stored will decrease to 10 to 20 percent of the normal capacity. For example, if the field tester normally stores 500 tests, it may only hold 50 tests.

Backup. Backup protection is another important issue. A simple strategy could be to print test data every few hours. This might be suitable for some small jobs, but too time consuming for most. A better strategy is to download to a computer every four hours and make duplicates on floppy disks.

When applicable, make sure an extra set of charged batteries is available.

Delay skew—ANSI/TIA/EIA-568-A-1. Pairs within a 4-pair cable each use a different twist length to improve NEXT performance. This introduces a physical difference in length between pairs. Additionally, the velocity of propagation may also vary between pairs if different insulating materials are in use for one or more pairs. These cables are referred to as 1 + 3, 2 + 2, and 3 + 1 cables. A number of field testers have settings to account for this difference. These differences should be within the operating limits of current cabling and applications standards.

At this time, these variations between pairs should not be a major source of concern when checking the quality of the installation. However, delay skew can become a significant factor in the performance of such recently introduced LAN technologies as 100VGAnyLAN and 100BASE-T2/T4. A value of 45 ns maximum delay skew has been established under Addenda 1, ANSI/TIA/EIA-568-A for cable.

Noise. Some measuring equipment manuals cite the interference effect of portable radios; be sure to coordinate such usage. Noise is an issue that concerns most customers but remains nonstandardized and difficult to assess. Some field testers offer a noise diagnostic test. If noise is suspected or if a test result appears to be incorrect, then two successive tests of the same cabling link should be performed.

Maintenance and upgrades. Software upgrades are usually obtained directly by downloading new executable files into a computer and transferring these to the field tester.

There are a number of ways to obtain the latest firmware for a particular tester:

- A diskette from the manufacturer
- Access to downloadable software from an electronic bulletin board service

- Access to software and other information at worldwide Web sites on the Internet

It is important to upgrade to the latest firmware and maintain the current documentation (including "readme" files) of the testers to ensure up-to-date operational procedures. In addition to these upgrades, it may be necessary to return the field tester to the manufacturer on an annual basis for factory calibration.

Batteries. All of these units rely on batteries, in most cases, by using NICAD packs. Some testers charge while the pack is in the tester, others offer a separate charging unit. Charging time, operating time, and battery life vary between units. Some units also allow for tests while using ac power or while charging.

Battery status is an important issue to consider because the many hours of required testing might depend on a reliable battery charge. Continuous battery status in the main display window is a useful feature. Alternatively, warning messages with beeps to alert users about a low battery condition may be sufficient for most users. In any case, suitable battery management strategies should be outlined by each manufacturer.

Most batteries require exercising to prevent battery memory. It is very important to let the batteries completely drain themselves on a regular basis and then be fully recharged. Failure to drain them will result in the creation of a memory that will not allow the battery to drain below the memory voltage and will shorten the battery's life expectancy.

There is a small disk battery in the field testers to maintain the volatile memory while batteries are switched. This battery is changed at the factory when the unit is sent in for its annual calibration.

Measurement problems

In the event that a link fails to meet specifications, it is time to stop and understand what may be the cause of the failure. Note that a failure is a measurement failure and could be caused by the cabling, components, test equipment, or poor installation practices.

Follow the procedures below to investigate a failure:

- Ensure that the tester is working properly by checking it against the calibration link mentioned earlier. If this link measures within the accuracy limit of its original measurement (maximum 1.5 dB for NEXT and 1 dB for attenuation), the field tester appears to be functioning properly.
- Follow the troubleshooting guidelines in Table 8.1 for various problems.

- If the trouble has been identified, take corrective action to fix the cabling.
- Retest to ensure that the corrective action has worked effectively.

TABLE 8.1 Troubleshooting

Problems	Solutions
Cannot turn on tester.	■ Recharge or change batteries.
Tester cannot perform or fails remote calibration.	■ Be sure both units are turned on and batteries are charged. ■ Replace tester cord.
Tester set for incorrect cable type.	■ Reset tester parameters for proper cable type, and calibrate NVP.
Tester set for incorrect link configuration.	■ Reset tester to basic link or channel as required.
Link/Channel fails autotest. Check area(s) of failure a) Fails wire map	■ Check that tester is set to match existing pin wiring. Examine both connections for open, split, or crossed pairs.
b) Fails length	■ Check and recalibrate NVP with known cable. ■ Check total length of patch/equipment cords.
c) Fails attenuation	■ Check all cables in link/channel for suitable category rating.
d) Fails NEXT	■ Check all components for suitable category rating. ■ Check quality of all wire terminations.
Tester will not run an autotest.	■ Check settings on control knob or menu. ■ Check to see if unit has failed calibration. ■ Check tester connections at both ends of link.
Tester will not store an autotest result.	■ Check that a unique name is selected for test results. ■ Check amount of free memory available.
Tester will not print stored results.	■ Check that serial interface to autotest printer and field tester are set to same parameters/emulation. ■ Check that results are selected for printout.

Backbone Cable

Introduction

Backbone cabling is defined as that cabling used to connect the main cross-connect in the equipment room with either the intermediate or horizontal cross-connect located in the telecommunications closet. It may be any of the cables utilized in the horizontal cabling system or it may be the high-pair-count cable used for voice distribution. ANSI/TIA/EIA-568-A recommends a length limitation for UTP, ScTP, Category 3, 4, and 5; and STP-A of 90 m (295 ft). The testing of these units is the same as that utilized for the horizontal runs. Therefore, the balance of this section will discuss the testing of the high pair-count cable utilized for voice distribution.

Continuity testing

Continuity testing may be accomplished by using a multimeter or any of the available wire map scanners. Several of these scanners are available that will test either a 4-pair group or a 25-pair binder group at one time. If the scanner does not provide length verification and verification is required, this may be accomplished utilizing a TDR.

Step	Utilizing a Pair Scanner (Continuity Tests)
1	Disconnect any existing cross-connect wires.
2	Attach pair scanner to one end of cabling.
3	Attach the pair scanner's remote device to the opposite end.
4	If the pair scanner is unable to locate the distant end, locate the break in the cable utilizing a TDR or the TDR function of a certification field tester.
5	Diagnose and fix faulty cable pairs.
6	Maintain a record of the test results for turnover to the customer.

Length

Step	Determining the Length of the Cable Utilizing the Multimeter
1	Set the multimeter to the ohms position.
2	Calibrate (zero out) the meter.
3	Place a shorting pin across one pair of the cable.
4	With the multimeter, read the dc loop resistance of the cable segment. ■ The nominal dc resistance of 24 AWG wire is 57.2 Ω per 305-m (1000-ft) segment or 0.0572 Ω per 300 mm (12 in). ■ The nominal dc resistance of 22 AWG wire is 36.0 Ω per 305-m (1000-ft) segment or 0.036 Ω per 300 mm (12 in).
5	Divide the loop resistance reading by .0572 or .036 to find the approximate length of the cable.

Coaxial Cable—Data

Introduction

Unlike the testing of a balanced cable, when testing an unbalanced cable (such as coax), the cabling installer is primarily concerned with only two parameters: continuity and length. Continuity and length may be tested by using the multimeter and a shorting plug. The certification field testers also have this capability, which will simplify the process. There are two primary coaxial networks in use today. They are the IEEE 802.3 Ethernet network (50 Ω) and the 75 Ω video network. The Ethernet network is composed of the cable segments, T connectors, barrel connectors, and terminators. The following figure is a schematic representation of an Ethernet network.

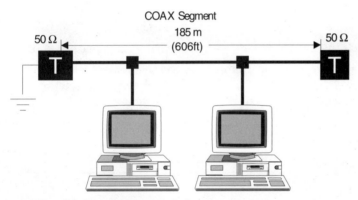

Figure 8.8 Ethernet segment—typical

As shown in the schematic, the coaxial network is composed of two components: the 50 Ω terminator located on each end of the cable and up to 185 m (606 ft) of RG58 IEEE 802.3 Thinnet cable in between.

Note. RG58 cable is not necessarily Ethernet cable. RG-58 should be Ethernet cable, but always look for the IEEE 802.3 to be sure. It will work, but it may not support the distance or connected device requirements of Ethernet. To test this network, it is necessary to test each of the coax segments as well as the terminators. The dc resistance of the terminator is 50 Ω ± 1 percent. One terminator must be bonded to ground. The nominal dc loop resistance value of the 185 m (606 ft) segment is 10 Ω. For this discussion, testing of the 50 Ω coax cable to the individual link will be considered.

50-Ohm Coaxial Cable

Continuity test

This test requires the use of the multimeter and a shorting plug. A shorting plug can be purchased or quickly fabricated from a BNC connector with its center pin shorted to the outer housing. (In some cases, the terminator may be used for this purpose. Be sure to measure and subtract the resistance value of the terminator in any further calculations.)

Observe the following procedures if using the multimeter to test cable:

Testing the cable:

- Remove all equipment and the 50 Ω terminator from each end of the cable.
- Set the multimeter for ac and then dc volts while checking the cable for unwanted signals that would damage the meter when testing resistance.
- Remove any equipment that may have been generating unwanted signals.
- Set the multimeter to the ohms position
- Calibrate (zero out) the meter
- With the multimeter, ensure that there is no continuity with the far end open.
- With the multimeter, read the dc resistance of the cable segment.

    Note. The maximum should not exceed 10 Ω.

Length determination

Testing the terminator:

- Measure the dc resistance between the center pin and the connector housing (ground). This dc resistance should be 50 Ω ± one percent (49.5–50.5 Ω).
- Diagnose and fix faulty cable segments and connectors.
- Maintain a record of the test results for the customer's records.

Using the dc resistance reading previously taken, the length of the segment can be calculated as follows:

- In accordance with the IEEE 802.3 standard, the maximum loop resistance of a segment cannot exceed 10 Ω. To determine maximum loop resistance per foot, divide the maximum loop resistance (10 Ω) by the maximum length of the segment 185 m (606 ft). The loop resistance value is approximately 0.054 Ω per m (0.016 Ω per ft).

- Unlike UTP, the resistance values of the core and the shield are not the same. Using the manufacturer's data, add the nominal dc resistance of the core to the nominal dc resistance of the shield (resisters in series).

- As this specification is given in either resistance per 1000-foot or 1-kilometer segment, divide this number by 1000 to provide the loop resistance per foot or meter.

- Divide the reading received during the continuity test by dc loop resistance per foot derived from the manufacturer's data. (If the test was made with the terminator in place, subtract the measured value of the terminator.)

- The result will be the length of the cable in meters (feet).

For example:

The loop resistance value we determined in the continuity test was 58 Ω. The measured value of the terminator was 50.2 Ω. To determine the length of the segment:

- Subtract the value of the terminator (50.2 Ω) from the total (58 Ω) for a value of 7.8 Ω.

- From the manufacturer's data, it was determined that the nominal dc resistance of the core is 8.8 Ω per 1000 foot segment and the nominal dc resistance of the shield is 5.8 Ω per 1000 foot segment for a loop resistance of 14.6 Ω per 1000 feet.

- Divide the loop resistance of 14.6 Ω by 1000 to determine a dc loop resistance per foot of .0146 Ω per foot.

- Divide the dc loop resistance (7.8 Ω) by the nominal dc loop resistance per foot (.0146 Ω). The approximate overall length of the segment as 534 feet.

If performing the length test with the TDR or other field tester, perform the test in accordance with the field tester manufacturer's instructions. For the reading to be accurate, the cabling installer must either refer to the cable manufacturer's specifications for the NVP, or determine the NVP utilizing the field tester and the field tester manufacturer's instructions.

75-Ohm Coaxial Cable

Continuity test

This test requires the use of the multimeter and a shorting plug. Unlike testing Thinnet cable, where the entire system could be tested end to end, this system requires the removal of any amplifiers and splitters. The system may have to be tested one segment at a time. Observe the following procedures if using the multimeter to test the cable:

- Set the multimeter for ac and then dc volts while checking the cable for unwanted signals that would damage the meter when testing resistance.
- Remove any equipment that may have been generating unwanted signals.
- Set the multimeter to the ohms position.
- Calibrate (zero out) the meter.
- With the multimeter, ensure that there is no continuity with the far end open.
- Install a shorting plug on one end of the cable.
- Using the multimeter, read the dc resistance of the cable segment.
- Calculate the dc resistance from the manufacturer's data for the cable.
- Compare the calculated value with the reading from the multimeter.
- Diagnose and fix faulty cable pairs.
- Maintain a record of the test results for turnover to the customer.

Length determination

Using the dc resistance reading previously taken, we find that the length of the segment can be calculated as follows:

- As the resistance values of the core and the shield are not the same, using the manufacturer's data, add the nominal dc resistance of the core to the nominal dc resistance of the shield (resistors in series).
- As this specification is given in either resistance per 1000-foot or 1-kilometer segment, divide this number by 1000 to provide the loop resistance per foot or meter.

- Divide the reading received during the continuity test by the dc resistance per foot derived from the manufacturer's data.
- The result will be the length of the cable in meters (feet).

Optical Fiber Cable

Overview

The testing of an optical fiber link is conducted for the following reasons:

- Preinstallation testing
- Acceptance testing
- Preventive maintenance testing
- Troubleshooting

Preinstallation testing. Preinstallation testing is performed on a reel of optical fiber cable after it is received from the supplier and before it is installed on the job. The benefits of preinstallation testing are verification of the following:

- Cable has not been damaged during shipment.
- Cable contains no factory defects.
- Attenuation of the fiber matches the factory test report which should be shipped with each reel of cable (if shipped by the manufacturer).

If a defect is found during the preinstallation testing, the shipper, supplier, or manufacturer should be contacted for replacement. If the cabling installer does not carry out preinstallation testing and then finds a defect after installation, there is no way to prove that the installation crew did not cause the defect. A few hours spent on preinstallation testing may save thousands of dollars in replacing defective cable.

Preinstallation testing can be performed with an optical fiber flashlight, depending on the length of fiber, a light source and power meter, or an optical time domain reflectometer (OTDR). An optical fiber cable on a spool or reel should be ordered wound so that both ends of the cable are accessible for testing. To use a light source and power meter, connectors or reusable mechanical splices must be installed on each end of every fiber strand.

A light source and power meter can only measure the attenuation (end to end signal loss) of the fiber. If a value is not within the expected range, the power meter cannot determine the nature or location of the defect.

Preinstallation testing utilizing an OTDR requires access to only one end of the cable. An OTDR can measure the following:

- End-to-end attenuation. This should be compared to the test report on the reel, being careful of the test method used by the manufacturer.
- Distance to a point of high attenuation. This could be a defect in the fiber strand, a fiber break, or it could be the end of the cable. An example of the need for this form of testing would be to identify the location where a forklift damaged a cable during shipment. The OTDR will show the distance from the beginning of the reel to the damaged area in either meters or feet. The footage markers on the cable sheath can be used to assist in locating the damaged section.
- Optical loss per unit of measure (dB/km).
- Continuity testing end to end of each fiber.

Acceptance testing. The following standards are applicable to premises optical fiber cable testing:

- ANSI/TIA/EIA-455-171A Currently Standards Proposal No. 3017, Proposed Revision of EIA-455-171 FOTP-171 "Attenuation by Substitution Measurement for Short Length Multimode and Graded Index and Singlemode Optical Fiber Cable Assemblies."
- ANSI/EIA/TIA-455-61 Currently Standard Proposal No. 2837-B, Proposed Revision of EIA/TIA-455-61, FOTP-61 "Measurement of Fiber or Cable Attenuation Using as OTDR."
- ANSI/TIA/EIA-526-7 Currently Standard Proposal No. 2974-B, Proposed New OFSTP-7 "Measurement of Optical Power Loss of Installed Singlemode Fiber Cable Plant."
- EIA/TIA-526-14A Currently Standard Proposal No. 2981, Proposed Revision of EIA/TIA-526-14 OFSTP-14 "Optical Power Loss Measurements of Installed Multimode Fiber Cable Plant."

An optical fiber link is a path consisting of one strand of fiber that has a connector on both ends. Optical fiber links normally begin and end in administration housings called fiber distribution units. Links are connected to electronic devices such as hubs, multiplexers, and routers with jumpers. Multiple links can be connected in the fiber distribution units by using jumpers to create circuits.

An optical fiber link or circuit should be tested before it is put into service and at other times, such as during troubleshooting. A light source and power meter or an OTDR can be used to measure the attenuation of the optical fiber link or circuit.

The following are the benefits of acceptance testing:

- Verifies that the total attenuation of all passive components in the link is within the design parameters (loss budget).
- Verifies that the passive components were installed properly.
- Minimizes downtime due to maintenance on improperly installed passive components.
- Establishes accountability when circuits are configured with multiple links connected together and installed by more than one vendor.
- Provides a benchmark for comparing future measurements.

Acceptance testing of intrabuilding fiber is most often performed with a light source and power meter. Two people may be required to perform this test. They may need some form of communications such as a fiber talk set, portable two-way radios, or telephones. The light source and power meter are capable of a more accurate insertion loss attenuation measurement than an OTDR.

Acceptance testing of a new installation employing a light source and power meter should be performed on individual links and not on a circuit that has one or more jumpers. This is because the circuit may be reconfigured in the future and the attenuation value for each link will be needed by the customer or technician to properly design the new configuration. The results of each measurement should be recorded on the optical fiber link attenuation record.

Preventive maintenance testing. Preventive maintenance testing is the periodic testing of a link to monitor its attenuation and then compare that attenuation to the original acceptance test values. This is a preventive maintenance measure to find potential problems before they cause an out-of-service condition. Preventive maintenance testing is normally performed at night or on weekends to minimize disruption to the users. The benefits of preventive maintenance testing are

- Minimize out-of-service downtime.
- Planned testing time instead of unplanned out-of-service.
- Greater network reliability.

Preventive maintenance testing may be performed with a light source and power meter or an OTDR. Testing circuits rather than individual links is acceptable for preventive maintenance testing. If a discrepancy is found while testing a circuit, the individual links must be tested to identify the problem area.

Light Source and Power Meter Testing—Channel or Link

Overview

This section covers procedures for the testing of optical fiber cables using a light source and power meter. Typical applications include pre-installation testing, acceptance testing, preventive maintenance testing, and troubleshooting.

Tools and equipment

Light source, multimode. Multimode light sources use a light-emitting diode (LED) as the light source. The standard wavelengths used to test multimode fiber links are 850 nm and 1300 nm. Most multimode light sources have an output port for both the 850 nm and 1300 nm windows.

Figure 8.9 Multimode light source

Light source, singlemode. Singlemode light sources typically use a laser diode as the light source. The standard wavelengths used to test singlemode fiber links are 1310 nm and 1550 nm. Most singlemode light sources have an output port for both the 1310 nm and 1550 nm windows.

Figure 8.10 Singlemode light source

Power meter. Power meters use detectors which measure a wide range of wavelengths. The wavelength selector on the power meter selects the appropriate calibration for the detector at the selected wavelength.

Figure 8.11 Power meter

Power meters are available, which are calibrated for only the multimode wavelengths (850 nm and 1300 nm) or singlemode wavelengths (1310 nm and 1550 nm). Power meters are also available, which are calibrated for both multimode and singlemode wavelengths. A basic power meter will only show a reading in dBm. A meter with more features will have the capability to set a zero reference, provide a direct reading in dB, and record the data to memory for later download to a PC. New to the marketplace are power meters which will test one at both windows in either multimode or singlemode at the same time.

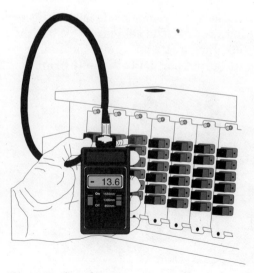

Figure 8.12 Sample power meter reading

Reference adapter. A reference adapter should be tested by the cabling installer and verified that it is known to be of good quality. Its purpose is to couple the test jumpers from the light source and power meter to determine the reference reading. Since the ferrule alignment sleeve has a tighter tolerance, it is good practice to use a singlemode ceramic ferrule adapter for the reference adapter for singlemode or multimode fiber strands.

Figure 8.13 Reference adapter

Test jumpers. The connector type (ST, SC) on one end of the test jumpers must match the connector type on the light source and power meter. The connector type on the other end of the test jumpers must match the connector type used on the optical fiber link to be tested. It is not necessary for the connector type of the light source and power meter to match that of the optical fiber link to be tested. The jumpers shall be 1 to 5 m (3.3 to 16 ft) in length. It is recommended that test jumpers and reference adapters with a variety of connector types be carried in the field tester case to facilitate any connector type found at a job site.

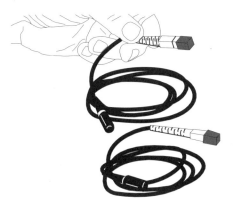

Figure 8.14 Test jumpers

Two-way radios. Two-way radios, a fiber talk set, or some other form of two-way communication is essential for the efficient use of the cabling installer's time. The two-way radios should be tested during the power meter reference reading procedure to ensure that they do not interfere with the power meter measurements.

Steps—Administrative

Step	Pretest Administrative Activities
1	Determine whether the fiber link to be tested is multimode or singlemode. If the fiber is multimode, verify that the size of the core is 50 or 62.5 µm. This information may be obtained from the work order or the cabling installer's supervisor. Select jumpers of the same fiber, core, type, and size.
2	Determine the wavelengths that should be measured. Multimode fiber is normally measured at 850 nm or 1300 nm for horizontal links and both wavelengths for backbone links. The work order should detail that one or both wavelengths be tested. If both wavelengths are to be tested, the cabling installer should test all of the links at one wavelength, change the settings to the other wavelength, reference the setup for the new wavelength, and test all links again at the new wavelength.
Note.	This is applicable only if manually recording test results on paper.
3	Singlemode fiber is normally measured at 1310 nm and 1550 nm. The work order should detail this information.
4	Determine whether the test setup will be in accordance with OFSTP-14A for multimode or OFSTP-7 for singlemode fiber Method A or Method B. (Refer to ANSI/TIA/EIA-568-A, Appendix H.) ANSI/TIA/EIA-568-A recommends Method B. Determine whether the measurements will be single direction or bidirectional. OFSTP-14A states "Bidirectional testing is a default requirement of this document as it is the most conservative. Depending on the size and complexity of the cable plant and if OTDR testing is required, the specifier may delete the requirement for bidirectional measurements." The recommendation of the ANSI/TIA/EIA-568-A and the current edition of the BICSI *Telecommunications Distribution Methods Manual (TDMM)* is single direction only. The work order should tell the cabling installer whether the testing will be in one direction or bidirectional.

Note. The recorded attenuation for a link with bidirectional testing is the average of the attenuation in each direction.

5   Identify the designed link attenuation from the work order or the job supervisor. An estimate of the attenuation can be made by following the worksheet in Attachment A. This will be compared to the measured attenuation of each link as a checkpoint to validate the measured attenuation reading.

Steps—Connector cleaning

Step	Connector Cleaning Procedure
1	Apply about four drops of isopropyl alcohol (99 percent pure) to a clean lint-free cloth.
2	With moderate pressure, wipe the endface of the connector in one direction only with the alcohol-soaked area of the lint-free cloth for about five seconds.
3	With a light pressure, wipe the endface of the connector in one direction only with a dry area of the lint-free cloth.

Steps—Adapter cleaning

Step	Adapter Cleaning Procedure
1	Apply about eight drops of isopropyl alcohol (99 percent pure) to one end of a clean swab.
	Note. Do not use a pipe cleaner, which is wire cord, and would damage the sleeve. This would have an adverse effect on composite or ceramic inserts.
2	Slide the alcohol-soaked end of the swab into the adapter.
3	Move the swab back and forth with one-inch strokes five times so that the alcohol-soaked area cleans the alignment sleeve in the adapter.
4	Remove the swab from the adapter and insert the dry end of the swab into the adapter.
5	Move the swab back and forth five times to dry the alignment sleeve.
6	Remove the swab from the adapter.
7	Use compressed air to blow any lint from within the adapter.

Steps—Link attenuation

Step	Calculating the Measured Link Attenuation
1	The displayed reading on a basic power meter is a number that represents an absolute power level. The unit of measure for this reading is dBm. The reference level for dBm is 1 milliwatt of power. One milliwatt can also be expressed as 0 dBm. Any measurement greater than 1 milliwatt is expressed as a positive number such as +3 dBm. Often, the (+) sign is not shown. Any measurement less than 1 milliwatt is expressed as a negative number such as –3 dBm. A typical multimode light source may have an output in the range of –11 dBm.
2	To determine the attenuation associated with an optical fiber link, the light source and power meter are first connected with test jumpers and a reference adapter. The reading on the meter will be less than the actual output of the light source because of the attenuation in the test jumpers (e.g., –14 dBm). Remember, –14 dBm is less power than –13 dBm. The –14-dBm value is called the reference reading (see Attachment B) and is recorded on the optical fiber link attenuation record. This is actually the loss values for the connectors, adapters, and the fiber.

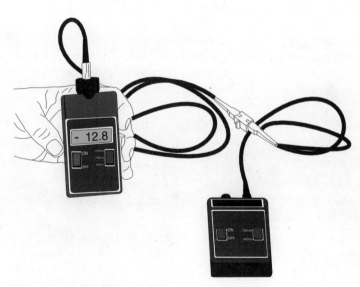

Figure 8.15 Calibration of meters

3   The test jumpers are disconnected from the reference adapter and are connected to the optical fiber link to be tested. The observed reading on the power meter is subtracted from the reference reading (−15.5 dBm subtracted from −14 dBm = 1.5 dB). This is the actual attenuation of the optical fiber link being tested. The unit of measure associated with this number is dB because it is the difference between two dBm measurements. The negative sign is shown because it denotes less power than the reference.

Note. Some power meters provide the capability to be zeroed and, thus, are direct reading.

Safety

Always perform a thorough visual inspection of work areas. Look for the following potential hazards:

- Open electrical panels and exposed wiring
- Trip or fall hazards on the floor
- Overhead hazards, such as cable trays, which could cause head injuries
- Hazardous chemicals storage

The alcohol used for cleaning connectors is flammable. Do not expose the fumes to an open flame or electrical arc. The fumes can also be toxic when used in confined spaces. Ensure that there is adequate ventilation when using alcohol.

Never look at the end of a connector, especially with a magnifying device, without first ensuring that there is no light source on the other end. Laser sources are potentially dangerous because the output level can be very high and the wavelength is out of the visible range of the human eye.

LEDs are safer than lasers because of their lower output power. Regardless of the light source, it is a wise practice to always verify that the distant end is disconnected from any light source.

When installing connectors or splicing fibers, there will be short pieces of fiber that need to be disposed of properly. The cabling installer should have a container with a lid to store the pieces of fiber. The container should be labeled properly and, when it is full, the closed container should be disposed of in a place that will not cause injury to anyone. A soft drink can or styrofoam cup is not acceptable for this purpose. Electrical tape or masking tape formed into a loop with the sticky side out is also not an acceptable method of disposing of fiber ends. However, a short piece of tape used to pick up a small piece of fiber from a work surface is acceptable. Fiber ends are not considered hazardous waste, but common sense should govern their disposal.

Steps—Calibration

Step	Field Tester Preparation and Calibration
1	Remove the light source, power meter, test jumpers, and reference adapter from the field tester carrying case.

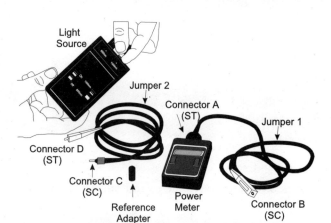

Figure 8.16 Field tester

2	Remove the dust cap from both connectors of test jumper 1 (ends A and B).

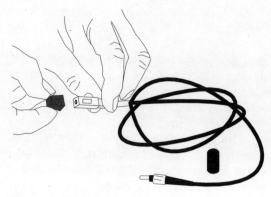

Figure 8.17 Dust cap removal

Testing Cable 443

3. Clean the connectors by performing the optical fiber connector cleaning procedure.
4. Remove the dust cap from the power meter.
5. Attach connector A of test jumper 1 to the power meter.

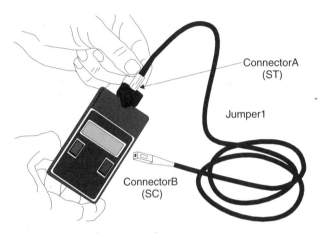

Figure 8.18 Attach connector A

6. Remove the dust cap from both connectors on test jumper 2 (ends C and D).

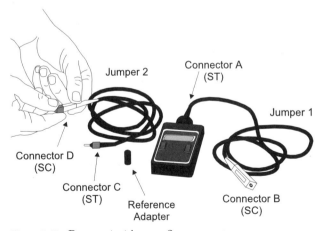

Figure 8.19 Prepare test jumper 2

7   Clean the connectors by performing the optical fiber connector cleaning procedure.

8   Remove the dust cap from the light source.

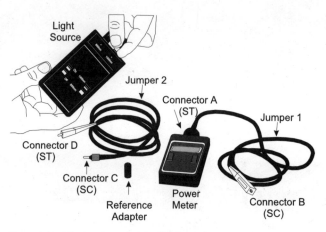

Figure 8.20   Dust cap removal—light source

9   Attach connector C of test jumper 2 to the light source.

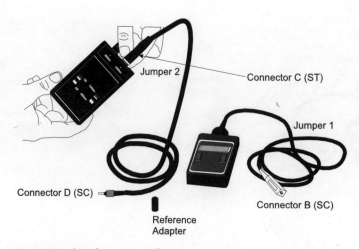

Figure 8.21   Attach connector C

Testing Cable   445

10   Remove the dust cap from both ends of the reference adapter.

Figure 8.22   Reference adapter

11   Clean the reference adapter using the adapter cleaning procedure.

12   Attach connector B to one side of the adapter.

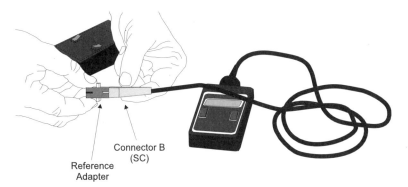

Figure 8.23   Attach connector B

13   Attach connector D (SC) to the other side of the adapter.

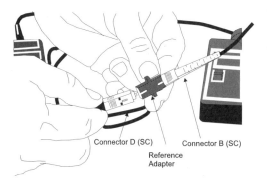

Figure 8.24   Attach connector D (SC)

Steps—Reference reading

Step	Power Meter Reference Reading Procedure
1	Turn on the light source a minimum of three minutes or utilize the time duration recommended by the field tester manufacturer prior to use. Verify that there is a display and the battery does not need to be replaced or changed.
2	Set the wavelength selector to the appropriate wavelength.
3	Turn on the power meter.
4	Set the wavelength selector to the same wavelength.
5	Observe the reading on the power meter. This reading is the power level expressed in dBm.
6	Record this reading on the optical fiber link attenuation record as the reference power level. Do not turn off the power meter or light source until all measurements have been recorded. If either unit is turned off, the reference procedure should be performed again.
7	Disconnect connector B (SC) and connector D (SC) from the adapter.
8	Install dust caps on connectors B and D.
9	Install dust caps on both ends of the reference adapter and store the adapter in the field tester carrying case.

Steps—Link measurements

Step	Link Measurements
1	Cabling installer 1 will go to the main cross-connect (MC), or one end of the link to be tested, and take the following items: - Power meter with test jumper 1 connected - Optical fiber link attenuation record - Fiber talk set or two-way radio **Note.** When the fiber talk set is used, predetermine which fiber will be utilized for the initial (meet me) talk path. Normally, this is the last strand in the cables. - Supplies for cleaning connectors and adapters
2	Cabling installer 2 will go to the telecommunications closet (TC), or other end of the link, and take the following items: - Light source with test jumper 2 - Fiber talk set or two-way radio - Supplies for cleaning connectors and adapters

3. Cabling installer 1 is responsible for determining which fibers will be tested and in what sequence. This information can be obtained from the work order or the job supervisor.

4. Cabling installer 1 connects test jumper 1 to the adapter of the first fiber to be tested.

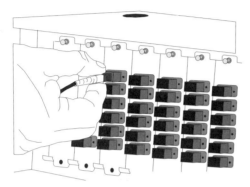

Figure 8.25 Connect test jumper 1

5. Cabling installer 2 connects test jumper 2 to the adapter of the first fiber to be tested.

6. Cabling installer 1 contacts cabling installer 2 using the fiber talk set or two-way radio to confirm that the connection has been made.

7. Cabling installer 1 observes the reading on the power meter.

Figure 8.26 Meter reading

8. Cabling installer 1 subtracts the reference reading from the observed reading. This is called the measured link attenuation. If the attenuation is not within the expected range, the cabling installer should check all of the connections and retest the fiber. If the same results are obtained, try a second fiber and perform the troubleshooting procedures outlined in the troubleshooting section of this manual.
9. The link attenuation should be recorded on the optical fiber link attenuation record.
10. Cabling installer 1 notifies cabling installer 2 that the measurement is complete for that fiber and it is time to move the test jumpers to the next adapter.
11. Cabling installer 2 moves the test jumper to the next fiber adapter to be tested.
12. Cabling installer 1 moves the test jumper to the next fiber adapter to be tested.
13. This procedure should be repeated for each of the optical fiber links to be tested.
14. Rereference to verify that the setup has not changed and affected any readings.
15. When all measurements have been performed and recorded, the light source and power meter should be turned off.
16. The test jumpers should be removed from the light source and power meter.
17. Dust caps should be installed on all test jumper connectors.
18. Dust caps should be installed on the light source and power meter ports.
19. Store the light source, power meter, test jumpers, and reference adapters in the field tester carrying case.
20. Cabling installer 1 should attach the optical fiber link attenuation report to the completed work order form and return it to the appropriate supervisor.

Test measurement documentation

Optical fiber link attenuation record. This is a paper document or computer database where the link attenuation measurements are recorded and the link attenuation calculations are performed. This is a permanent record and should be retained for future reference. See Attachment B. For those field testers that record the test data in digital format, download and save the test data according to the field tester manufacturer's procedures.

Light Source and Power Meter Testing— Patch Cables

Overview

TIA draft standard FOTP-171 (SP-3017) contains four different methods, each having three optional procedures for the testing of optical fiber patch cords. Method B is typically used for factory testing and is, therefore, considered to be most appropriate for the testing of components to be used in the premises network. The procedure uses a two-jumper substitution method with a mode filter. The two reference quality components are tested to zero the meter. Then the jumper to be tested is inserted and the new reading is the fiber attenuation. In all other respects, the setup is identical to that used in testing either the channel or link configuration.

Reference components

The reference components (jumpers) must be composed of the same fiber as the cable to be tested. They must be long enough (typically 2 m [6.5 ft]) to have a mode filter formed within its length and have one meter or more extending past the mode filter. The reference jumper selection process is defined in SP-3017, Annex A.

Mode filter

The mode filter is fabricated by wrapping one of the reference jumpers five times around a smooth, round mandrel creating a fiber loop diameter of 20 mm (0.8 in). For example, a cable with a 2.5 mm (0.1 in) diameter would require a 17.5 mm (0.7 in) mandrel to support a fiber loop diameter of 20 mm (0.8 in). The mode filter is included in the launch cable to produce restricted launch conditions. In other words, the input signal appears to the test jumper as though it were attached to a LAN backbone cable.

Test configuration

The following figure shows the test setup for the testing of an optical fiber jumper.

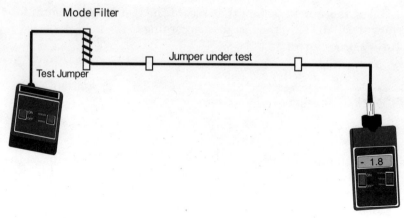

Figure 8.27 Test setup for the testing of an optical fiber jumper

Optical Time Domain Reflectometer

Overview

This section covers the testing of optical fiber cables using an optical time domain reflectometer. Typical applications include acceptance testing, preventive maintenance testing, preinstallation testing, and troubleshooting. Procedures utilized are those specified in TIA draft standard SP-2837B.

Tools and equipment

An optical time domain reflectometer (OTDR) is an electronic-optical instrument that is used to characterize optical fibers. It locates defects and faults, and determines the amount of signal loss at any point in an optical fiber.

An OTDR takes thousands of measurements along a fiber. The measurement data points can vary in distance between them. The data points are displayed on the screen as a line sloping down, from left to right, with distance along the horizontal scale and a reduced signal level on the vertical scale. By selecting any two data points via movable cursors, the operator can read the distance and relative signal level between them.

Applications

OTDRs are widely used in all phases of an optical fiber system's life, from construction to maintenance to fault locating and restoration. An OTDR is used to:

- Measure overall (end-to-end) loss for system acceptance and commissioning.
- Measure section loss (individual fiber reels) for incoming inspection and verification of specifications.
- Measure splice loss (both fusion and mechanical splices) during installation, construction, and restoration operations.
- Measure reflectance of connectors and mechanical splices for CATV, SONET, and other analog or high-speed digital systems where reflections must be kept low.
- Locate fiber breaks and defects.
- Indicate optimum optical alignment of fibers in splicing operations.
- Detect the gradual or sudden degradation of fiber by making comparisons to previously documented fiber tests.

Theory of operation

The optical time domain reflectometer (OTDR) uses the effects of rayleigh (pronounced RAY-lay) scattering and fresnel reflection to measure the characteristics of an optical fiber. By sending a pulse of light (the optical in OTDR) into a fiber and measuring the travel time (the time domain in OTDR) and strength of its reflections (the reflectometer in OTDR) from points inside the fiber, it produces a characteristic trace, or profile, of the length versus returned signal level on a display screen.

The trace can be analyzed on the spot, printed out immediately for documentation of the system, or saved to a computer disk for later analysis. A trained operator can accurately locate the end of the fiber, determine the location and loss of splice, and estimate the overall loss of the fiber.

Rayleigh scattering. When a pulse of light is sent down a fiber, part of the pulse gets blocked by microscopic particles (called dopants) in the glass and scatters in all directions. This is called rayleigh scattering. Some of the light (about 0.0001%) is scattered back in the opposite direction of the pulse (back toward the source) and is called the backscatter. Since dopants in optical fiber are uniformly distributed throughout the fiber due to the manufacturing process, this scattering effect occurs along its entire length.

Rayleigh scattering is the major loss factor in fiber. Longer wavelengths of light exhibit less scattering than shorter wavelengths. For example, light at 1550 nm loses 0.2 to 0.3 dB per kilometer (dB/km) of fiber length due to rayleigh scattering; light at 850 nm loses 0.4 to 0.6

dB/km from scattering. A higher density of dopants in a fiber will also create more scattering and thus higher levels of attenuation per kilometer. An OTDR can measure the levels of backscattering accurately and use it to measure small variations in the characteristics of fiber at any point along its length.

The rayleigh scattering effect is like shining a flashlight in a fog at night: the light beam is diffused, or scattered, by the particles of moisture. A thick fog will scatter more of the light because there are more particles to obstruct it. The fog can be seen because the particles of moisture reflect small amounts of the light back. The light beam may go a long way into the fog if it is not very thick but, in a dense fog, the light attenuates quickly due to this scattering effect.

Fresnel reflection. Whenever light traveling in a material (such as an optical fiber) encounters a different density material (such as air), some of the light (up to four percent) is reflected back towards the light source while the rest continues in the material. These sudden changes in density occur at ends of fibers, at fiber breaks, and (sometimes) at splice points. The amount of the reflection depends on the magnitude of change in material density (described by the index of refraction [IOR]—larger IORs mean higher densities) and the angle that the light strikes the interface between the two materials. This type of reflection is called a fresnel (pronounced freh-NELL) reflection. It is used by the OTDR to precisely determine the location of fiber breaks.

Backscatter level vs. transmission loss. Although the OTDR measures only the backscatter level and not the level of the transmitted light, there is a close correlation between the backscatter level and the transmitted pulse level. The backscatter is a fixed percentage of the transmitted light. If the amount of transmitted light drops suddenly from Point A to Point B (caused by a tight bend, a splice between two fibers, or by a defect), then the corresponding backscatter from Point A to Point B will drop by the same amount. The same loss factors that reduce the levels of a transmitted pulse will show up as a reduced backscatter level from the pulse. The ratio of backscattered light to transmitted light is also known as the backscatter coefficient.

The OTDR

The OTDR consists of a laser light source, an optical sensor, a coupler/splitter, a display section, and a controller section.

Laser light source. The laser sends out pulses of light on command from the controller. The duration of the pulse (the pulse width) can be selected by the operator for different measuring conditions. The light goes through the coupler-splitter and into the fiber under test. Some OTDRs have two lasers to allow for testing fibers at two different wavelengths. Only one laser is used at a time. The operator can easily switch between the two with the press of a button.

Coupler/splitter. The coupler/splitter has three ports—one each for the source, the fiber under test, and the sensor. It is a device that allows light to travel only in specific directions: from the laser source to the fiber under test, and from the fiber under test to the sensor. Light is not allowed to go directly from the source to the sensor. Thus, pulses from the source go out into the fiber under test. The returning backscatter and fresnel reflections are routed to the sensor.

Optical sensor section. The sensor is a photodetector that measures the power level of the light coming in from the fiber under test. It converts the optical power in the light to a corresponding electrical level—the higher the optical power, the higher the electrical level output. OTDR sensors are specially designed to measure the extremely low levels of backscattered light. The sensor section includes an electrical amplifier to further boost the electrical signal level.

The power of fresnel reflection is about 40,000 times higher than that of backscatter and is greater than the sensor can measure—it overloads the sensor, driving it into saturation. The electrical output level is subsequently clipped at the sensor's maximum output level. Therefore, whenever a test pulse encounters the end of a fiber (whether at a mechanical splice or at the end of the fiber), it causes the sensor to be blinded for as long as the pulse occurs. This blind period is known as the dead zone.

Controller section. The controller is the brain of the OTDR and performs the following:

- Tells the laser when to pulse
- Gets the power levels from the sensor
- Calculates the distance to scattering and reflecting points in the fiber
- Stores the individual data points
- Sends the information to the display section

A major component of the controller section is a very accurate clock circuit, which is used to precisely measure the time difference between when the laser pulses and when the sensor detects returning light. By multiplying this round-trip pulse travel time by the speed of light in fiber (which is the speed of light in free space corrected by the index of refraction), the round-trip distance is calculated. The distance from the OTDR to the point (one-way distance) is simply half of the round-trip distance.

Since backscattering occurs all along a fiber, there is a continuous flow of light back into the OTDR. The controller samples the level measured by the sensor at regular time intervals to get its data points. Each data point is described by its sequence time (which relates to distance from the OTDR) and power level. Because the original pulse gets weaker as it travels down the fiber (due to rayleigh scattering induced loss), the corresponding returned backscatter level gets weaker further down the fiber. Therefore, the data points generally have decreasing power levels from start to end. When a fresnel reflection occurs, the power level of the corresponding data point for that location suddenly goes up to its maximum level—way above the level of the backscatter just prior to it.

When the controller has gathered all its data points, it plots the information on the display screen. The first data point is displayed at the left edge of the graph as the starting point of the fiber. Its vertical position is based on its power level—a higher power is plotted higher up on the graph. Subsequent data points are placed to the right. The resultant trace is a sloping line that runs from the upper left towards

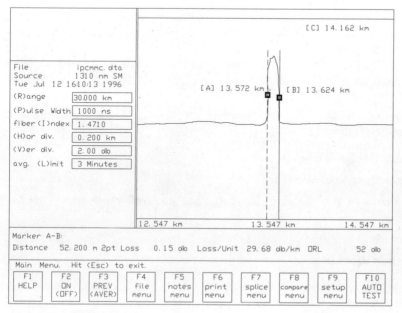

Figure 8.28 OTDR trace

the lower right. The slope of the line indicates its loss-per-unit-distance (dB/km) value. Steep slopes mean larger dB/km values. Data points corresponding to backscatter level make up the line. Fresnel reflections look like spikes coming up from the backscatter level. A sudden shift of the backscatter level indicates a point loss, which may indicate either a fusion splice or a stress point in the fiber where light is escaping.

Display section. The display section is typically an LCD screen that shows the data points, which make up the fiber trace and displays the OTDR setup conditions and measurements. Most OTDR displays connect the data points with a line to give a better look at the overall trace. The operator can manipulate cursors on the screen to select any data point on the fiber trace. When the cursor is on a data point, the distance to that point is displayed on the screen. An OTDR with two cursors will display the distances to each cursor and the difference in backscatter levels between them in decibels (dB). The operator can choose the type of measurement being made with the cursors, such as two-point loss, dB/km, splice loss, and reflectance. The measurement results are shown on the display.

OTDR specifications

Dynamic range. The dynamic range of an OTDR determines the length of fiber that can be measured. It is listed as a dB value—larger values mean longer distance measurement capability. A test pulse needs to be strong enough to get to the end of the fiber to be tested. The sensor has to be good enough to measure the weakest backscatter signals that come from the end of a long fiber. The combination of the total pulse power of the laser source and the sensitivity of the sensor determines the dynamic range. A very powerful source and a sensitive sensor will provide a large dynamic range; a weak source and an average sensor will yield a low dynamic range.

Dynamic range for an OTDR is determined by taking the difference between the backscatter level at the near end of the fiber and the upper level of the average noise floor at or after the fiber end. A large dynamic range will produce a clear and smooth indication of the backscatter level at the far end of the fiber. A small dynamic range will produce a noisy trace at the far end—the data points that make up the trace backscatter level will not form a smooth line, but will vary up and down from one to the next. It is difficult to distinguish details in a noisy trace.

Increasing the total pulse output power of a laser source can be accomplished in two ways: increase the absolute amount of light emitted or increase the pulse duration (pulse width). There are limits to each of these procedures:

- A laser diode has a natural maximum output level that cannot be exceeded. Also, a higher output level means a shorter component life—the laser might burn out faster.

- When the pulse width is increased, other performance characteristics, such as the dead zone, are affected. Longer pulse widths produce longer dead zones.

- Sensors also have natural limitations to their ability to measure low light levels. At some point, the electrical level sent out by the sensor (which corresponds to the optical power level detected) becomes lost in the electrical noise and sensor measurements. Electrical shielding within an OTDR is critical in order to decrease the adverse effects of ambient electrical noise in the instrument. Additionally, when a sensor is operating at its peak sensitivity, its level accuracy is decreased. To improve accuracy at lower light levels, an OTDR will use averaging techniques to combine the measurements from thousands of pulses. The use of averaging will improve the sensitivity of a sensor and can therefore improve an OTDR's dynamic range.

Dead zone. Dead zone refers to the space on a fiber trace following a fresnel reflection in which the high return level of the reflection covers up the lower level of backscatter.

An OTDR's sensor is designed to measure the low backscatter levels from a fiber and becomes blinded when a larger fresnel reflection hits it. This blind period lasts as long as the pulse duration. When the sensor receives the high levels of backscatter, which may follow immediately after a reflective event, the dead zone includes the duration of the reflection plus the recovery time for the sensor to readjust to its maximum sensitivity. High quality sensors recover quicker than cheaper ones and thus achieve shorter dead zones.

The dead zone effect can be illustrated by considering what could happen while you are looking at a starlit sky: with no other lights around, the eyes become sensitive and you are able to see very dim stars (like backscatter). If someone then shines a flashlight in your eyes, the overpowering light (like a fresnel reflection) blinds you and you are no longer able to see the stars. You will not be able to see anything, but the bright light for as long as it is in your eyes (pulse duration). After the light is removed, your eyes slowly readjust to the darkness, becoming more sensitive, and you are able to see the low light level of the stars again. The OTDR sensor acts very much like our eyes in this example. The period of blindness is the dead zone.

Since the dead zone is directly related to the pulse width, it can be reduced by decreasing the pulse width. But decreasing the pulse width lowers the dynamic range. An OTDR designer must make a compromise between these two characteristics. Likewise, the OTDR user must choose a pulse width depending on whether it is more important to see closely spaced events or see farther out in a fiber.

Importance of dead zone. Dead zones occur in a fiber trace wherever there is a fiber connector and at some defects (such as cracks) in the fiber. There is always at least one dead zone in every fiber—where it is connected to the OTDR. This means that there is a space starting at the beginning of the fiber under test where no measurement can be made. This space is directly related to the pulse width of the laser source. Typical pulse widths in OTDRs range from 30 ns (nanoseconds—billionths of a second) to 10,000 ns. In distance, this works out to be from 6 m (20 ft) to over 1600 m (5280 ft) long. If you need to characterize the part of the fiber that is close to the near end, you will need to select the shortest pulse width possible. A second point is the end of the fiber. By the time the receiver has recovered, there is nothing to reference, but the noise level of the field tester. Therefore, it becomes imperative that a method is established to overcome this limitation.

Dead zones are characterized as an event dead zone or an attenuation dead zone. An event dead zone is the distance after a fresnel reflection before another fresnel reflection can be detected. An attenuation dead zone is the distance after a fresnel reflection until the backscatter

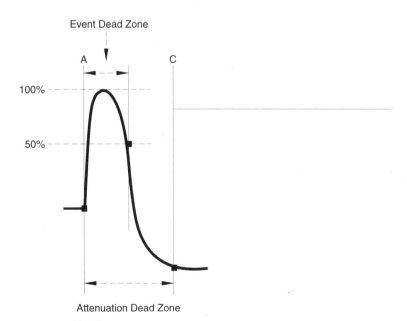

Figure 8.29 Dead zone measurements

level can be detected. Event dead zones are always shorter than attenuation dead zones. The following figure shows how the two dead zones are measured.

Resolution.   There are two resolution specifications:

- *Loss resolution.*   Loss resolution is the ability of the sensor to distinguish between levels of power it receives. Most OTDR sensors can display down to 1/100th of a decibel (0.01 dB) difference in backscatter level. This specification must not be confused with level accuracy, which is discussed later.

- *Spatial resolution.*   Spatial resolution is how close the individual data points, which make up a trace, are spaced in time (and corresponding distance). The OTDR controller samples the sensor at regular time intervals to obtain data points closely spaced, enabling the OTDR to detect events closely spaced. The ability of the OTDR to locate the end of the fiber is affected by the spatial resolution: if it only takes data points every 8 m (26.2 ft), then it can only locate a fiber end within ±8 m (26.2 ft). See "Distance accuracy" under "Accuracy and Linearity."

    The operator is able to select and measure the distance and loss between any two data points. Those that are closely spaced will provide more detail. An OTDR displays the fiber trace as a line that connects the data points and allows the operator to place a cursor between points as well as on points. This interpolation of the information produces a better display resolution than the actual spatial (or data point) resolution.

    Spatial resolution is reduced in certain areas by a dead zone. Valid measurements of fiber attenuation are only made from backscatter level to backscatter level. Data points, which were taken while the sensor was in saturation due to a fresnel reflection, cannot be used to make loss measurements, since the sensor was not able to make an accurate level measurement at that time. Therefore, the spatial resolution around a fresnel reflection is worse (lower resolution), because the only usable points occur before and after the dead zone.

Accuracy and linearity.   A number of factors contribute to accuracy and linearity:

- *Level accuracy and linearity.*   The level accuracy of the OTDR sensor is measured in the same way as optical power meters and photodetectors of any kind. The accuracy of any optical sensor depends on how closely the electrical current output corresponds to the input optical power. Most optical sensors convert incoming optical power to a corresponding electrical current level evenly across its operating range, but the electrical output is extremely low. All sensors use electrical amplifiers to boost the very low electrical output level. All amplifiers introduce some amount of distortion to the signal. High-quality amplifiers are able to boost both high and low levels by the same amount. In other words, they have a very linear response to an input over most of the operating range.

Lower-quality amplifiers introduce significant distortion into the amplified signal at either high or low input levels—they become nonlinear at the extremes of operation. The linearity inherent in an optical sensor and its amplifier will determine how accurately the incoming optical power is converted to an amplified electrical event.

Level accuracy for many optical sensors is stated either as a flat plus-or-minus (+/–) dB amount (if its measurement range is small), such as +/–0.10 dB, or a percentage of the power level, such as two percent. For OTDRs, a better representation of the accuracy is the linearity, stated as a +/–dB amount per dB of power measured over certain ranges of measurement, such as +/–0.10 dB/dB in the 10 to 20 dB range. OTDRs are expected to maintain reasonable accuracy over very wide ranges of measurements (some spanning over 35 dB of backscatter levels) and thus require good linearity over the entire optical input range of the sensor. Linearity problems in an OTDR often show up as rolling off or bumpy appearances of the displayed fiber trace.

Fresnel reflections are generally outside the measurement range of the sensor and are not considered in the linearity specifications. However, the trace display during the recovery period after a reflection often exhibits characteristics of nonlinearity as the incoming power level transitions from extremely high to very low.

- *Distance accuracy.* There are three components to distance accuracy of an OTDR:
  - Clock stability
  - Data point spacing
  - Index of refraction uncertainty.

The accuracy of distance measurements depends on the stability and accuracy of the clock circuit which times the pulses going out and the interval between sampling of the sensor output. Clock accuracy is stated as a percentage which relates to percentage of distance measured. For instance, an accuracy of 0.01 percent of distance means that if the distance to the end of the fiber is measured as 20,000 feet, then the accuracy of this measurement is +/–2 feet (20,000 × 0.0001). If the clock runs too fast or too slow, then the time measurements will be shorter or longer than the actual value.

An OTDR can only make accurate distance measurements based on the actual data points it takes. The closer the data points are spaced, the more likely one of them will fall close to, or on, a fault in the fiber.

Distance by using an OTDR is calculated from the speed of light in the fiber. The speed of light in fiber is calculated from the speed of light in free space (a constant value) divided by the index of refraction (IOR). This means that the user-setable IOR is critical in accurate measurement of distance. If the IOR is wrong, then the distance will be wrong. The characteristics of a fiber can change along its length, producing slight variations in its IOR and, therefore, cause distance inaccuracies. Thus, fiber distance is due to the variance of IOR within the same fiber and between two or more fibers spliced together.

In most cases, use the fiber manufacturer's recommended IOR setting for the fiber type and wavelength to be tested. Check with the fiber manufacturer if there are any questions about the IOR.

Wavelength

Optical fiber is used and tested at the following wavelength bands: 850 nm, 1310 nm, and 1550 nm. Multimode fibers work in the 850 nm and 1300 nm bands. Singlemode fibers work in the 1310 nm and 1550 nm bands. OTDR measurements are made at the same wavelength at which the fiber system will be operated.

The measuring wavelength of an OTDR is listed as its central wavelength with a certain spectral width. Linewidth is the spread of wavelengths around the central wavelength of the laser source. For example, a laser with a central wavelength of 1300 nm and a linewidth of 20 nm will include wavelengths from 1290 nm (1300 − 10) to 1310 nm (1300 + 10). Lasers with narrow linewidths are more expensive than those with wide linewidths. Central wavelengths are also normally specified as being within a certain tolerance, such as +/−30 nm.

Loss in fiber is wavelength-dependent. It is important to test fiber at about the same wavelength at which it will be operated. Optical transmitters (lasers and LEDs) are generally specified as to their wavelength band, that is, 850, 1300, or 1550 nm. The specific central wavelength and linewidth is not always clearly listed. In some cases, if a test for attenuation is made at one end of a wavelength band (at 1320 nm, for instance), then the test signal will be attenuated at a slightly different amount than the operating signal. In long fiber runs, this could lead to unexpected problems at the receive end of the system.

Operation

The following discussion will assist in understanding the operation of an OTDR.

Configuration. Choosing the configuration of an OTDR depends on the fiber to be tested. OTDRs can measure only one fiber type at a time—singlemode or multimode. Either a single or dual wavelength measuring capability for each type of fiber can be selected. Thus, OTDRs are available to test 850 nm and/or 1300 nm multimode or 1300 nm and/or 1550 nm singlemode.

Mainframe and optical card. In the modular design, the OTDR mainframe contains the controller, display, operator controls, and optional equipment (such as printers, plotters, external interfaces, and modems). The optical card, consisting of the laser source and optical sensor sections, is plugged into the mainframe and can be changed to allow testing at various wavelength and fiber type combinations.

Note. All OTDRs are not set up as modular.

Fiber type. An optical network interface card (NIC) is limited to working with either singlemode or multimode fiber. The primary difference between these two fiber types is the diameter of the fiber core; multimode cores are at least five times larger than singlemode. Since an OTDR must both send and receive light, it cannot efficiently connect with both fiber types. For example, a module designed for singlemode fiber will couple light into both singlemode and multimode fibers easily. When the light returns, most of the backscatter from the multimode fiber will be lost as it tries to couple into the smaller core of the singlemode fiber going to the detector.

Singlemode fiber cores are all about the same size: between 8–9 µm in diameter. A singlemode optical module is optimized for best light adapter in this diameter range.

Multimode fiber core diameters are 50 µm and 62.5 µm. A multimode optical card is optimized at only one of these core sizes internally, although it will be able to measure the other two size cores without difficulty.

Wavelength. The test wavelength is one of the important specifications of an OTDR. It is critical to test a fiber system at the same wavelength at which it will operate. However, it may also be useful to test the system at the other wavelength.

Shorter wavelengths of light exhibit greater amounts of attenuation in the same fiber due to their higher sensitivity to rayleigh scattering.

Longer wavelengths are more sensitive to bending loss and will allow light to exit out of the fiber core more than shorter wavelengths. This means that a fiber that has been stressed by bending will show a greater loss at the point of bending when tested at 1550 nm than it will at 1300 nm, even though the overall end-to-end attenuation will be lower at 1550 nm than at 1300 nm. The sensitivity of different wavelengths of light to different loss mechanisms in fiber can be a very important tool in troubleshooting a fiber cable.

Connector. Fiber systems are always connectorized before they are put into use. Either a connector is field-installed on the fiber or a preconnectorized fiber pigtail cable is spliced onto the fiber end. The connector on the OTDR should match the connector of the fiber system for best results. Some OTDRs offer universal-type connectors that change in the field. Other OTDRs have fixed, nonchangeable connectors.

When making a connection to a nonconnectorized, bare fiber, it is necessary to use a pigtail. This is the same as a jumper, but with only one end connectorized. The other end is bare fiber so a temporary splice to the bare fiber end to be tested can be made.

Measurement parameters. Once an OTDR has been properly configured for the fiber system to be measured, the test can be made. There are few decisions that must be made in determining the instrument setup conditions to get the best results. Many of these measurement parameters only need to be set once and will remain in the instrument's memory. If trace data can be stored in the OTDR's memory or on a diskette, the setup information is also normally recalled when the trace is recalled back to the screen.

Distance range. Distance range is also known as display range. It limits the amount of fiber that will be displayed on the screen. The distance range must be longer than the fiber to be tested. The distance range affects test accuracy and the time required to complete a test.

Since an OTDR must send out one test pulse at a time and allow all returns from the pulse to get back to the detector before sending out another pulse, the distance range determines the rate at which test pulses are sent out. This is known as the pulse repetition rate (PRR). The faster the rate, the quicker the averaging time for a given number of averages.

Since a longer fiber requires longer pulse transit times, the overall averaging is slower for longer distance ranges because the PRR is slower. If a long fiber is tested using a shorter distance range, then there is a possibility that a new test pulse will be sent into the fiber before all the return signals from the previous pulse are received by the OTDR detector. The resultant multiple received signal levels will produce unpredictable results on the OTDR display and will affect level measurements. It may also produce ghosts in the fiber trace.

Resolution. Measurement resolution (the spacing between data points) can be selected on some OTDR configurations. Higher resolutions (closer data points) will provide more detail about a fiber, but a test will usually take longer than one made at a lower resolution.

Higher resolution can provide more accurate location of an event. For instance, if an OTDR takes measurements every 8 m (26.2 ft) along a fiber, it is likely that a break could occur 7 m (23 ft) after a data point. The resulting fresnel reflection would appear to start at the following point, with the last backscatter level occurring at the previous point. Since distance to a break is always located at the last backscatter level prior to a fresnel reflection, the actual location of the break (reflection) would be off by 7 m (23 ft). If the data point resolution were shortened to 0.5 m (1.6 ft), then the reflection location would be more accurately located—to within about 0.3 m (1 ft).

Resolution should not be confused with the horizontal display scale. Also, the cursor resolution (how short a distance the cursor can be moved on the screen) has nothing to do with data point spacing. Most cursors can be placed between data points and thus appear to offer a better resolution.

Pulse width. The duration of the laser pulse can be changed by the operator. By selecting a long, media, or short pulse width, the operator can control the amount of backscatter level coming back and the dead zone size. A long pulse width will inject the highest amount of optical power into the fiber and, therefore, will travel farther down the fiber and produce stronger backscatter levels. It will also produce the longest dead zones. Conversely, a short pulse width will give the shortest dead zones but will send back weaker backscatter.

Long pulse widths give the maximum dynamic range for an OTDR and are used to quickly find defects and breaks in a fiber. Because the backscatter levels are higher, shorter averaging times are required to get a clean trace.

Short pulse widths are used to look at the part of the fiber that is closest to the OTDR. Because of the shorter dead zone, it can detect details in the fiber backscatter just past a fresnel reflection. Because of the lower levels of backscatter, longer averaging times are needed.

The rule of thumb for setting the pulse width is "long pulse to look long; short pulse to look short."

Averaging. The data points obtained from a single measurement pulse may vary in level from one to the next even though there is little change in the pulse they came from. The resulting trace looks noisy or fuzzy. To get a more reliable and smoother-looking trace, OTDRs send out thousands of measurement pulses every second. Every pulse provides a set of data points which are then averaged together with the data from previous pulses in order to improve the signal-to-noise ratio of the trace. Averaging takes time. Usually, a lot of averaging is required when a long fiber is being tested and/or when a short pulse width is being used. The operator can preset the amount of averaging that takes place so that test results will be consistent.

After a fiber has been scanned and the resulting trace is displayed on the screen, the operator must interpret the trace. Cursors are used to select the end points of measurements. The numerical results are displayed on the screen.

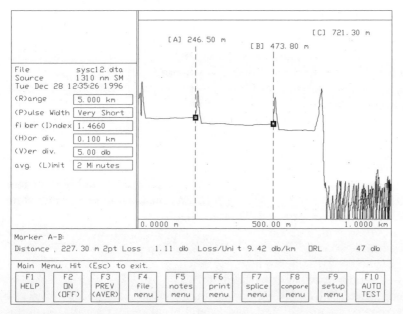

Figure 8.30 Interpret fiber trace

Fault location. The most critical measurement that is made with an OTDR is the location of a defect or break in the fiber. In order to repair a fault, the precise location must be located.

A fresnel reflection occurs at all fiber faults. This appears as either a sudden spike or drop off of the fiber trace, indicating that the OTDR pulse has encountered a sudden change in the density of the glass; that is, it encountered air at the end of the fiber. The distance to this reflection on an OTDR trace is the point at which the trace spikes up. If the trace recovers to the backscatter level after the reflection, then the fiber is not completely broken. The amount of shift in the backscatter level from before the reflection to after the reflection indicates how much light is lost at the fault.

Some mechanical splices produce a fresnel reflection. It is important to know where mechanical splices are located in a fiber, in order to keep from mistaking them for faults.

Distance measurements. The distance to a cursor is displayed on the screen. By simply moving a cursor to any point on the trace, the operator can read the distance to that point from the OTDR. Units of measurement can usually be selected to display distance in meters, feet, kilometers, or miles.

The measured distance to an event in a fiber, such as a mechanical splice, fusion splice, or fiber end depends on where the cursor is placed. To get the most accurate distance measurements, always place the cursor on the last point of backscatter just prior to an event. That is, place the cursor at the point just before a reflection so that the cursor does not ride up on the spike. For a nonreflective event (such as fusion splice or fiber macrobend), place the cursor at the point just before the trace drops (or rises).

When two cursors are used, the OTDR will display the distance to each cursor from the OTDR as well as the distance between the two cursors. This feature is used to isolate sections of a fiber.

Loss measurements. Loss measurements are indicated between two or more cursors. Loss can only be accurately measured from backscatter level to backscatter level. This means that both cursors must not be on a fresnel reflection or in a dead zone.

- *Overall loss.* The end-to-end attenuation of a fiber can be estimated by placing one cursor just to the right of the near-end dead zone and another cursor just to the left of the far-end fresnel reflection. The loss from cursor to cursor is read on the screen.

    Note. This means of measurement is not as accurate as an end-to-end attenuation measurement made with a light source/power meter combination, because it does not count the connection.

- *Section loss.* The loss in a section of fiber is measured by simply placing two cursors on the ends of the section to be measured and reading the level difference between the two.

- *Splice loss.* A splice is identified as a shift in the backscatter level. It can appear as a drop off or even a rise or gain. There may also be a fresnel reflection if it is a mechanical splice. A splice loss occurs at a single point in the fiber. A bend or point of stress may also show up as a point loss. The loss at a single point can be measured in two ways:

- *Two-point method.* The two-point method is identical to the way section loss is measured, except that the two cursors are placed as close together as possible, with the left cursor set just to the point to be measured and the right cursor set as close as possible on the right side of the loss event, but still on the backscatter. The backscatter level at a point loss does not shift directly down, but rather, has a slight roll-off to it. The length of the roll-off region is related to the pulse width (just like the distance taken up by a fresnel reflection).

  Since valid loss measurements can only be taken from backscatter to backscatter, the right cursor cannot use any part of this roll-off portion. This forces the right cursor to be placed farther out on the fiber than the next data point after the loss occurs. Consequently, the level measurement between the two points includes the loss occurring at the point plus an amount of loss that would normally occur over such a distance, although this is usually minimal.

- *LSA splice loss method.* The LSA splice loss method uses a mathematical technique called least squares averaging to eliminate the distance-induced loss found in the two-point method. The operator simply places a cursor on the point to be measured and selects LSA splice loss from the control panel. The OTDR then determines what the loss would be if the drop in backscatter level display was straight down instead of rolling off.

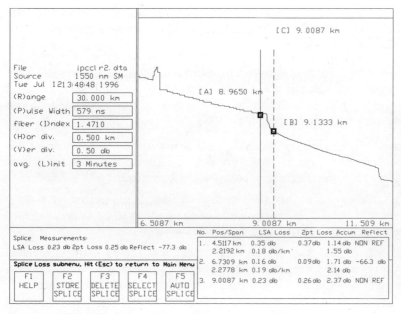

Figure 8.31 LSA splice loss

Quality factor (dB/km). The measurement of loss per unit distance is a common method for determining a fiber's quality. Less loss per unit length means a stronger signal at the receiver. The distance unit of measurement is usually kilometers (km). Fiber cable is ordered based on its type (singlemode/multimode) and its loss per km at a specified wavelength. Typical values for singlemode fiber at 1310 nm is 0.3 to 1.0 dB/km. At 1550 nm, this drops to 0.2 to 0.50 dB/km. For multimode fibers, the values range from 1.0 to 6.0 dB/km.

The quality factor is automatically calculated for a fiber when two cursors are set on the trace and the dB/km measurement is selected from the control panel. The OTDR simply displays the distance and loss between the two cursors and calculates the dB/km value by dividing the distance into the loss.

Reflectance. The amount of reflection at the location of a connector, fiber break, or mechanical splice depends on a connector's polish, the angle of cleave or break, and how much the index of refraction changes when the light leaves the fiber. Most mechanical splices use an index-matching gel or fluid to reduce the amount of change. Smaller changes in the index of refraction produce smaller reflections.

Some OTDRs can measure the amount of reflecting light automatically by placing one cursor just in front of the reflection and pressing the appropriate button on the control panel. Reflectance is measured in –dB (negative decibels) with a small negative value, indicating a larger reflection than a large negative value. That is, a reflectance of –33 dB is larger than a reflectance of –60 dB. The larger reflectance will show up as a higher spike on the trace. This connector reflectance is known as return loss.

By knowing the level of reflectance at a connector, it is possible to determine whether a problem is occurring very close to the connector—perhaps at the pigtail splice or in the connector strain-relief part. If the OTDR measures the distance to the far connector to be the correct length, but the reflectance is much worse than it was previously, the fiber might be fractured just inches short of the connector endface—thus causing a lower reflection due to the ragged edges of the shattered glass.

Since the connectors are the most handled part of the fiber, it is easy to damage the fiber right next to the connector and not realize it. An OTDR can be used to isolate this type of problem by using its reflectance-measuring capabilities.

Measurement problems

Even a trained and experienced OTDR operator may have difficulty interpreting a fiber trace at times. There are a few cases where it is almost impossible to get an exact distance or loss determination based on one measurement. In some extraordinary circumstances, it may be necessary to test a fiber with different setup conditions or from both ends in order to get meaningful results.

Nonreflective break. When a fiber is cut or broken, the end may shatter such that the angle at which the light hitting the end may not reflect at all. Also, the end of the fiber may become immersed in a liquid, which may also eliminate the fresnel reflection. When this happens, the trace will suddenly fall off into the noise level. There may be a rounding-off of the backscatter where it falls off so that it may be difficult to judge where the fall-off point is. The best method to determine the break point is to use the two-point loss method to detect at which point the backscatter level drops off by 0.5 dB.

Place the left cursor as near to the end as possible, but still on the backscatter. Then move the right cursor in towards the left cursor until the loss between the two reads 0.5 dB. The actual end of the fiber should be very close to the point measured by the right cursor. To increase your confidence in this location, take the OTDR to the other end of the fiber and test back to the break from the other side. It is possible that the other side of the break will reflect some light. (Keep in mind that the fiber could be broken at more than one point.)

Gain splice. Sometimes, when two fibers are spliced together, the backscatter level at the splice point shifts up instead of down. At first glance, this would appear to be a gain in power at the splice. The OTDR may even indicate a negative splice loss. What has happened is that the two fibers were mismatched: the second fiber has a higher backscattering coefficient than the first and more light is scattered back by it. The OTDR sensor reads this as a higher level than the end of the first fiber and plots the corresponding data points higher up on the screen.

If the same splice is tested from the opposite direction, the OTDR would indicate a higher normal loss than the amount of the negative loss. In this case, the true splice loss value is the average of the difference between the two readings. That is, if the gainer reads 0.25 dB and the opposite direction reads 0.45 dB, then the actual splice loss is 0.1 dB ([0.45 − 0.25]/2).

The following figure shows what a gain splice looks like on an OTDR display in comparison to a normal splice. Note that the slopes of the two fiber traces are different. The second fiber has a steeper slope than the first fiber, which indicates a higher backscatter level throughout the fiber. It would normally appear higher on the screen than the first fiber because it returns more light to the OTDR. A difference in index of refraction can produce different backscatter levels and, thus, different slopes of the trace. The other possible cause of gain is that the mode field diameter (which is related to the fiber's core size) is different for the two fibers. This causes more backscattering to come back from the second fiber. When a gain splice occurs, it is because the two fibers being spliced together are mismatched in some way.

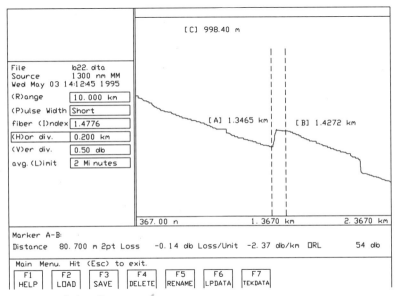

Figure 8.32 Gain splice

The average of all splices in a fiber span (a span is one or more fibers spliced together to make a continuous fiber link from one connector end to the other) is usually the benchmark used in construction of a system. If the average is equal to or better than the goal, then the overall loss budget planned for will be met. Gain splices can be confusing in determining splice loss averages, since they usually are displayed as a negative loss on the OTDR. In order to determine the average splice loss value for a string of splices in a fiber span, include the gain splice values along with the normal loss values. That is, use both the positive and negative values as displayed by the OTDR in summing the total of all splice loss values.

The most accurate method of determining the average splice loss values in a fiber span is to make a bidirectional measurement of each splice—that is, measure the splice of fiber A to fiber B, first from the end of fiber A, then from the end of fiber B. This method is time consuming and can usually only be done after the entire system has been spliced. The next best method is to take the one-direction average of all splices in a span using the splice values measured from the same direction only.

Normally, when a gain occurs, the next splice will be a higher-than-normal loss. This is because the fiber with the higher backscatter level causing the gain will also cause a higher measured loss going to the next fiber. The effects of the two splice measurements will cancel out. Avoid calculating splice loss averages using one-directional splice loss values when the individual splice values along the span were measured from different directions.

Ghost reflection. Sometimes a fresnel reflection can be seen where it is not expected—usually after the end of the fiber. This usually happens when a large reflection occurs in a short fiber. The reflected light actually bounces back and forth within the fiber, causing one or more false reflections to show up at multiple distances from the initial large (true) reflection. That is, if a large reflection occurs at 400 m (1312 ft), and there is an unexpected reflection at 800 m (2624 ft) (twice the distance to the first) and another at 1200 m (3936 ft) (three times the distance to the first), then it is likely the second and third reflections are ghosts. Here are a few techniques to use to determine if ghosts are occurring and to eliminate them:

- Measure the distance to the suspect reflection. Place a cursor half this distance on the fiber. If an expected reflection is at the half-way mark, then the suspect is probably a ghost.
- Suppress or reduce the known (true) reflection. By making the amount of returned power smaller, the ghost will also be reduced (or eliminated). To reduce the reflection, use index-matching gel at the reflection, or reduce the amount of power going to the reflective point by selecting a shorter pulse width or adding attenuation in the fiber before the reflection.
- Change the distance range (display range) of the OTDR. In some OTDRs, a ghost is caused when the distance range is too short.
- If a ghost seems to occur in the fiber, then measure the loss across the suspected reflection. A ghost will show no loss across it when doing a splice loss measurement.

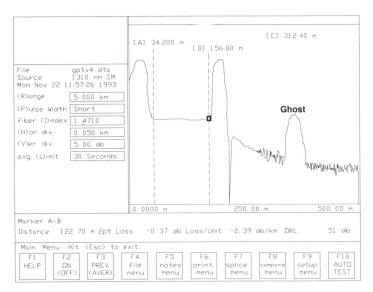

Figure 8.33 Ghost reflection

Testing procedures

This section will discuss the setup and testing procedure used with the OTDR.

Considerations. The testing setup required is dependent on the fiber system under consideration. The first and probably the most frequently used OTDR test is the determination of the overall attenuation and characterization of the fiber link. The following figure shows the components to be tested and their relationship to the OTDR screen display. The cursors are positioned, to measure the overall attenuation of the link under test.

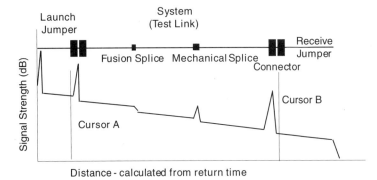

Figure 8.34 OTDR test configuration—end to end attenuation

The components required to complete this test are the OTDR, connecting jumpers, and the test link. To complete this test correctly, it is imperative that the correct launch and receive jumpers are used. As stated in the discussion of the dead zone, there are two points at which a measurement cannot be made. This is the beginning and end point of the fiber. One method of overcoming this limitation is the selection of the launch and receive jumpers. Remember, to make any loss measurement in fiber, you must see backscatter at both sides of the event. Since the dead zone at the OTDR may be from a few feet to over a mile in length, select the launch jumper that has enough length to allow the system to recover prior to the first connector of the fiber to be measured. As a rule of thumb, when testing multimode fiber in the LAN, the minimum length of the launch jumpers should be at least 100 m (328 ft) in length. The jumper at the receive end must also be of sufficient length to allow the system to recover and again show sufficient backscatter to allow a measurement. For singlemode, the launch jumper must be longer than the dead zone of the longest pulse width of the OTDR.

Once this field test setup has been established, it is now possible to test all of the parameters necessary to completely characterize the fiber link. By positioning the cursors on each side of the event, determine the loss of each connector and splice, as well as the loss per kilometer of each fiber segment within the link. Refer to the "Measurement Parameters" section for additional details on interpreting the test results. The only major problem that remains is the built-in problem of reading a scope. Accuracy is dependent on how well the cursors are positioned when the readings are taken. The view angle of the screen when placing the cursor makes a difference. With the addition of more computing power built into today's field testers, comes some additional capability which makes this difficult part of the cabling installer's job a lot easier to accomplish accurately and quickly.

Automatic measurements

The latest-model OTDRs have the additional capability to automatically configure themselves and perform standard measurements. These autoranging and autoanalysis features allow each operator to make the same precise, consistent measurements on any fiber.

Autorange feature. Autoranging, also known as Automode or Auto-Setup, sets the variable distance range (display range) of the OTDR, as well as the pulse width and resolution, to the best settings for the fiber under test.

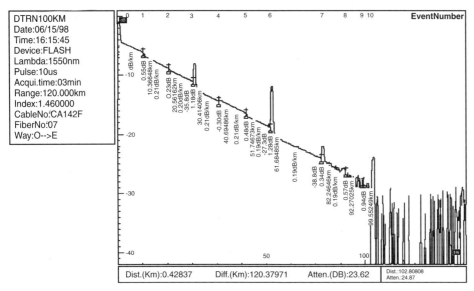

Figure 8.35 OTDR trace with automatic event selection

Fiber analysis software. One of the most important tasks in using an OTDR is being able to interpret the fiber trace correctly. The fiber analysis feature can perform this task by using the computer part of the OTDR to scan and analyze the digital test results obtained from the fiber. It will look for shifts in the backscatter level, which indicate splice losses or defects. It will look for sudden spikes up from the backscatter, which indicate Fresnel reflections (usually mechanical splices, breaks or fiber terminations). It will measure the loss in each of these events that it finds. The results are summarized in an event table, which lists each event, its location, loss, and reflectance (if any).

Some OTDRs will link the event table to the actual trace, so it can be seen where each event is situated in relation to the others. Events may also be deleted and added. Text comments may be added for each event. This capability will allow complete documentation for each fiber for use in maintenance or for providing detailed records to customers. Most OTDR manufacturers who use the auto-analysis feature also have a similar standalone program available for use on a desktop or laptop computer. If the OTDR has a floppy disk drive for storing test data, it can either load a previous test into the OTDR or into the computer and run the analysis software. This standalone software is a good choice if using an older OTDR that does not have built-in analysis, but does have trace data storage.

	Distance km	Attenuation	Reflectance	Rel. dist.	Slope	Link budget	Uncertainty
1	10.36648	0.55	10.36648	0.20	2.03		
2	20.56162	0.23		10.19531	0.21	4.22	
3	30.41406	1.18	−35.8	9.85244	0.20	6.21	
4	40.69486	−0.30		10.28081	0.21	8.32	
5	51.74673	0.48		11.05187	0.21	10.64	
6	61.68485	1.28	−27.3	9.93811	0.19	12.57	
7	82.24646	0.34	−38.8	20.56162	0.19	16.53	
8	92.27025	0.57		10.02379	0.19	18.46	
9	99.55249	0.94		7.28224	0.25	20.27	Two Pt.
10	102.20836	25.96	>−32.2	2.65588			

Figure 8.36 Event table linked to figure 8.35

Automatic fiber analysis routines do not always find all the events in a fiber and sometimes will find events that are not there. If splices are very good and it is difficult to find the splice on the trace, then the software will probably also have trouble finding and measuring them. The ability of the software to locate an event depends on the thresholds set up, the length of the fiber, and the amount of averaging done on the trace. There is the option of setting the sensitivity of the analysis routine so that it will only find events that meet or exceed a certain threshold. If the threshold is set to 0.2 dB, then the software will not mark splices with lower losses. On the other hand, when measuring very long fibers, the far end of the trace may be very noisy, with the individual data points fluctuating up and down from one to the next. This fluctuation may cause the software to find a false event that is not there. The noise in the trace may be decreased by increasing the averaging time (scan time). The number of false events may be reduced by keeping the loss threshold above 0.1 dB.

Some fiber analysis software will allow preset of the location of the splice points into a template or master trace file. All the other fibers in the cable can be run against this master to only get the measurements on the splices known to be there. By only measuring specific known points, an overall cable loss table can be created by exporting the analysis results to a spreadsheet or database. This is an excellent way to manage fibers and to determine if the splices or other sections of the fiber are getting worse over time.

Field test equipment selection

The field test equipment represents a major investment. As such, there are several factors which must be considered.

Basic test equipment includes such items as the multimeter, inductive amplifier, wire map testers, station identifiers, fiber flashlight, etc. Each of these represents a comparatively small investment. On the other hand, the purchase of field test equipment such as a TDR, certification field tester, light source and power meter, OTDR, etc., are high-cost items and require a great deal more research prior to purchasing. The first question to be answered is what type of media will be tested? Will it be copper or fiber? If copper, which of the several types will be primary? If fiber, is it multimode or singlemode? Is the primary usage for new installations or for troubleshooting? These are some of the questions to ask before making a major purchase.

Copper media. In addition to the basic field testers listed earlier, probably the most versatile tester is the certification field tester. Each of the field testers available will accomplish the basic tests required by ANSI/TIA/EIA-568-A. The certification field testers have the ability to accomplish all required tests at one time by the use of an autotest function. If a large number of tests are to be performed at one time, a unit may be required that will download each test directly to a computer as it is being performed. This may save substantial time in the documentation process. A unit may be required to save the Pass/Fail criteria and the traces. In addition to the certification tests, the unit may also be used for troubleshooting. If so, a listing of the type of tests required will need to be determined. Tests such as impedance, capacitance, loop resistance, and noise are available on most units. If the TDR function is required, a unit that will provide the trace on a continuous real time basis may be needed. With this function, a change to the circuit can be made and the result seen as it happens. If active circuits need to be tested, a field tester will require the ability to monitor a data circuit and provide real time analysis. Make sure the certification field tester chosen can do all of these easily, quickly, and accurately. Controls should be easy to operate and easily accessible without going through layers of menus. Remember that each function added has a cost.

Fiber media. There are two primary field test equipments for the fiber media. The first is the light source and power meter. These are available in many different configurations and price ranges. If a great deal of fiber testing will be done, consider a standalone version. Another option is a unit that connects directly to the certification field tester. One advantage to this unit is that the tests may be saved the same as the copper tests and then downloaded with the other test results in the same format. New to the market is field test equipment that will test fiber in pairs. Each fiber is tested at one wavelength (850) in one direction and the other wavelength (1300) from the other direction at the same time. This alleviates the need for multiple trips to each location to test both wavelengths, and, in the case of bidirectional testing, each direction. If testing beyond the capability of the light source and power meter, the next option is the OTDR. Of primary importance is the kind of measurements to be made—loss, distance, reflectance, splice alignment. Be sure the OTDR selected can do all of these easily, quickly, and accurately. If live tests (hot cut—splicing of fibers in a working cable) are needed, an OTDR that can do an active splice loss measurement in real time will be required. Look for a simple front-panel control layout as well as a display that is easy to interpret. If the instrument will be used only when a problem occurs in the fiber, or for semiannual maintenance evaluations, then a simple unit with few front-panel controls that has a built-in HELP feature is adequate.

Training. Considerations for training curriculum can include

- The number of operators required for each piece of equipment.
- The frequency of use by each operator.
- Whether the test result is open to interpretation by the operator.
- How to make interpretations consistent.
- The current training level of the cabling installer.

It may be that only supplemental training is required.

Documentation. It is normally required that a permanent record of the test data is kept on file. Therefore, choose field test equipment with a printer, plotter, and/or disk storage capability. Also, consider the range and type of external printers and plotters supported by the unit. Determine if the unit is compatible with the currently owned equipment. The field test equipment selected should have the ability to store trace information on a disk, print it out on an internal printer/plotter, and/or print/plot it from a computer program using the computer's printer.

Test data analysis. A very efficient means of keeping trace data on file is to store the information in a computer. In the case of the certification field tester, does it have the ability to store the trace data along with the pass/fail data? These traces can become a valuable troubleshooting aid as it has the ability to do a direct comparison of the original test with the one taken today. This allows the traces to be recalled at any time for comparison and analysis. Most OTDR manufacturers have programs that emulate an OTDR on a computer; the same loss and distance measurements can be made on a laptop computer or office computer just like on an OTDR. The software should also be able to analyze the trace data to produce fiber event tables and cable loss tables for complete documentation. If this option seems beneficial, make sure the program will operate on the type of computer in use. Some programs run under MS-DOS®, Windows®, or both.

Manufacturer/vendor support. The type and amount of vendor support may be an important consideration when selecting a test set. Testing can get confusing at times and some fast answers from someone who has seen the problem before may be needed. Check to see if the supplier has a help line or some way to provide technical information. Verify that the local representative is technically qualified and experienced enough to talk the user through problems. Check the warranty repair turnaround policy. How long will it take to have the field test equipment repaired? It is always good to know the manufacturer's reputation in the industry.

Another consideration is whether or not the field test equipment can be upgraded to add new features or improve operations. While some field test equipment must be taken out of service and sent to the factory for upgrades, most can be upgraded in the field simply by uploading the new software from a computer or placing a disk with the new operating software in the unit. Software upgrades can provide completely new features and improved performance without losing the use of the unit.

Compatibility. Will other field testers be purchased? If compatibility between models is important, then determine if the data from a manufacturer's previous models are compatible with the current (and future) models. Verify that it will be easy to compare traces with data taken years ago on a different model. Some field test equipment manufacturers completely change their models and do not provide backward compatibility in features, operation, or data. If models are changed, it may be necessary to retrain the operators. However, most ODTR manufacturers have proprietary software programs that cannot be used with other OTDR software programs.

Appendix A   Fiber Performance Calculations Worksheet

A. Calculate the Passive Cable System Attenuation			
1	Calculate Fiber Loss at Operating Wavelength	Cable Distance × Individual Fiber Loss	1.5 km × 1.5 dB/km
		Total Fiber Loss	2.25 dB
2	Calculate Connector Loss (Exclude Tx and Rx Connectors)	Individual Connector Loss × Number of Connector Pairs	0.75 dB × 4
		Total Connector Loss	3.0 dB
3	Calculate Splice Loss	Individual Splice Loss × Number of Splices	0.3 dB × 3
		Total Splice Loss	0.9 dB
4	Calculate Other Components Loss	Total Components (None)	0.0 dB
5	Calculate Total Passive Cable System Attenuation	Total Fiber Loss + Total Connector Loss + Total Splice Loss + Total Components	2.3 dB + 3.0 dB + 0.9 dB + 0.0 dB
		Total System Attenuation	6.2 dB
B. Calculate Link Loss Budget			
	Example Manufacturer's Electronic Specifications	System Wavelength	1300 nm
		Fiber Type	62.5/125 μm multimode
		Average Transmitter Output	− 18.0 dBm
		Receiver Sensitivity (10^9 BER)	− 31.0 dBm
		Receiver Dynamic Range	11.0 dB
6	Calculate System Gain	Average Transmitter Power − Receiver Sensitivity	− 18.0 dBm − −31.0 dBm
		System Gain	13.0 dB
7	Determine Power Penalties	Operating Margin (none stated) + Receiver Power Penalties (none stated) + Repair Margin (2 fusion splices at 0.3 dB each)	2.0 dB + 0.0 dB + 0.6 dB
		Total Power Penalties	2.6 dB
8	Calculate Link Loss Budget	System Gain − Power Penalties	13.0 dB − 2.6 dB
		Total Link Loss Budget	10.4 dB
C. Verify Performance			
9	Calculate System Performance Margin to Verify Adequate Power	Link Loss Budget	10.4 dB
		− Passive Cable System Attenuation	− 6.2 dB
		System Performance Margin	4.2 dB

Note: 4.2 > 0. Therefore, the system will operate as installed.

Appendix B Fiber Link Attenuation Record

Fiber Optic Link Attenuation Record

Location: _____   Technicians: _____

Date: _____

Wavelength: 850 nm   1300 nm   1550 nm

Buffer Tube Color:

Reference Reading:

Fiber Number	A to B Reading	A to B Attenuation	B to A Reading	B to A Attenuation	Average Attenuation	Remarks
1						
2						
3						
4						
5						
6						
7						
8						
9						
10						
11						
12						

Notes:

Chapter 9

Troubleshooting

General Reference	481
Copper Cable	489
Optical Fiber Cable	505
Attachment A—Typical OTDR Fault Presentations	516

General Reference

Overview

The following section lists general topics that a telecommunications cabling installer should be familiar with in order to complete troubleshooting tasks.

Test equipment. The troubleshooter should have appropriate test equipment available and know how to use it. Depending on the type of cabling system in question, this could include the following:

- Wire map tester
- Handheld cable tester
- Certification test set
- Optical light source and power meter
- Optical time-domain reflectometer (OTDR)
- Optical fiber flashlight
- Infrared conversion card
- Low-intensity laser

- Multimeter
- Electromagnetic field strength meter
- Toner/Wand
- Telephone test set (butt set)
- Appropriate test adapters, leads, and cables

Wire map tester. Wire map testers are low-cost cable testers which can test for opens, shorts, crossed pairs, and improper wiring. These units are good for quick, basic tests, but lack the sophisticated diagnostic capabilities of more expensive testers. Most testers in this category are designed for UTP.

Handheld cable tester. Testers in this category usually provide, at a minimum, the functions of a wire map tester, plus that of a time-domain reflectometer (TDR) function to measure length, as well as functions to measure near-end crosstalk (NEXT), attenuation, and induced noise.

Some will also include tools for identifying multiple cables and tracing cables in the wall or ceiling. Testers in this category are generally designed to test at frequencies up to 20 MHz and, therefore, are not suitable for certifying Category 5 installations. Many are designed to work with UTP, ScTP, STP-A, and coaxial cable. More expensive units in this category will store test results and allow results to be printed or exported to the database.

Certification test set. These units include all of the functions of the handheld testers described above. They also have additional functions that allow them to verify whether a cabling system meets the transmission performance requirements of TIA/EIA TSB-67.

> **Note.** TIA/EIA TSB-67 is the Telecommunications Systems Bulletin which specifies the testing parameters for Categories 3, 4, and 5. It is expected that TIA/EIA TSB-67 will be incorporated into the next revision of ANSI/TIA/EIA-568-A.

Certification tests units will test a UTP and ScTP cabling system to at least 100 MHz, and will measure and record the following parameters:

- Wiremap
- Length
- Attenuation
- NEXT

The autotest feature compares the actual measured values with required values for Category 3, 4, or 5 performance and displays pass or fail for the entire battery of tests. The certification test equipment will also display pass or fail and the actual measured values for each test individually. These units are capable of other tests, including impedance, capacitance, and loop resistance. Certification testers can store data and export it to a database or output it to a printer.

Cable tracer. A cable tracer is an accessory for certain cable testers and certification test sets. It consists of a signal generator and a signal receiver. The generator is connected to one end of the cable and the receiver is then used to follow the path of the cable—through the wall, floor, or ceiling.

Cable-end locator kit. Sometimes called an office locator kit, this is a set of numbered 8-pin modular plugs, each of which can be uniquely identified by the locator. The standard practice is to insert the plug into an outlet in the work area, then search a patch panel for the unique identifying signal.

Optical light source and power meter. An optical light source and power meter are useful in performing loss (attenuation) measurements for optical fiber cable.

Optical time-domain reflectometer (OTDR). An OTDR may be employed to carry out certain measurements and functions for optical fiber cable.

> **Caution.** Effective use of an OTDR requires experience and skill. Results often require interpretation.

Optical fiber flashlight. It is used to determine continuity of optical fiber strands. It generates a safe light that can be checked with the naked eye.

> **Warning.** Never look at optical fiber strands unless you personally confirm that they are disconnected from potential laser light sources. Such sources can result in permanent eye damage.

Infrared conversion card. An infrared conversion card allows the troubleshooter to visually detect an infrared signal when such a source is directed upon the card's phosphorus material.

Low-intensity laser. Also known as the hot red light, the low-intensity laser operates in the visible light range. The laser is used to identify individual fibers within a cable by passing a red light down the fiber. When used as a troubleshooting tool, the cable will glow red at the break point.

Multimeter. A multimeter is a combination electrical meter which at a minimum measures voltage, current, and resistance. A multimeter is used for a variety of functions—including finding opens, shorts, and high-resistance connections in cables and connectors.

Electromagnetic field strength meter. An electromagnetic field strength meter measures the presence of electromagnetic interference. It is used when a problem is believed to be due to electromagnetic interference.

Toner/Wand. A toner is often used to trace and to identify cables by sending an audio tone which can be heard through a telephone test set or identified by means of a wand.

Telephone test set. Sometimes referred to as a butt set, a telephone test set is used for the following voice circuit functions:

- Simulate the user's telephone equipment
- Identify circuits

Test adapters, leads, and cables. The cabling installer may require the proper adapters to connect the troubleshooting equipment to the cable under test.

Other equipment. Other equipment that may be required includes:

- Ladder.
- Two-way radio.
- Fiber or copper talk sets.
- Alcohol and wipes.
- Compressed air.
- Notebook.
- Inspection microscope.

Ladder. When troubleshooting cabling problems, it may be necessary to access difficult-to-reach areas, such as the space above a drop ceiling.

Two-way radio. Many troubleshooting tasks require two cabling installers—one positioned at each end of a cable link or data channel. A two-way radio can be helpful when installers at opposite ends of a cable need to communicate with each other.

Note. Interference from such radios has been known to affect the operation and accuracy of some handheld cable testers.

In addition, the use of two-way radios may not be allowed in some facilities for safety or security reasons. Check with the job supervisor before using this equipment. Additionally, be aware that two-way radios do not function properly in all environments.

Some light source and power meter equipment includes a communications function. This is only useful, however, when both parties are connected to opposite ends of the same cable.

Alcohol and wipes. Isopropyl alcohol (99 percent pure) and lint-free wipes, including cotton swabs, can be used to clean dirty electrical and optical connections.

Compressed air. Compressed air is used for removing loose dust and other debris from electrical and optical connections. Ensure that the air is manufacturer approved as suitable for cleaning fiber terminations (e.g., without Freon™).

Notebook. A notebook will serve for documenting the data if the test equipment does not store these results.

Inspection microscope. An inspection microscope is used to examine in detail the condition of the end of a fiber cable or a fiber connection.

Documentation. The process of troubleshooting can be greatly eased when appropriate documentation is available. This should include:

- Cabling diagrams.
- Description and functioning of the equipment attached to the cabling system.
- Certification test data for the network.

Communication skills. A troubleshooter must:

- Be able to read technical manuals, instructions, catalogs, etc.
- Verbally communicate with customers, coworkers, contractors, and other support personnel.
- Understand blueprints and drawings that relate to cabling installation.

Ideally, a cabling installer will have computer skills to add, look up, update, and interpret cabling database information, view electronically stored documents, and obtain support information via online resources.

Industry and standards knowledge. It is recommended the troubleshooter be familiar with:

- *National Electrical Code* (NEC).
- ANSI/TIA/EIA standards.
- Basic electrical and electronic principles.
- Basic optical fiber and light theory.
- Basic troubleshooting techniques.
- Test equipment operation.
- Voice and data network basics.

Safety standards and procedures. The troubleshooter should be familiar with the rules of safety that pertain to the job, as well as any safety requirements of the facility in which the work is being performed. These rules include, but are not limited to, the following:

- *National Electrical Code* (NEC)
- Common safety practices
- Confined-space safety
- Electrical safety
- Ladder safety
- Laser safety
- OSHA standards
- Personal protective equipment
- Safe optical fiber handling practices
- Tool safety

Troubleshooting tips for isolation. Becoming an effective troubleshooter takes practice and experience. Fortunately, modern cable testers can make the job easier. Here are a few troubleshooting tips:

- Determine the nature of the problem and make sure it is cabling related. Many network problems can appear to be cabling related. Someone without the proper diagnostic skills and experience can easily misdiagnose a defective network interface card as a cabling problem. Conversely, many cabling problems can appear to have other causes. For example, out-of-specification cabling can result in transmission errors, which will often be manifested as poor performance. The first thing to do is talk to the user(s). Determine the following:

- Has the basic link or channel in question ever worked?
- Has the basic link or channel in question been tested previously?
- When and how did the problem begin?
- Is there some specific event that correlates with the beginning of the problem? Remember, correlation does not necessarily mean causation. Do not make assumptions. Sometimes, equipment moves, construction work, or other events can be related and may indicate possible causes of problems.
- Why is the problem believed to be cabling related? Again, assume nothing. Just because someone believes a problem is cabling related does not mean that it is. In either case, the users' response can often help locate the source of the problem.
- Does it affect all activity or specific applications? If only specific applications are affected on a network link, chances are the problem is not cabling related. Some applications are more sensitive to cabling variations than others; some cabling problems may only be associated with specific applications.
- What is the scope of the problem? Is it one connection or many? If the problem is affecting a single connection, the chances are it is within that specific channel or the equipment directly connected to it. If the problem affects multiple links, it could be related to backbone cabling, a link to a shared service (such as a network file server), or with such equipment as a hub or a server.

- Look around for anything unusual. Check the work area for anything that suggests intervention. Examine the telecommunications closet(s) for clues, such as sloppy wiring or evidence of recent changes. Be alert for signs of recent construction or electrical work.

- Use a cable tester to check suspected cabling segments. The results will often provide a good indication of possible causes as well as the location of a problem. If the tester has an autotest function, be sure to look at the results of the individual tests. (See "Using Test Equipment for Troubleshooting.")

- Break the problem into smaller parts. If performing a basic link test, examine each component (the channel and each patch cable) separately. If troubleshooting a 10BASE-2 bus, break the bus at the midpoint and then test each half. If the problem is isolated to one half or the other, using this method break the problem section in half again and repeat the process until the defect is located.

- Look for defective terminations or improperly installed connectors. Loose, improperly installed, or damaged connectors are a major cause of cabling problems. Check for intermittent problems by moving or bending the cable under test near the point of termination.

- Ensure that the proper outlets and other connections are employed. For example, outlets designed for Category 3 use will probably not meet Category 5 specifications.

- Patch cables and plugs are another potential trouble source. Although the standards call for the use of stranded wire for UTP patch cords, many people construct their own using solid (nonstranded) wire. This practice is not recommended.

- Check to make sure the appropriate type of cable for the application has been installed.

- Direct attention to the installation practices. Cabling installers who do not follow proper installation practices can create a multitude of problems—most commonly not maintaining twists close enough to the point of termination and splitting pairs. This is usually due to using the old USOC pinouts instead of T568A or T568B pinouts.

- Equipment and patch cords should be examined carefully, especially in the case of Category 5 installations. Patch cords are easily damaged, especially at or near the point of termination. Patch cords also may not be properly made. It is difficult, for example, to maintain the pair twist in close proximity to the modular connector when making Category 5 patch cords; therefore, it is recommended that they be purchased from a reputable manufacturer rather than made by the cabling installer.

- Look for missing terminators on 10BASE-2 systems. Terminators are often not properly reinstalled when equipment is moved or added.

- Has the cable been damaged? If an equipment cable is laid across the floor, it can easily be damaged. In addition, improper installation practices, building construction, or remodeling can damage installed cable. Also, furniture pushed against a wall plate may create a bend radius smaller than the standards allow.

- Look for user intervention. This is a major problem with 10BASE-2 systems. Users have been known to add sections of cable to a 10BASE-2 bus. Two problems are common: adding a drop cable between a 10BASE-2 T-fitting and a network interface card, or extending the bus by using an incorrect cable (for example, cables for IBM 3270 terminals employ the same BNC connectors used by 10BASE-2 but are of a different impedance). A user may have removed one or more computers from the network with the 10BASE-2 T-fitting still attached, thereby breaking the bus.

- Inspect for sources of interference. For example, cable running over fluorescent lights, down elevator shafts, or in parallel with and in close proximity to ac power lines may cause problems.
- Check for excessive cabling distances. This can be caused by using extra-long patch cords. This is a common problem with 10BASE-2 systems—which seem to grow in length until they just quit working.

Copper Cable

Overview

Effective troubleshooting requires experience, knowledge, patience, and skill. The effective troubleshooter:

- Has good communications skills.
- Is able to gain useful information by asking questions in a manner that does not alienate the person being questioned.
- Has learned how to observe objects and events in order to extract information about the possible causes of problems.

Additionally, a good troubleshooter knows where and how to locate information: how to get to the right people on vendor technical support lines, how to use online resources, and how to use manuals and technical documents. A good troubleshooter understands the application for each of the available tools and keeps them in proper condition.

Troubleshooting is an investigative process—a troubleshooter looks for clues and follows leads. A troubleshooter must avoid making assumptions about the cause of problems—things are not always as they first appear and false assumptions may cause hours of wasted time. Finally, a troubleshooter must know when to take a different approach or get outside help. Sometimes, the most obvious things are difficult to identify because the telecommunications cabling installer is so close to the problem.

Although there is no set of rules to cover all situations, the cabling installer can use some general guidelines as a starting point. Gather relevant information; this includes locating cabling diagrams and documentation, asking questions of users, and observing the situation. Specifics may include:

- Type of system (LAN, PBX, etc.).
- Device experiencing problems (PC, telephone, etc.).
- Type of cable (UTP, coaxial, optical fiber).
- Location of closets and equipment rooms that house the problem circuit connections.
- Location of the telecommunications equipment.
- Location of user equipment.
- Details of the circuit.
- Cabling diagrams.
- Certification test data.
- Previous service records.
- Other information (components, etc.).

Contact the end user before beginning work and ask the following questions:

- What is the specific problem? This initiates the troubleshooting process and helps to narrow the possible problem.
- When did the problem begin?
- How often does this problem occur?
- What else was happening in the area at that time? The problem could be related to the installation of other equipment or to another event which could help determine the cause of the problem.
- What other actions can be taken that will help in narrowing the cause? For example, if the user's PC is working properly but cannot connect to the network, have the user check the connections. If the connections are good, ask the user to check the faceplate. If the user has a "Data 1" and "Data 2" connection, have the user change the connection and try to access the network. If the trouble is a telephone set, again have the user check the connections. If the connections are good, determine if talk battery is available. If so, have the user disconnect the set for 10 minutes. Sometimes, especially after an electrical storm, the PBX fault-detection program will correct the problem.

Perform a visual inspection. Check out the overall site, as well as the specific problem area. Look for evidence of equipment movement, construction, previous sloppy work, incorrect cable or components, and anything else that can aid in determining the cause of the problem.

Complete all actions that the troubleshooter would normally have the user do when talking over the telephone.

If the problem has not been resolved, make sure all equipment is removed from the cable in question and perform the appropriate diagnostic tests. Refer to "Interpreting test results" in the "Using Test Equipment for Troubleshooting" paragraphs of this section. Troubleshooting often involves two cabling installers working together to test cables, one at each end of a line. A set of two-way radios, or the test equipment's talk set (copper or fiber), is often required for efficient testing. Turn off two-way radios during testing to eliminate the possibility of inducing unwanted noise.

If the problem has not been identified using the above procedures, then break the problem into smaller parts and test those parts separately. Refer to "Troubleshooting tips for isolation" in the "General Reference" paragraphs of this chapter.

It may be necessary to obtain assistance in locating a problem. Modern networks are complex, and one person cannot be expected to know everything.

> **Note.** Asking for help is not a sign of failure or incompetence. An expert in a specific area, such as interpreting the scope pattern on an OTDR may be needed. In other cases, someone with a fresh viewpoint looking at the situation may identify the problem.

Reconnect the equipment to the cable and make sure everything is working properly.

Using test equipment for troubleshooting

A good cable tester should help locate the sources of cabling problems. Looking at the results of the individual tests can provide a wealth of information.

Interpreting test results. When reading results, be aware that there may be more than one potential cause for a particular test result; in some cases there may be multiple contributing causes. The following paragraphs list test results and possible associated causes.

High attenuation (copper). Attenuation is the loss in power of a transmitted signal as it travels along a cable. If attenuation is too high, transmission reliability will be affected. Attenuation increases with temperature, frequency, and cable length.

High attenuation can be caused by:

- High temperature.
  Cables installed in areas that may be subjected to high temperature may exhibit an increase in attenuation.
- Wrong grade or category of cable.
  The cable under test may be unsuitable for the data rate at which it is being used. In this case, it may be necessary to replace the cable.
- Cable is too long.
  The cable length may exceed the specified maximum length. Check cable lengths to make sure. Some cabling installers, afraid of waste, will leave excess cable coiled in the ceiling or telecommunications closet. Also, excessive lengths of patch and equipment cables can be a problem.
- Incorrect equipment patch cable.
  Make sure equipment and patch cables meet requirements for the desired installation (i.e., Category 3, Category 4, and Category 5).
- Improper terminations.
  Ensure that all connections are terminated properly. Make sure that connecting hardware including work area outlets, patch panels, and termination blocks meet the transmission performance requirements of the desired category.

Attenuation-to-crosstalk ratio (ACR). ACR is the result of the formula NEXT minus attentuation. The values for each of these parameters are recorded at a specific point of frequency and recorded in decibels. For example, if the worst-case NEXT value at 100 MHz was 30 dB and the worst-case attenuation value at 100 MHz was 20 dB, then the ACR value at 100 MHz would be 10 dB. A positive ACR value is desirable.

Incorrect capacitance. For twisted-pair copper cables, capacitance is measured between the two wires of a pair. For coaxial cable it is measured between the center conductor and the shield.

Possible causes for an incorrect capacitance value include:

- Broken conductor in a cable; check continuity.
- Split pairs; investigate the pair terminations and patch cables.
- Wrong cable type for application; determine the cable type and replace if incorrect.
- Excessive noise (refer to Noise in this chapter for more information).
- Shorted conductors; wire map, length, or scan tests should help locate this kind of fault.

If this capacitance value appears erratic, investigate for intermittent connections or noise.

Excessive near-end crosstalk (NEXT). Excessive NEXT may be attributed to:

- Wrong grade of cable.
  Make sure the grade of cable being used is correct for the application.
- Improper termination practices.
  If proper twisting is not maintained when terminating twisted-pair cable, NEXT may increase. Ensure the twisting is maintained as specified for the category of cable being used.
- Split pairs.
  Proper pairing must be maintained. The twisting of the pairs tends to minimize interference radiating from the cable pair and that received by the cable pair. By splitting pairs (using wires from two different pairs as a signal circuit), the beneficial effects of pair twisting are lost.
- Incorrect or substandard components.
  The quality of the basic link or channel will be determined by the weakest component. If a component is poorly made or does not meet the requirements for the desired installation category, the component should be replaced.
- Incorrect or defective cables and test adapters.
  Make sure that proper cables are utilized. It is not uncommon, for example, for an end user to install flat, untwisted, silver satin telephone station wire as network patch and equipment cables.

Incorrect cable length. Due to the twisting of the pairs, the actual physical length of a twisted-pair cable will appear slightly shorter than the electrically measured length.

- Length too short.
  A length test that displays a significantly shorter-than-actual length, may indicate an open or short in the cable. It can also indicate a poor termination at an intermediate connection, such as a punch-down block, patch panel, or wall outlet.
- Length too long.
  - Check for excess cable coiled in wall, ceiling, or telecommunications closet.
  - Verify cable length runs from blueprints.
  - Replace long patch cords with shorter cords where possible.
  - Notify job supervisor or customer as required.

- Inaccurate length measurement.
  Make sure that the NVP setting of the cable tester is correct.

The following applies to 10BASE-2 coaxial cable:

- If a length appears different from that expected, retest from the opposite end. If these two readings are different, there may be a problem with the cable construction or its installation.
- If a length measurement indicates the bus is too long, remove part of the bus and retest. It is not uncommon for 10BASE-2 networks to grow beyond the length considered maximum at the time of installation. If the bus is too long, a repeater may be required.
- If performing a length test or scanning the cable with the far end of the cable open, and the length test indicates the bus is shorter than expected, the coax may be open somewhere in the cable. Install a shorting plug on the far end of the cable and retest. If there is still an indication of an open, then the coax is probably open.

Excessive loop resistance. Loop resistance may be determined by shorting both conductors of a copper pair at one end of a cable and measuring the total resistance of both conductors from the opposite end. Excessive resistance can be caused by:

- Excessive cable length.
  Make sure the length of the cable is within specifications for the application.
- Poor connection at termination point.
  Improperly punched-down terminations or damaged wall outlets, connection blocks, and patch panels could be to blame.
- Defective shorting plug, test cable, test adapter, equipment cable, or patch cable.
- Test each component separately.

Incorrect connections. This includes crossed pairs, split pairs, opens, and shorts. When testing a channel, investigate each cabling component separately. Possible causes are:

- Incorrectly installed or damaged cable or termination
- Wrong patch cables
  Patch cables wired to older USOC specifications will create split pairs for ANSI/TIA/EIA-568-A applications.
- Mismatched terminations
  The wiring scheme for the T568A and T568B connectors is different. Do not use a T568A connector on one end of a cable and a T568B connector on the other.

Incorrect impedance. Impedance mismatches are associated with:

- Poor installation techniques.
- Incorrect or defective cable.

Noise. Noise is unwanted electrical activity that interferes with the desired signal transmission. Noise can be induced directly by a connected signal source, such as a computer with a defective, noise-generating component. Also, it can be radiated from some external source, such as an electric motor, copier, or electrical power cable in close proximity to a data cable. There are two types of noise to be addressed—impulse noise and continuous noise.

Impulse noise refers to discrete noise spikes which occur on a regular or irregular basis. These spikes can be as high as 350 volts. In some instances, impulse noise can damage equipment but more often it will cause machine lock up (network file servers, network hubs, etc.) or corrupt data. Impulse noise can be caused by an electric motor, copier, laser printer, elevator, air conditioner, or any device which produces a large power disruption upon startup or shutdown.

Continuous noise is usually below three volts. While it can affect transmission on a cable, it will generally not result in hardware damage. Continuous noise can be caused by a fluorescent light, computer, ac power line in close proximity to a data cable, and many other sources.

Most cable troubleshooting equipment will measure for continuous noise. Some can also monitor for impulse noise. A field strength meter can also be used to monitor for impulse noise.

The frequency of continuous noise can provide a clue as to its source:

- *Under 150 kHz.* Noise in this range is usually due to sources such as ac power lines, fluorescent lighting, and machinery.
- *150 kHz–20 MHz.* In this range, noise is often attributed to light dimmers, medical equipment, computers, copiers, and laser printers.
- *Above 20 MHz.* Such noise may result from radios, cellular phones, wireless phones, television sets, microwave ovens, and broadcast equipment.

Here are a few things the cabling installer can do about noise:

- Move cables away from possible sources of interference. ANSI/TIA/EIA-569-A, *Commercial Building Standard for Telecommunications Pathways and Spaces,* provides guidelines for separating data cables from common sources of noise. Ensure that these guidelines are followed.

- Move interference sources away from cabling. Change the location of microwave ovens, copiers, etc., if possible.
- Install a better grade of cable. If using UTP, install Category 5, even if the application only requires Category 3.
- Maintain cable pair twist up to the point of termination. Category 5, for example, requires that twists be maintained to within 13 mm (0.5 in) of the point of termination.
- Make sure there are no split pairs.
- Maintain separate cross-connect fields for voice and data.

Locating cables. If documentation is not complete, it may be necessary to follow the path of a cable to locate the ends. Many cable testers have accessories available to help with these tasks.

Cable tracer. A cable tracer consists of a signal generator and a signal receiver. The signal generator is connected to one end of the cable, and the receiver is then used to follow the path of the cable—through the wall, floor, or ceiling.

Cable-end locator kit. Sometimes called an office locator kit, this is a set of numbered 8-pin modular plugs, each of which can be uniquely identified by the locator. The standard practice is to insert each of the plugs into outlets in the work area, then search a patch panel for the unique identifying signal. This allows the matching of a patch panel connection with an outlet in the work area.

Copper media

Step	Troubleshooting twisted-pair
1	Follow the general guidelines outlined in the overview of this section.
2	Complete the visual inspection to determine that the: ■ Location has been correctly labeled. ■ Correct patch cords/equipment cords have been used. ■ Correct cable and termination devices have been installed in accordance with the applicable standard.
3	Having completed the verification phase and determined that the trouble is within the cabling network, the fault-isolation phase begins.

Isolation. The troubleshooting flow chart (Figure 9.3) provides a comprehensive, organized approach to cable troubleshooting. However, before beginning this phase, wire properties should be reviewed.

To most cabling installers, a wire pair appears as:

Figure 9.1 Twisted-pair wire.

However, to an electron, this same wire pair appears as a transmission line.

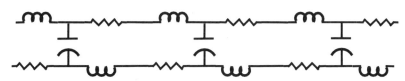

Figure 9.2 Schematic of a transmission line.

As seen in the above diagram, a transmission line is composed of resistors, inductors, and capacitors. When a second pair is added, the influence of these components is felt on the second pair. Crosstalk occurs when a signal traveling down a pair exceeds the proper limits and induces a current into an adjacent pair.

If the cable is stretched, the capacitance (the geometry of the cable pairs within the cable sheath) is changed; if stretched hard enough, the resistance (a change in wire gauge) is also changed. If a coil is placed in the link an inductor is added. By placing one or more cables together for an extended length, there is the chance of an increase in crosstalk. If cable ties are placed too tightly, the jacket of the cable is deformed, resulting in a change in capacitance and, therefore, a change in NEXT and return loss.

Note. When dealing with ScTP or STP-A, it is also important that the ground as well as the individual pairs be verified.

498 Chapter 9

The following figure provides the guidelines for successful problem isolation.

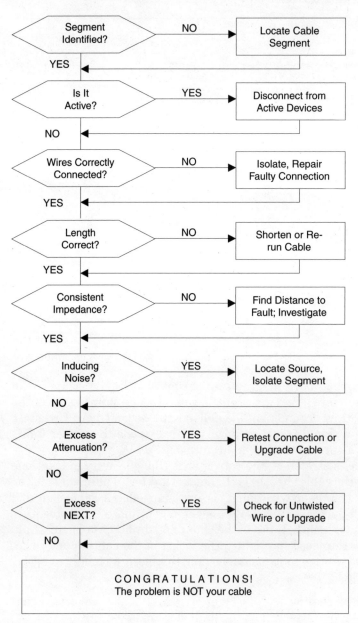

Figure 9.3 Troubleshooting flow chart. *(Printed with permission of Wavetek.)*

After developing the guidelines, the next objective is to match the test equipment with the situation.

TABLE 9.1 Test-Set Selection

Step	Media type			
	UTP	ScTP	STP-A	Coax
Segment Identified?	Documentation Toner/Wand	Documentation Toner/Wand	Documentation Toner/Wand	Documentation Toner/Wand
Active?	Telephone Test Set	Telephone Test Set	Telephone Test Set	Telephone Test Set
Wires Correctly Connected?	Wire Map Tester	Wire Map Tester	Wire Map Tester	Multimeter
Length Correct?	Multimeter TDR	Multimeter TDR	Multimeter TDR	Multimeter TDR
Consistent Impedance?	Certification Test Set	Certification Test Set	Certification Test Set	Certification Test Set
Inducing Noise?	Certification Test Set	Certification Test Set	Certification Test Set	Certification Test Set
Excess Attenuation?	Certification Test Set	Certification Test Set	Certification Test Set	Certification Test Set
Excess NEXT?	Certification Test Set	Certification Test Set	Certification Test Set	N/A

Up to this point, obvious faults have been discussed. What about the types that are more difficult to isolate? Two of the most common are discussed below.

Symptom. Certification test set cannot locate the remote end.

Type of fault. Obvious.

Possible cause. Assuming that both the test set and injector are attached to the correct circuit, the most probable cause is an open in all four pairs of the cable.

Isolation. Switch the test set to the TDR function and it will display the cable length. If the length is shorter than anticipated, check at that point for an open.

One of the more difficult faults is an intermittent fault. It is in a category all by itself. An example is:

Symptom. Unable to log onto the network, sluggish performance, disk errors, station freezes, reboots.

Type of fault. Intermittent

Possible cause. If the cable has been checked and eliminated as the problem, there is a good possibility that there is an intermittent noise problem. Noise is being induced into the system by an outside source such as an electric motor, (e.g., electric pencil sharpener) lighting, ac power lines, heaters, etc. This effect only occurs when the device is activated and puts a "spike" onto the line.

> **Note.** Check for the location of the electric pencil sharpener.

Isolation. Using the certification test set, first check the average noise on the circuit. [Make sure your two-way radios are turned off. Check for impulse noise (the spike)]. If these levels are high and the test set has an active time-domain crosstalk capability, it should be able to determine where the noise spike is entering the line. If it does not, the line will have to be hand traced to locate the possible equipment, which may be inducing the noise. If these efforts prove negative, another potential source of the problem may be in the grounding, bonding, or electrical system.

In this case, perform a visual inspection, beginning with the connection between the bonding conductor for telecommunications and the TMGB of the telecommunications grounding and bonding network. If all connections are correct, use a multimeter to test the connections for a maximum value of 0.1 Ω. If these efforts prove negative, a strip-chart recorder may be placed on the circuit to the device. If power fluctuations are evident, the branch circuit serving the equipment may need to be moved to another phase of the three-phase power feeding the building or the unit be placed on a UPS system.

Once the problem has been successfully isolated, the final phase is to repair and test the circuit. In the case of a noise-interference problem, it may not be possible to clear the problem by moving the cable or the interfering source. If this is true, consideration must be given to placing a Coupled Bonding Conductor (CBC). The CBC provides equalization like the TBB and also provides protection through the electromagnetic coupling with the telecommunications cables.

> **Note.** Refer to ANSI/TIA/EIA-607, Annex-B, for more specific information.

Another possibility is to replace the UTP cable with either ScTP or STP-A to correct the situation. Once the repair has been completed, the circuit should be retested in accordance with the testing procedures provided in Chapter 8, "Testing Cable."

Step	10BASE-2 (thin ethernet) coaxial cable
1	Follow the general guidelines outlined in the Overview of this section.
2	Perform a visual inspection to make sure that the correct cable has been installed, the correct terminator is in place at each end of the bus, and proper installation practices have been followed. During the physical inspection look for: • Proper connector installation—Make sure that connectors appear to be properly crimped onto the cables. • Correct terminators—A 50 Ω terminator must be used at both ends of the 10BASE-2 bus. • Correct connectors—Are the connectors the correct type for the cable? • Correct cable—Does the cable meet 10BASE-2 or thin-ethernet requirements? Both RG-58A/U and RG-58C/U coaxial cable meet these specifications if designated IEEE 802.3 compliant on the jacket. • Drop cables—Drop cables between the T connector and the network interface are not allowed. They are only permitted between external transceivers and 15-pin AUI connectors at network interfaces. • Damaged or kinked cable—Replace or repair damaged cable segments. • Proper grounding—A 10BASE-2 bus should be grounded at one point only.
3	Test the bus using appropriate equipment. At a minimum, measure the following: • *Length.* With terminators removed, attach the test equipment to one end of the cable and perform a length test. Note the results. Next, install a shorting plug at the open end of the cable and retest. If the results of the two tests vary by more than 0.6 m (2 ft), then evaluate every cable segment, T-fitting, and cable connector individually. The maximum length for a 10BASE-2 bus is 185 m (606 ft). • *Loop resistance.* Remove the terminators and place a shorting plug at one end of the cable. Using a multimeter, measure the resistance between the center pin and the shield at the open end of the cable. Loop resistance should be no more than 5 Ω per 100 m (328 ft) at 20 °C (68 °F) Refer to Chapter 8, "Testing Cable," for detailed instructions.

- *Characteristic impedance.* Determine the characteristic impedance, remove the terminators from both ends of the cable, attach a cable tester to one end and perform the impedance test. The average impedance should be 50 Ω.

In the case of a coaxial network, a phenomenon, called a standing wave, exists. When this occurs, one workstation on the network will fail to connect to the network. Stations ahead or behind work fine. Replacing the unit with one that worked at another location will not clear the problem. This condition is created because the cable portion is just long enough to coincide with the wavelength or a harmonic of the wavelength that is carrying the signal. The solution is to either shorten or lengthen the cable feeding the workstation.

4. If problems are indicated, investigate each cabling component separately, including segments of cable, T connectors, and terminators. Test each terminator for proper resistance (50 Ω). T connectors are examined for opens, shorts, and resistance between the housing and each center connector.

Steps—UTP cable

Step	UTP cable troubleshooting
1	Disconnect active equipment so that it will not interfere with the testing. For an existing system, ensure that users have been contacted and a convenient time has been arranged to remove equipment from service. - Determine which equipment is connected to the line being tested. - Disconnect the equipment.
2	Connect the cable tester and carry out the appropriate measurements. Some test sets require daily calibration. Calibrate according to the manufacturer's specifications. - Set the tester for the correct type of cable. - Conduct the appropriate measurements.
3	If the tester is not the type to return a pass/fail reading, compare the results with the specifications for the cable. Verify that the readings fall into the acceptable range for the cable. If the original cable certification test results are available, compare the current reading/test result with the result of the original test.

4   Record the results.

5   Identify failing measurement values.
    - Determine which results do not fall into the normal range or are different from the original test results. If the results of a test are near the limit (other than length) for troubleshooting purposes, consider that test to indicate a failure. Properly installed and functioning cable should be well within limits, and marginal results indicate potential problems.
    - Record any failing results.

6   Use measurement results to determine the source of the problem.
    - Inspect hardware and connections.
    - Visually inspect the cable for damage.
    - Check the cable type.
    - Verify test settings.
    - Inspect the terminations.
    - Examine the bend radius, the pair twist, and the sheath.
    - Observe the color codes.
    - Verify pin configuration.

7   Test specific channel components, the basic link, and each patch or equipment cord separately.

8   Perform additional tests as necessary. See "Interpreting test results" in the "Using test equipment for troubleshooting" and the "Isolation" paragraphs in this section.

9   Take corrective action for any observed conditions which do not meet cabling standards.

10 Retest the cable to verify that it passes all relevant tests.

11 Record the:
    - Circuit identification number and location.
    - Customer name and phone.
    - Service type.
    - Test results.
    - Repairs made.
    - Passing test results.

12 Perform housekeeping as needed.
    - Replace any cables moved during the testing.
    - Ensure that all cables are properly routed.
    - Cables must maintain at least the minimum bend radius.
    - Ensure that all cables have the appropriate slack.
    - Leave a clean space when the installation is complete.

The following procedure assumes the use of a cable tester or certification test set. If a multimeter is used, refer to Chapter 8, "Testing Cable," for the procedures.

Steps—Coaxial cable

Step	10BASE-2 coaxial cable troubleshooting
1	Disconnect active electronic equipment from the cable under test. Contact the user before disconnecting active equipment.
2	Adjust the test sets' autotest parameters for the cable to be investigated. By way of an example: • 10BASE-2 or thin-ethernet cable • Specific settings for a particular brand and model of cable (e.g., settings for PVC and FEP cable may be different). Refer to tester's documentation for specific instructions. If testing a type or brand of cable not defined in the database, it may be necessary to calculate the nominal velocity of propagation (NVP) for that cable.
3	Remove the terminators from both ends of the cable.
4	Using the loop resistance test (or a multimeter), measure the resistance of each terminator. It should be 50 Ω, ± 1 percent.
5	Connect the test set to one end of the cable.
6	Perform the following measurements as described in 10BASE-2 (thin ethernet) coaxial cable in previous paragraphs in this section: • Resistance • Length • Characteristic impedance Some tests require the use of a shorting plug at the far end of the cable.
7	Record the results.
8	Identify failing measured values. • Determine the results that do not fall into the normal range. If the original cable certification test results are available, compare the current reading/test result with the result of the original test. If the results of a test are near the limit (other than length) for troubleshooting purposes, consider that test to indicate a failure. Properly installed and functioning cable should be well within limits, and marginal results indicate potential problems. • Record any failing results.

9   Use measurement results to determine the source of the problem.
    - Inspect hardware and connections.
    - Visually inspect the cable for damage.
    - Check the cable type.
    - Verify test settings.
    - Inspect connector crimps.

10 If the measured value does not meet requirements, examine the details to find out which particular test failed. If necessary, test each component as described earlier. Locate and repair the problem and retest.

11 Perform additional tests as necessary. See "Interpreting test results" in the "Using test equipment for troubleshooting" and the "Isolation" paragraphs in this section.

12 Take corrective action for any observed conditions which do not meet cabling standards.

13 Retest the cable to verify it passes all relevant tests.

14 Record the:
    - Circuit identification number and location.
    - Customer name and phone.
    - Service type.
    - Test results.
    - Repairs made.
    - Passing test results.

15 Perform housekeeping as needed.
    - Disconnect test equipment.
    - Reinstall terminators and reconnect equipment.
    - Replace any cables moved during the testing.
    - Ensure that all cables are properly routed.
    - Cables must maintain at least the minimum bend radius.
    - Ensure that all cables have the appropriate slack.
    - Leave a clean space when the installation is complete.

Optical Fiber Cable

Overview

Troubleshooting is a process for locating a decreased performance condition (problem) that is not acceptable to the user of an optical fiber circuit. This condition may be minor in nature, an intermittent condition, or a total outage. Restoration is the process of returning the circuit to an operational condition that is acceptable to the user.

The troubleshooting process for optical fiber cable is identical to that used for copper cable. The four steps of verifying, isolating, repairing, and testing are applicable regardless of the media.

Verify

Here again, the questions asked of the user remain basically the same. It may also be helpful to take the user through the same telephonic verification as performed for copper cable.

- Answers to the following questions are needed:
  - Who is affected by the condition?
  - What is the nature, scope, and severity of the condition?
  - Where are the possible areas that the condition could occur?
  - When did the condition occur? Does it correlate with any other activities or abnormal conditions?
  - Why does this condition exist?
  - How could this condition happen—electrical outage, weather conditions, people working in the area?
- Upon arrival at the site, assess the situation. The goal is to determine the priority of the condition and the resources available to correct the problem. The following items should be considered:
  - What is the priority of the restoration? A circuit serving one computer is not as critical as a backbone cable in a hospital.
  - What test equipment is available?
  - What trained manpower is available?
  - What method or procedure will be used to find the problem?

Isolate

This is the most interesting part of the troubleshooting process. The two common approaches for isolating a problem are:

- Start at the beginning of the circuit and work toward the other end. This is appropriate when there is only one link or one that is relatively short in length (the horizontal).
- Start in the middle of the circuit and determine in which direction the problem lies. Move to the middle of an identified defective section and again determine which section is defective. Continue this process until the defective link is found (the backbone).

A thorough visual inspection of all areas is necessary. It is important that each connector be thoroughly cleaned before inspection. A dust particle on the end of a connector becomes a very efficient attenuator. Refer to Chapter 8 for detailed cleaning procedures.

Most problems in an existing circuit can be located in this manner, such as a:

- Disconnected jumper.
- Jumper which has been incorrectly changed due to move, add, or change (MAC) activity.
- Jumper to a hub or router that has been unplugged.
- Hub or router that has been turned off. Look for the power on indicator.
- Broken fiber inside a fiber distribution unit. The fiber inside a 900-µm buffer can be pinched or broken by a door being closed or a drawer being moved. The plastic buffer on the fiber may or may not show some damage.

  Note. The damaged area can usually be felt with your fingers, because any compression severe enough to damage the fiber will also distort the plastic buffer.

- Damaged cable inside a telecommunications closet (TC), equipment room (ER), or entrance facility (EF).

  Note. A main cross-connection (MC) may be found in a TC, ER, or EF. Cabling installers performing other functions in these areas may have rearranged the cable to accomplish their task. Most likely, they will not be trained in the proper procedures for handling optical fiber cables. (This is a good example of why optical fiber cables should be clearly marked.)

Until this point, all previously listed conditions would result in a dead fiber; that is, no light exits. There are other conditions, such as higher attenuation at 1300 nm than 850 nm or a greater loss in one direction than in the other. If the system had been tested according to ANSI/TIA/EIA-568-A (at one frequency and in one direction), the majority of these conditions could have existed when the fiber was originally placed but did not become evident until the fiber was activated.

ANSI/TIA/EIA-526-14A, currently standard Proposal 2981, states, "Bi-directional testing is a default requirement of this document as it is the most conservative." Some examples are:

- At the time of the original testing, the fiber was tested from the TC to the workstation in the 850 nm window. The fiber is now to be activated and is to operate in the 1300 nm range. The loss reading at 1300 nm is higher than at 850 nm. Possible causes of the problem are:
  - Excessive bending at the connector.
  - Excessive bending along the route.
  - In the case of a backbone fiber, excessive bending at a splice point.

- Lower wavelengths normally have a greater loss than higher wavelengths because of a higher scattering loss at the lower wavelength. On the other hand, the higher wavelengths tend to leak out easier because of bending in the fiber.
- The second condition is a greater loss at both frequencies when tested from the other direction. In the original test, the light was transmitted from connector 2 to connector 1. In that case, all of the light traveled into connector 1.

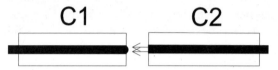

Figure 9.4 Connector loss C2 to C1.

When the transmit and receive directions are reversed because of the domed-shape polish on connector 1, part of the light is allowed to escape, causing an increase in attenuation. This can be corrected by repolishing the connector.

Figure 9.5 Connector loss C1 to C2.

- Another possible cause may be small asymmetrical nicks in the fiber. When light is transmitted in one direction, the light is reflected into the fiber, while transmission from the opposite direction causes the light to be reflected out of the fiber.

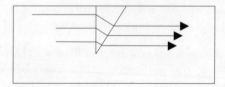

Figure 9.6 Reflective loss.

This condition causes some light to be reflected outside the fiber and lost. When the light is transmitted from the opposite end of the fiber, all light is reflected into the fiber.

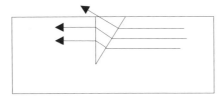

Figure 9.7 Reflective loss.

Repair

Once the problem has been isolated, the next action is to correct the problem and restore the circuit to its original working condition. This may involve the following:

- Cleaning a connector.
- Reseating a connector.
- Replacing a connector.
- Installing a splice on a broken backbone fiber.
- Replacing a broken splice.
- Replacing a broken jumper.
- Replacing a broken pigtail.
- Installing a replacement section of either horizontal or backbone cable.

Safety. Always perform a thorough visual inspection of areas where working. Look for the following potential hazards:

- Open electrical panels and exposed wiring
- Trip or fall hazards on the floor
- Overhead hazards, such as cable trays, that could cause head injuries
- Hazardous chemicals storage

The alcohol used for cleaning connectors is flammable. Do not expose the fumes to an open flame or electrical arc. The fumes can also be toxic when used in confined spaces. Ensure that there is adequate ventilation when using alcohol.

Warning. Never look at the end of a connector without first ensuring that there is no dangerous light source on the other end. Laser sources are potentially dangerous because the output level can be very high and the wavelength is out of the visible range of a person's eyes.

LEDs are generally safe because of their lower output power. Regardless of the light source, it is a wise practice to always verify that the distant end is disconnected from any light source.

When installing connectors or splicing fibers, there will be short pieces of fiber which need to be disposed of properly. The technician should have a container with a lid to store the pieces of fiber. The container should be labeled properly, and when it is full, the closed container should be disposed of in a place which will not cause injury to anyone. A soft-drink can or styrofoam cup is not acceptable for this purpose. Electrical tape or masking tape formed into a loop with the sticky side out is also not an acceptable method of disposing fiber ends. A short piece of tape used to pick up a small piece of fiber from a work surface is acceptable; however, fiber ends are not considered hazardous waste, but common sense should govern their disposal.

Test

Once repairs are finished, the final step is the testing and documentation of the repaired circuit. Refer to Chapter 8, "Testing Cable," for more information.

Test equipment

There are many types of test equipment available to assist the technician in the search for the problem location. The test equipment includes:

- Optical fiber flashlight—A small pocket flashlight equipped with a fitting which will mate with a connector. The flashlight injects light into the core of the fiber that can be seen at the other end on a good circuit. It works well for fiber identification also, since the color code does not always uniquely identify a single fiber.
- Flashlight advantages:
  - Inexpensive.
  - Easy to use.
  - Minimum training required.
  - No calibration or referencing is required.
- Flashlight disadvantages:
  - Short range.
  - No quantitative measurement.
  - Not useful in locating broken fibers.
  - Cannot be used on working circuits.
- Visible LED light source—A device similar in size to a small flashlight featuring a light-emitting diode as the light source. It is equipped with a fitting that will mate with a connector. The light source injects light into the core of the fiber which can be seen at the

other end on a good circuit. It works well for fiber identification also, since the color code does not always uniquely identify a single fiber.
- LED advantages:
  - Inexpensive.
  - Easy to use.
  - Minimum training required.
  - Useful in locating broken fibers in 900-µm buffered fiber and translucent connectors.
  - No calibration or referencing is required.
- LED disadvantages:
  - Media range.
  - No quantitative measurement.
  - Cannot be used on working circuits.
- Visible laser light source—A device containing a laser diode as the light source. The wavelength of the light from the laser diode is usually 660 nm, which is in the visible light range. The connection method is the same as the flashlight and visible LED sources.
- Laser advantages:
  - Easy to use.
  - Minimum training required.
  - No calibration or referencing is required.
  - Useful in locating broken fibers in 900-µm buffered fiber and translucent connectors.
  - Long range.
- Laser disadvantages:
  - Somewhat expensive.
  - No quantitative measurement.
  - Cannot be used on working circuits.
- Light source and power meter—A multimode light source depends on a light-emitting diode that provides an output at either 850 nm or 1300 nm. A singlemode light source relies upon a laser diode providing an output at either 1310 nm or 1550 nm. The light source electronic circuitry makes the output very stable so that precise measurements can be made with the power meter.
- Power meters contain a photo detector that is sensitive to a wide range of wavelengths; a switch selects the operating wavelength. A power meter will usually be capable of measuring both multimode and singlemode wavelengths.
- Lightsource/Power-meter advantages:
  - Easy to use.
  - Some training required.

- Quantitative measurements (the most accurate of any method).
- Some meters offer internal data storage and download capabilities.
- Lightsource/Power-meter disadvantages:
  - Somewhat expensive.
  - May require two technicians, depending on the number of tests to be completed.
  - Displays the amount of loss in the circuit but does not give the location of the loss.
  - Requires access to both ends of the fiber and both ends must have connectors installed.
  - Cannot be used on working circuits.
- Fiber identifier—This is an electronic device which uses clip-on technology to determine if there is live traffic or a test signal on a fiber. The identifier will indicate not only the presence of live traffic but also the direction of flow. A test signal, such as a 1004-Hz tone, can be placed on the fiber by a light source equipped to do so. Fiber identification can then be obtained with a very high degree of certainty. The fiber identifier can also be used to verify continuity of a fiber.
- Fiber identifier advantages:
  - Easy to use.
  - Some training required.
  - Can be used by one technician.
  - Does not require access to either end of the fiber.
  - Can be used on working circuits.
- Fiber identifier disadvantages:
  - Somewhat expensive.
  - Requires access to one fiber at a time, not multiple fibers in a sheath.
- OTDR—A sophisticated electronic testing device which sends out a light pulse on a fiber and then measures the time and amplitude of the reflected signal. The results are displayed as an X–Y plot on a screen. It is used as a troubleshooting and acceptance testing measurement tool.
- OTDR advantages:
  - Will provide quantitative attenuation measurements but with less accuracy than a light source and power meter.
  - Will furnish distance measurements to events (fault, splice, connector, etc.) and to the end of the fiber.
  - Visual display of all events.
  - Print and storage capability of all displays.
  - May be accessed from either end of the fiber.

- OTDR disadvantages:
  - Expensive.
  - Requires considerable training to properly operate.
  - May require two technicians, depending on the number of fibers to be tested.
- Fault finder—Similar to the OTDR, but does not have an X–Y screen. Results are displayed one at a time as alphanumeric characters. Used as a troubleshooting tool, but not an acceptance-test tool.
- Fault-finder advantages:
  - Will provide quantitative measurements but are not as accurate as a light source and power meter.
  - Will furnish distance measurements to events and to the end of the fiber.
  - Can be used by one technician.
  - Requires access to only one end of the fiber.
  - Less expensive than the OTDR.
  - Requires minimum training to properly operate.
- Fault-finder disadvantages:
  - Alphanumeric display instead of an X–Y display.

Steps—Fault isolation

Step	Isolating a fault
1	Perform a visual inspection of the fiber and connectors.
2	If physical damage is evident (chipped fiber at the connector, broken fiber at the panel or faceplate, etc.), make the necessary repair and proceed to Step 5. If physical damage is not evident, proceed to Step 3.
3	Clean the connectors by performing the optical fiber connector cleaning procedure detailed in Chapter 8.
4	Check the fiber with the optical fiber flashlight. If light is visible, proceed with Step 5. If light is not visible, proceed to Step 12.
5	Set up and calibrate the light source and power meter.
6	Test the fiber from the origin to the extreme.
7	Document results.
8	Test the fiber from the extreme to the origin.
9	Document the result.
10	Perform the mathematical averaging of the two readings (R1 + R2)/2.

11 If the results are within the standard (2 dB for the horizontal or 11 dB for the backbone), document the results and place the fiber back into service. If not, continue with Step 12.

   Note. The maximum loss for horizontal (2dB) and backbone (11 dB) is based on worst-case maximum length. These values may not be acceptable based on a link loss budget.

   At this point of the fault isolation procedure, the fault is one of the following conditions:
   - A broken fiber that was not evident during the initial visual inspection (Steps 12 through 15).
   - High attenuation from an unknown cause (Steps 16 through 20).

12 If looking for a no-light condition, connect the visual fault locator (Hot Red Light) to one end of the fiber.

13 Perform a visual inspection of the fiber path. A red glow should be evident at any point where light is escaping from the fiber. Mark each location on the sheath. Note that there may be more than one fault in the cable. If no light is detected, proceed to Step 16.

14 If a broken cable is detected in the horizontal, replace the cable. If a broken cable is detected in the backbone, pull slack from the service loop and splice.

15 Once the repair has been completed, complete Steps 5 through 11.

   At this point, there are two options: the fault locator or the OTDR. Each will find the cause of the no-light or the high-attenuation condition. For the remainder of this checklist, the steps used will be those for the OTDR test set. Prior to setting up the OTDR, obtain a copy of the original OTDR tests for the fiber in question. These will be invaluable in the interpretation of the new test results.

16 Set up the OTDR. If the original traces are available, take special care to set the refractive index, pulse width, pulse duration, sampling rate, etc. to that original trace.

17 Connect the test set and perform the test. Refer to Chapter 8 paragraphs on fault isolation, distance measurement, and loss measurement as needed. As our interest is in the total cable, it is imperative that both a launch and receive jumper of proper length be used for the test. Representative samples of each type of fiber fault are contained in Attachment A.

18   Compare the new test with the original. Note any point of difference between the two tests. These differences are the faults, either broken, partially broken, or high-impedance locations.

19   Make repairs as necessary.

20   Once the repairs have been completed, complete Steps 5 through 11.

Appendix A Typical OTDR Fault Presentations

Reflective Break

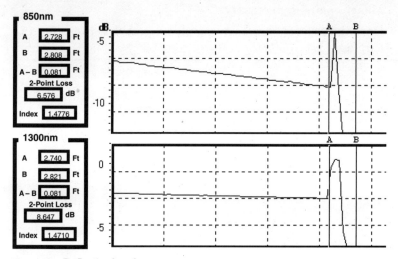

Figure 9.8 Reflective break.

A reflective break is most commonly observed under the following conditions:

- Unconnected fiber—As the loss is greater at 1300 nm than it is at 850 nm, this is an indication that the fiber is being influenced by microbending. Therefore, the fault is not the connector but with excessive bending in the cable prior to the connector.
- Broken or bad connector.
- Broken fiber.

Nonreflective Break

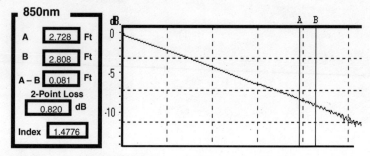

Figure 9.9 Nonreflective break.

When a fiber is cut or broken, the end may shatter such that the angle at which the light hitting the end may not reflect at all. The end of the fiber may become immersed in oil or grease, which may also eliminate the fresnel reflection. When this happens, the trace will suddenly fall into the noise level. There may be a rounding of the backscatter where it falls so that it may be difficult to judge where the fall point occurs. The best method to determine the break point is to use the two-point loss method that indicates the point the backscatter level drops by 0.5 dB.

Place the left cursor as near to the end as possible but still on the backscatter. Then move the right cursor in towards the left cursor until the loss between the two reads 0.5 dB. The actual end of the fiber should be very close to the point measured by the right cursor. To increase confidence in this location, take the OTDR to the other end of the fiber and test back to the break from the other side. It is possible that the other side of the break will reflect some light. Keep in mind that the fiber could be broken at more than one point.

This indication is most often seen when testing fiber on a reel. The fiber at the distant end has been cut; therefore, all of the light is escaping into the air with little or no reflection. It becomes imperative to terminate both ends of the fiber to achieve an accurate length measurement.

Defective Splices

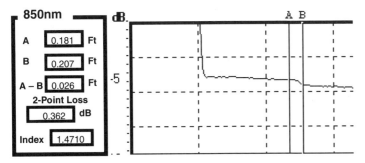

Figure 9.10 Defective fusion splice.

In the case of a defective fusion splice, the defect is noted by the two-point loss method. Measuring from the last flat spot on the trace before the splice (cursor A) to the first flat spot on the trace following the splice (cursor B), note that the loss is 0.362 dB. This loss is in excess of the 0.3 dB allowed by the standard.

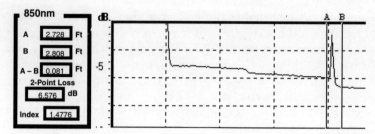

Figure 9.11 Defective mechanical splice.

As in the case of the fusion splice, the defect is determined by the two-point loss method. As the indicated loss is in excess of the 0.3 dB allowed by the standard, the splice is considered defective.

Chapter 10

Retrofit Installations

Retrofit Installations	519
System Cutover	545

Retrofit Installations

Overview

A telecommunications system's retrofit involves the replacement or augmentation of an existing installation. Performing a retrofit or extensive system upgrade will always be far more complicated and labor intensive than a new cabling installation in unoccupied spaces.

Planning for a retrofit installation involves a multitude of tasks. All of these tasks must be completed before any cable is removed or installed. When all of these planning tasks are completed, a comprehensive plan will have been developed and the telecommunications cabling installation team will be fully prepared to proceed with the project.

Determined by the scope and design of the cabling installation, each retrofit will require the cabling installer to address a unique set of circumstances. Some retrofits may involve a simple expansion of a phone system or a data network. Other retrofits may involve both voice and data systems, along with the security and fire alarm systems as well as an upgrade to the backbone or horizontal cabling infrastructure.

Good communications with the customer are essential when determining what part, if any, of the existing cable plant will be reused. If the customer wants to utilize portions of the present cables and equipment, the existing components will have to conform to the new requirements. In many cases, system upgrades require a new cabling configuration that will not allow the use of the current infrastructure.

Another concern is grandfathered systems. Grandfathered systems consist of nonplenum-rated cables that were installed in a plenum prior to the mandated use of plenum cables. These systems are said to be acceptable because they were installed prior to the code. When replacing these types of systems, the cabling installer may be required to replace all of the cables. Many localities have rules of percentage. This means that if a substantial percentage of the cable is being replaced, it will be necessary to bring the entire installation up to code. Check with the local inspectors to determine the rules of percentage for the area.

Prior to installing a new cable plant in a retrofit installation, it is necessary to determine whether the existing cable system or any portion of it can be reused. Obtain as much information as possible from the customer and verify that the information is accurate. If the existing documentation is questionable or does not exist, it may be necessary to trace and document the current system. In addition, the cabling installer must test the existing cable plant to verify the level of service it can support.

In ideal situations the retrofit will take place in unoccupied spaces. This allows for the removal of all equipment and cables that will not be part of the new installation. With these items removed, valuable pathway and closet space is freed up for the installation of new system components.

In most cases, the new system or components to be integrated into the existing working system must be installed with minimal disruption to the customer. This may involve installing the new system parallel to the existing system, while working nights and weekends. After the new cables and equipment have been installed and tested, the transfer to the new system may take place.

When transferring to the new system, a detailed plan must be developed to cut the new components into the final working configuration. This final cutover from the old to new system is one of the most detailed and critical steps in a retrofit. To help ensure a quality transition and to expedite the cutover, cutsheets are used. Cutsheets are the documentation that contain the existing cross-connect terminations and show the changes that must be performed to execute the cutover. A poorly planned cutover could cause unnecessary interruptions to the customer's communications.

TABLE 10.1 Detailed Cutover Plan

Cutsheet		Project: BICSI, Tampa				
Existing		Temporary cabling		New system		Description
KSU-4-12	TB-7-17	KSU-4-12	Temp-3-17	PBX-2-21	TB-9-1	WA-102 phone
KSU-4-13	TB-7-21	KSU-4-13	Temp-3-21	PBX-2-22	TB-9-5	WA-103 phone
KSU-4-14	TB-8-1	KSU-4-14	Temp-5-5	PBX-3-1	TB-11-9	WA-122 phone

Legend
KSU—Existing Key System TB
PBX—New Phone System TB
Temp—Temporary Cabling TB
TB—Terminal Block
WA—Work Area

Another way of documenting each circuit is with a circuit layout record (CLR). CLRs provide a quick graphical representation of how the circuit is laid out. It shows all equipment, cables, cross-connects, and their termination locations. CLRs are extremely beneficial when troubleshooting.

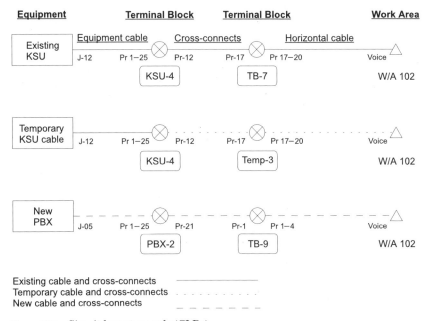

Figure 10.1 Circuit layout records (CLRs).

The first CLR shows:

- The existing key service unit (KSU) has a 25-pair cable plugged into its J-12 jack and the other end is terminated on block KSU 4, pairs 1–25.
- The voice cable for work area 102 terminates on terminal block (TB) 7, pairs 17–20.
- A cross-connect is terminated between block KSU 4, pair 12 and terminal block 7, pair 17.

The second CLR shows:

- The existing key service unit has a 25-pair cable plugged into its J-12 jack and the other end is terminated on block KSU 4, pairs 1–25.
- A temporary voice cable for work area 102 terminates on terminal block temporary 3, pair 17.
- A cross-connect is terminated between block KSU 4, pair 12 and terminal block 3, pair 17.

The third CLR shows:

- The new PBX has a 25-pair cable plugged into its J-05 jack and the other end is terminated on block PBX 2, pairs 1–25.
- The new voice cable for work area 102 terminates on terminal block 9, pairs 1–4.
- A cross-connect is terminated between block PBX 2, pair 21 and terminal block 9, pair 1.

Depending upon the size and scope of the retrofit, either a flash cut or a phased cutover may be necessary.

A flash cut transfers the old system to the newly installed system in one continuous process until 100 percent complete. The flash cut may be a hot cut where the equipment cables are unplugged from the existing system and are plugged into the new system. The customer is totally without communications during a hot-cut process. Another type of flash cut is the rolling cut. During a rolling cut, cross-connects are relocated one pair after another until complete.

In a phased cutover, portions of the old system are transferred to the new system in groups. As offices or buildings are rewired they are phased into the new system. Both the new and old systems will be working in parallel until the transfer is complete.

The removal of abandoned cable is an essential part of most retrofits. Abandoned cable is a fire hazard. If a fire were to occur, the old cable could become fuel, while providing a path to spread the fire throughout the building. They also take up valuable pathway space and place an unnecessary strain on the supporting hardware. Cable the customer does not plan to use should be removed and disposed of properly.

Remove cables carefully so as not to damage or disturb other cables. Ensure written permission has been obtained from the owner to remove the cable.

Regardless of the particular situations, every retrofit requires:

- Working with the customer to determine new system specifications.
- Obtaining, reviewing, and verifying existing documentation.
- Determining safety concerns and requirements for the project.
- Performing a site survey.
- Developing an installation or job plan.
- Installing in compliance to codes and standards.
- Developing cutover sheets.
- Performing cutover.

- Testing the installation.
- Performing final documentation.
- Removing abandoned cable and equipment.
- Firestopping.

Planning for a retrofit installation

When a retrofit installation is requested, the customer has already developed an idea of what they want. In most cases, you will have copies of the bid, request for proposal, formal contract, purchase order, or some other official document that conveys the customer's intentions. Most of these documents contain the drawings and specifications that determine which materials are to be used, when they are to be installed, and under what standards they will be installed. The development of a set of installation drawings, a list of materials, installation scope of work, and an overall project schedule will further enhance these documents. The larger the project, the more complex and important these documents become to a successful installation.

The cabling installer should obtain all the available documentation on the existing installation from the customer, including the cross-connect records and any information on the proposed system. Many times existing documentation only contains information from the initial installation and does not reflect the last 10 years of moves, adds, and changes (MACs). The cabling installer should plan on spending numerous man-hours or even days verifying the existing infrastructure.

A Registered Communications Distribution Designer (RCDD®) may have been employed by the customer, the architect, or the installation company to develop the required telecommunications drawings and specifications for the new retrofit installation. If not, a member of the installation team may have to develop them. Many organizations request that the design be reviewed and approved by an RCDD.

Separate drawings should be prepared for each type of telecommunications system to be installed. That is, separate drawings are prepared for copper cables, optical fiber cables, coaxial cables, and low-voltage cables. If the project is small enough, all of this information may be contained on a single drawing. It is still recommended that a single systems drawing be prepared.

Larger retrofit projects may require a series of drawings and complex project schedules. The first set of drawings include the existing closets and infrastructure. The second set might include temporary facilities needed to free up space for the final installation, and a third set of drawings could show the final configuration.

Elevation details of the closets and the arrangement of the various types of equipment should be part of these drawings. A detail of each wall and rack in each closet, as well as a plan view of the floor-mounted hardware, should be included. The drawings must also indicate the cable supporting structure.

Members of the selected installation team must conduct a site survey to identify all aspects of the installation and how each must be addressed by the team. The results of the site survey will be used to develop the project schedule and the installation job plan.

Developing an installation job plan must involve all aspects of the retrofit. An understanding of installation standards, national and local codes, BICSI manuals, manufacturer's specifications, and basic telephony and data communications principles will contribute to the success of the installation plan. The complexity of the plan is directly proportional to the size of the installation and number of systems being integrated.

In large retrofits that require extensive work area recabling, it is advantageous to have temporary office spaces where personnel can relocate until their section of the retrofit is complete. The personnel are then moved back into their recabled office spaces and another group relocated into the temporary office spaces. This practice is referred to as "swing-floor phasing," because groups of personnel are swung from one floor area to another as they are phased into service. This form of swing-floor phasing requires each group of users to relocate twice before they are in their final offices.

Another approach to swing-floor phasing is to work with the customer to develop a master plan of swapping office areas (i.e., the accounting department moves into temporary trailers, and the old accounting spaces are rewired). Then, the purchasing department relocates into the old accounting spaces and the vacant purchasing department spaces are rewired. Telemarketing then relocates into the rewired purchasing spaces, and so on, until all office spaces are rewired and the accounting department can move back into the last rewired space. This form of swing-floor phasing requires the accounting department to move twice while all other departments only moved once. This saves the customer time and money.

Additional factors to consider are whether the customer wants to utilize the present cable and equipment, and if so, whether the existing components meet the new requirements. In many cases, the upgrade of systems requires a new cabling configuration that does not allow the use of the current infrastructure. An example of this would be upgrading an existing 10BASE-2 coaxial system to an unshielded twisted-pair (UTP) Category 5 cabling system. This change in topology necessitates removal of the coaxial, daisy-chained network and the installation of a star-wired UTP cabling network. However, the customer may require that the coaxial system remain operational until the UTP system is completely installed, tested, and ready to be cutover.

In the case of voice services, the client may be upgrading from a key service unit (KSU) or an electronic key system (EKS) to a private branch exchange (PBX). In this event, each KSU telephone set would be using non-rated, 25-pair cable and the EKS could be using non-rated, 2-pair jacketed quad cable.

Quad cable is POTS cable originally used for voice service and consists of four conductors colored green, red, black, and yellow. Jacketed quad cable is usually not twisted.

Neither of these cables would meet the requirements of Category 3 cable and would have to be replaced to meet the standards. Again, this could require the present cable plant to remain in place until the cutover to the new voice system is complete. It may also require additional space, as both telephone systems will be operational until the new system is completed.

Prior to installing a new cable plant in a retrofit installation, it is necessary to determine whether the existing cabling system or any portion of it can be reused. The cabling installer should obtain as much information as possible from the customer and verify that the information obtained is accurate. If the existing documentation is questionable or does not exist, the installer may have to document the current system. In addition, the existing cable plant must be tested to verify the level of service it can support.

After the infrastructure has been verified and tested, the cabling installer can determine which pathways, cables, and equipment can be reused. The challenge is then to incorporate the new components into the ones that are to be reused.

If the telecommunications closet does not have adequate room for the old and new components, there are several options available to the cabling installer:

- Build additional closets.
- Expand the existing closet.
- Float the existing equipment and cables. In a small closet, the existing equipment and cables may need to be extended out into the hallway to make room for the new system. The new plywood and system components can be installed, tested, and prepared for cutover. The old system components can be removed after the cutover is complete.
- Install temporary cabling. Temporary cabling is exactly as its name implies. It provides service until the cutover is complete and then is removed.

Large office retrofits may require extensive work area recabling and closet renovations. Temporary cabling can be either backbone or horizontal cables. It may be easier to temporarily relocate a small telephone system from the closet into an office and run all the horizontal cables around the perimeter of the floor until the new system is installed. This allows the equipment and termination hardware to be installed in the closet while the new horizontal cables are installed to the workstations.

Scope of work

A scope of work for the project provides guidelines for the installation team. The scope of work is the document that lists all elements of the installation. It can be generated by the customer, the designer, or the installation company when bidding or responding to a customer's request for proposal (RFP). The scope of work should indicate:

- What work is to be performed.
- What materials are to be installed.
- What methodology is to be used.
- How the completed installation is to be tested.
- The identification, label, and documentation system to be used on the project.
- The test methods to be used.
- When and how the installation is to be turned over to the customer.
- Which cables and equipment are to be uninstalled and how they will be disposed of.
- Clarifications or understandings that elaborate on the various items involved in the installation.

Contract

The contract is a written document that states the entire understanding between the customer and the contractor. Some customers do not use a contract but simply generate a purchase order that refers to the other documents associated with the project. If a contract is available, the cabling installer should ensure that all documents listed therein are available. Contracts may also list penalties associated with not completing work or delays in completion. The cabling installer should pay particular attention to these liquidated damages, especially when performance bonds are required as part of the contract.

Project schedule

Companies use a variety of different project management styles and software. Various charts and graphs are used that allow the telecommunications installation team to track materials receipt and disbursements, labor items completed, and the overall status of the project on a day-to-day basis. Manually generated project schedules can be used, when the project is small and not complex. Examples of two project schedule documents are found as Attachments D and E to Chapter 2, "Planning."

Project log

The person in charge at the project site should maintain a daily project log. This log should reflect any work performed, whether the work is complete, and the plans for the following day. Sometimes, copies of this log are required to be turned over to the customer or general contractor on a periodic basis. Accuracy is critical to the credibility of this log. This log should be used to record all activities associated with the project, especially such occurrences that may result in changes in financial impact to the company. If an incident occurs with another contractor, it should be written in the log, initialed by the person in charge, as well as a representative of the other contractor. If the other contractor declines to initial the entry, indicate their refusal, as well as the date and time of the entry.

Site survey

While site surveys were discussed in great detail in Chapter 2, "Planning," this section will give an overview of the common practices and detail the steps taken when dealing with procedures unique to retrofit installations.

When all of the initial project documents are obtained, a site survey is performed. A retrofit installation's site survey will usually be far more involved than one for new construction. When dealing with new construction, the site survey is performed to verify that the proposed installation is sound and will go in as planned. In addition to this, site surveys for retrofit installations must verify that the existing documentation accurately depicts installed infrastructure. If the customer is not confident that their documentation reflects the actual installation, the cabling installer may have to trace and document the entire infrastructure. Also, the cabling installer may be required to test the existing system components to determine if they meet the requirements of the new equipment as well as current standards.

The process of verifying the existing documentation may be tedious but must be done thoroughly to establish the necessary framework to start the job plan. Once the existing infrastructure has been identified, labeled, documented, and tested, an informed decision can be made on what portions of the old infrastructure may be reused.

In most cases, the existing infrastructure in use must not be disturbed during the survey. This may mean that it is necessary to document some parts of the existing infrastructure after hours.

Retrofit installations usually require a team of cabling installers to visit the place of installation. While there the team should observe all of the various locations where installation work will be performed. When making this site survey they should carry the customer's and telecommunications designer's documents. This allows the team to identify specific locations related to the project and work to be performed. The documents may also indicate hidden obstacles not visible from floor level.

The cabling installer should use a checklist for each project during the site survey to ensure that all items of concern are addressed. When problems are found; plans can be formulated to overcome them while still on site, rather than having to return to the site. Examples of a checklist are shown in Chapter 2, "Planning."

All information gathered during the site survey should be placed into the project file. This information will become invaluable later, especially if new team members are assigned after installation begins.

The first stop at a job site should be the general contractor's site office. While there, the cabling installer can:

- Explain the work to be performed for the customer.
- Identify the other contractors working for the general contractor and the impact their work will have on the installation process.
- Obtain a copy of the general contractor's construction progress schedule.

   Note.   This document can be used to determine how the installation schedule can be coordinated with the contractors working on the project.

It is important for the cabling installer to review with the general contractor (customer on a small project) the responsibilities of the installation team. This allows the general contractor to fully understand the role the telecommunications company will play in the completion of the overall project.

The cabling installer should determine who is responsible for the construction of the pathways and spaces. The customer's documents should state whether they are being provided by the general contractor or by your company. Most of the time the pathways and spaces of a new building are the responsibility of the general contractor or their electrical subcontractor. Smaller retrofit installations often require the telecommunications contractor to assume responsibility for the pathways.

If the telecommunications contractor is responsible for the pathways, the cabling installer must determine how to install these pathways and what obstacles must be overcome to install them. More detailed information can be found in Chapter 3, "Installing Supporting Structures."

On larger jobs, it will be necessary to coordinate the installation team's schedule with other trades. The installation may be dependent on the efforts of the electrical contractor to install necessary pathways prior to cable pulling or the carpenter building the telecommunications closets. These are critical path items that can seriously affect the installation schedule.

The cabling installer should ensure that tools required to perform any work required for a thorough site survey are brought to the site. Some suggested tools might be personal protective equipment such as hardhat, safety glasses, dust mask, leather gloves, leather boots, and hearing protection. Additional items that may prove useful are a ladder, flashlight, measuring wheel, handheld tape recorder, digital camera, or video camera.

Another concern is firestopping the installation. In a retrofit installation, the installation team may be required to bring the entire installation, including the existing cable plant, up to code. The cabling installer should document the status of the existing firestopping and check with the local authorities and the customer to clarify responsibilities concerning the existing noncompliant firestopping.

During the site survey, the cabling installer should look for floors with a "Swiss-cheese" effect. This happens in installations using poorly designed poke-through distribution. Every time a user moves a desk, the cabling installer core drills another hole. Too many holes close together may result in a floor or entire building being condemned. It may be the cabling installer's first time in the building and after drilling one hole, the floor collapses. This could result in the installation company being held responsible for the damages.

Determine whether the floors are pre- or poststressed concrete. Both types are poured at the factory; however, each is constructed in a different manner and poses a great danger to the uninformed cabling installer. The danger is when a cabling installer core drills through one of these cables. The floor loses its midspan support, while the walls lose their stability. This could cause the floor, and possibly the building, to collapse.

530 Chapter 10

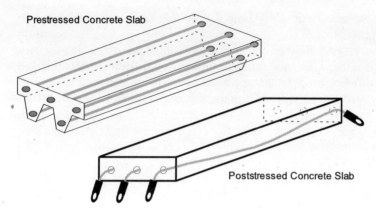

Figure 10.2 Pre- and poststressed concrete.

Prestressed concrete is made in a large form to produce a flat slab that will be used as part of a floor. It gets its name from the way it is constructed. Before the concrete is poured into the form, several large metal cables are placed in the form. The cables run lengthwise in the form and are pulled with tons of force from each end. The concrete is then poured into the form and allowed to cure. Once cured, the tension is released from the cables. The force from the cables now pulls both ends of the slab towards the center. This causes the slab to bow upward in the middle. When installed, the bow is on top. After office equipment and furniture are placed on top of the floor it levels out. This design allows the architect to use fewer columns in an open office design.

Poststressed concrete is the exact opposite of prestressed. The metal cables are placed in the form with their ends sticking out. Each cable end has a factory-crimped eye on it. The cables are free of any tension when the concrete is poured into the form. The concrete is cured and the slabs are sent to the construction site. When the slab is installed the cable eyes are attached to the vertical supporting structure of the building. The turnbuckles are tightened so the cables running through the slab are now holding the floor up and the walls in. The building is designed so the floors actually hold the walls in with less cross bracing.

Before drilling into a concrete floor, consult the building owner, engineer, or architect to determine how it is constructed. The floor may also be x-rayed to determine the location of rebar, conduits, and steel supporting strands. A bonded and insured coring firm that does this type of work on a daily basis may be hired.

When a cabling installer is working with an existing system, it is necessary to verify that the original installation meets current codes and standards. Check the cable pathways and distances, firestopping, and grounding systems. These are areas which frequently do not meet today's standards.

Retrofit projects require the identification of all existing pathways and spaces being used for telecommunications, their size, capacities, usability, congestion, and compliance with code.

When working with a retrofit project, some important considerations include:

- What is the size of the existing closets?
- How much space is available in them?
- Will new hardware fit within the confines of their spaces?
- Will additional equipment racks be required and will they fit?
- Is it necessary to expand any of the existing telecommunications closet?
- Are new telecommunications closets required?
- Are any existing pathways vacant?
- Does any pathway have usable space?
- Will pathway space be available after the abandoned cable is removed?
- Will the existing pathways support additional weight?
- Are new pathways required?
- Will any temporary cabling be required?
- Can existing facilities be utilized to assist installation of the new cable?
- Is the telecommunications grounding infrastructure installed?
- Does the infrastructure comply with ANSI/TIA/EIA-607 and the National Electrical Code?
- What, if any, are the requirements for maintaining dual service?
- Can any existing cable, hardware, or equipment be removed?
- Are the floors composed of prestressed or post-stressed concrete?
- Who is responsible for firestopping any existing noncompliant penetrations?
- Do any safety hazards exist?

The answers to these and other questions will determine how the cabling installer plans to implement the project.

Most of the answers to these questions should be available from the materials contained in the customer's and designer's documents. Do not leave anything to chance. Review all of the project's requirements before concluding additional pathways and spaces are not required. If additional pathways and spaces are required and have not been included in the original plans and specifications, job change orders may be required. At the very least, these changes may alter the cabling installer's approach to the installation methodology.

Verify existing infrastructure

The first step in verifying the existing infrastructure is to review the existing documentation with the customer. Copies of the blueprints should be made and one copy used for note taking. Items that should be noted on the blueprints include the locations of all:

- Active circuits.
- Low-voltage devices that will be involved in the retrofit, such as:
  - Computer terminals and their functions.
  - Telephones and facsimile machines and their phone numbers.
  - Credit-card terminals.
  - Modems and their phone numbers.
  - Cash registers.
  - Bar-code readers.
  - Security touch pads and intrusion detectors.
  - Fire alarms.
  - Pay phones.

Even though the cabling installer will probably not be dealing with all these devices, it is necessary to identify their cables to avoid disturbing them during the installation phase and the removal of abandoned cables.

Because an interoffice phone directory is usually kept more up to date than the cable records, obtain and use the interoffice phone directory as a reference to cross-check discrepancies in the cable plant documentation. The directory usually lists all the employee telephone numbers and may include the incoming central office numbers.

The next step is to perform a walk-through of all the office spaces that will be affected by the project to locate active devices. An active device is any equipment that is operational and plugged into the wall. When a computer is plugged into the wall it is a good indicator that it is active but it is not guaranteed. When users move to a new desk, the patch cord in the closet gets moved, but their old computer and telephone remain plugged into the wall even though they may not be operational. If a computer is not turned on, have someone test to see if the computer can log onto the network while testing for dial tone on the telephone. It is the

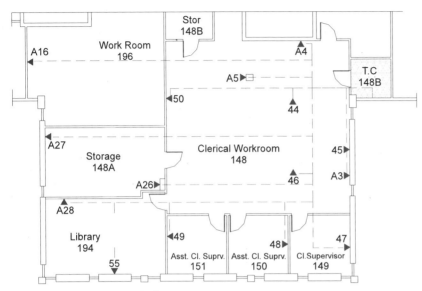

Figure 10.3 Typical "as builts."

cabling installer's job to gather all the clues and put the puzzle together. The installer should verify that each piece of equipment has been identified by the customer and is noted on the blueprints. Verify its workstation identifier and telephone number, if applicable.

The existing cross-connect records can be used to verify that each documented cross-connect is actually terminated where it should be. The fastest way to do this is with three cabling installers. One person reads aloud to the other two installers. One installer finds the first end of the cable and gently tugs at it with his/her fingers. If there is any slack in the wires, the wires will move. The third installer can trace the wires to their far point of termination. The person reading aloud will also note any corrections to the documentation.

Remove all cross-connects that are not in service. Performing a cutover when the cross-connects are filled with abandoned wires and patch cords can be extremely difficult. If a cross-connect is not being used, document its termination points and note that it was removed. If a later discovery indicates that it was in use, use the notes to reinstall it.

Locating copper cables. Another way to verify cable and cross-connect termination points is to tone them out. The tone generator induces a steady or warbled tone on a pair of copper conductors within the cable being traced. An inductive amplifier (probe or wand) is used to detect and amplify the tone at the far end of the cable.

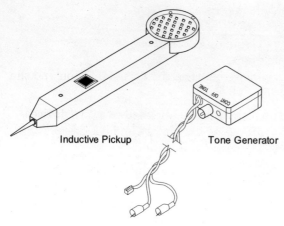

Figure 10.4 Signal generator and inductive amplifier.

When toning out cable and cross-connect termination points, disconnect the equipment at the workstation and inject the tone on a spare pair in the cable. If tone is placed on the cables as they leave the closet, every outlet on a floor would have to be checked until the correct one is found. By sending tone from the work area, an educated guess as to which closet the cable is in can save a great amount of time and walking.

Do not use the active pairs to transmit tone, because:

- Ringing current from the telephone switch may damage a toner.
- The tone will disrupt and possibly damage a digital phone circuit.
- The tone may damage data network equipment.
- The tone will disrupt network traffic on a computer circuit.

For additional information on testing, refer to Chapter 8, "Testing Cable."

Steps—Verify existing infrastructure

Step	Verify existing infrastructure and documentation
1	Request information from the customer. ■ Ask which equipment is active. ■ Ask which circuits are in use. ■ Ask for the cable plant documentation. ■ Ask for the location of telecommunications rooms, closets, and any other cable distribution points.

2   Check the existing documentation for accuracy.
    - Compare the existing documentation and any subsequent moves, adds, and changes (MACs) documentation to the installed cabling plant.
    - Verify all discrepancies between the documentation and the actual installation.

3   Visually determine if circuits are attached to active devices.
    - Perform a walk-through of the facility and document active devices connected to the installed infrastructure.

4   Check for labels and tags throughout the cable plant.
    - Inspect the cable pathways, cables, termination points, and any equipment for labels, tags, or other indications of the equipment attached to each circuit.
    - Label items as they are identified.

5   Inspect for terminations or jumpers.
    - Trace cables to their terminations and document findings.
    - Trace jumpers to determine their termination points and document findings.

6   Use appropriate test set for copper cable identification and tracing. Use the volt-ohmmeter (VOM) or tone generator and inductive amplifier on a pair of conductors not in use. The signal generated by the toner may interfere with other signals in the cable. The VOM test will not interfere with adjacent conductor signals.

    Tone generator
    - Attach tone generator to the cable pair at work area.
    - Use the inductive amplifier to detect the signal.
    - Trace and identify the cable from the detected signal.

    VOM
    - Check cable at the work area for stray voltages.
    - Set the VOM for resistance or continuity.
    - Attach VOM to the cable pair at work area.
    - Short the distant end of the pair under test.
    - Check for continuity.

    Office Locator Kit
    - Plug locator modules into workstation outlets.
    - Connect tester to suspected cables in the telecommunications closet.
    - Document the cable information when a locator module is detected.

Certification Test Set (if equipped with a tone generator)
- Attach tester with tone generator to the cable at work area.
- Use inductive amplifier to detect signal.
- Trace and identify cable from the signal detected.
- Connect the remote tester and perform auto test to verify cable performance.

7. Use appropriate testers for optical fiber identification and tracing.

   Warning. Never look into the end of an optical fiber. Always use an optical power meter to ensure that the cable is not energized. Looking into an energized fiber can cause serious damage to the eyes.

   Fiber flashlight
   - Ensure fiber is not energized.
   - Connect flashlight.
   - Look for colored distant fiber end.

   Power meter and light source
   - Inject light (at proper wavelength).
   - Measure and record results at distant end.
   - Test in opposite direction: record results.

   Fiber strand identifier
   - Clamp light detector around fiber.
   - Check for the presence of light.

   Visual fault finder
   - Connect the light source to fiber.
   - Look for the visible light at the remote end. It is not necessary to look directly into the fiber to see the light.

   Infrared conversion card
   - Place the optical fiber connect about 6 mm (0.25 in) away from the card and point the ferrule towards the card's surface.
   - Look for a luminescence across the phosphorus surface of the card.

Conversion Card

Figure 10.5 Conversion card.

8   Hand trace cables.
    - From any active equipment, trace the cable to its termination to determine the path of the active circuit.
9   Verify the backbone cable plant. Document the:
    - Path and termination points.
    - Number of conductors and fibers at each point of termination.
    - Splice locations and configurations.
    - Number of conductors and fibers in use (vacant or tagged bad).
10 Label both ends of the cables and all termination devices by:
    - Determining which circuits are active.
    - Labeling both ends of all circuits that are identified.
11 Test the remaining cables that are being considered for reuse but have not been tested.
12 Check the existing firestopping installations and document any that do not comply with applicable codes.

Evaluating the existing infrastructure

The site survey provided the opportunity to investigate the building's physical layout, verification of the installed infrastructure, and testing of the cable plant against today's standards. Now, the cabling installer must determine what portion, if any, can be re-used with the new system.

Gathered earlier in the site evaluation, the following information should be available to assist the cabling installer in evaluating the existing system's potential for reuse:

- List of all active devices:
    - Telephones
    - Computers
    - Fax machines, etc.
- Existing closets, pathways, and floor plans
- Installed equipment and functional block diagrams
- Verified existing cabling records and cross-connect sheets
- Initial system test results
- List of closets, pathways, cables, and equipment to be installed or modified
- Backbone cable routing and splicing sequences (if any), and the cable pairs to be extended to all telecommunications closets (TCs)
- The type of connecting hardware by size, quantity, and configuration (shown on the drawings)

- List of service providers with the provider's contact number, circuit identifiers, and circuit descriptions
- List of expected changes to the entrance facility

If the telecommunications closet does not have adequate room for the old and new components, there are several options available to the cabling installer:

- *Build additional closets.* This gives the needed extra space and shortens the length of the horizontal cables. Shorter cables use less material and are easier to install. Additional closets will also bring an older system that is out of specification within the standard's allowable cabling distances. Equipment and cables are easily installed in a new closet since there are no existing components to work around. However, new closets often require the customer to give up valuable office space.
- *Expand the existing closet.* This may be accomplished by knocking out one of the non-load-bearing walls and expanding into the customers office spaces. This can be more complicated than new closets, because equipment and termination devices on the wall may need to be removed, and the equipment must remain in service.

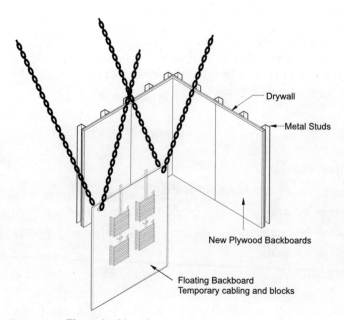

Figure 10.6 Float a backboard.

To remove the wall, the equipment and hardware must be "floated." To float a backboard, the cables and cross-connects are lengthened to allow the entire wall to be relocated a few feet. This practice gets its name because the plywood (or a large portion of the actual wall) is disconnected from the building structure and hung from the overhead supporting structure. The components are floating by their new temporary supports. The temporary relocation of an equipment rack can also be referred to as floating a rack.

Floating a backboard or rack allows the contractor to work around the existing system to rebuild the room and install and test the components of the new system prior to the cutover.

- *Float the existing equipment and cables.* In a small closet, the existing equipment and cables may need to be extended out into the hallway to make room for the new system. The new plywood and system components can be installed, tested, and prepared for cutover. The old system components can be removed after the cutover is complete.
- *Install temporary cabling.* Temporary cabling is used to provide service until the cutover is complete and then it is removed.

Temporary cabling is usually installed as quickly and cheaply as possible. It is often laid along the outside perimeter of the room or simply left hanging out of the ceilings. It is designed as a temporary solution and should never violate any fire or safety codes. Routing through plenum areas should be avoided due to the cost and complexity of the installation.

Large office retrofits may require extensive work area recabling and telecommunications closet renovations. Temporary cabling can be either backbone or horizontal cables. It may be easier to temporarily relocate a small telephone system from the closet into an office and run all the horizontal cables on the perimeter of the floor until the new system is installed. This allows the equipment and termination hardware to be installed in the closet while the new horizontal cables are installed out to the workstations.

Changes in the entrance facility will impact the space requirements within the closet. The cabling installer must be aware if the customer plans to add optical fiber cable or new T-1 circuits. These facilities require termination space and may need electronics that require valuable closet space. Contact the service providers and determine their requirements and clearly mark their designated areas within the closet.

Steps—Evaluating the existing infrastructure

Step	Evaluating the existing infrastructure
1	Determine planned use of systems. • Verify the customer's planned use of the data systems. Will it be Ethernet, token ring, FDDI, or a combination of these? • Is a new voice system or expansion to be installed or will the current system remain? • How many cable pairs does the voice system require? • List cables to be reused, if applicable. • List cables to be removed.
2	Backbone cabling • Verify whether a fiber backbone is required and, if so, the number of strands. • Will both multimode and singlemode fiber be required? • Determine whether there is a requirement for a copper backbone cable and, if so, what pair count and category.
3	Determine physical parameters. • Is riser-rated or plenum-rated cable required? • If the existing cable does not comply with the fire code, must it all be replaced even though it meets transmission performance specifications? • Confirm equipment rack configurations and space requirements. • Will firewalls be penetrated? If so, which firestop materials are acceptable to local authorities? • Will the existing penetrations need to be brought into compliance with the fire code? • Confirm the need for removal of existing cable from conduits and pathways prior to installation. • Confirm the status, if any, of asbestos within the facility. If present, has it been abated? If not, what planning is required? • Confirm that the system layout complies with ANSI/TIA/EIA-568-A distance limitations for both horizontal and backbone cables. • Verify that the grounding system is designed per ANSI/TIA/EIA-607.
4	Confirm findings with customer. • Meet with client to verify analysis. • Determine if any portions of existing system will be reused.

Job plan

Careful planning is required when performing a retrofit installation to ensure the customer is provided with the desired system upgrade while incurring a minimum of interruption to their business. A thorough review of the plan for implementation will eliminate any confusion and assist in making the transition to the new system as efficient as possible.

In many cases, the customer must continue to use existing equipment that will later be transferred to the new system. Therefore, the cutover must be scheduled to minimize downtime and disruption of customer operations. Normally, such activity is scheduled after business hours or on weekends; however, circumstances sometimes exist whereby the cutover takes place during normal office hours. Regardless of the time of cutover, the newly installed cable plant should be totally terminated, labeled, tested, and documented prior to the planned transition.

A good job plan is essential for the successful completion of the planned work. This plan should reflect all aspects of the work to be performed from a scheduling perspective. Job plans can be simple documents for a small project, such as a check list, to complex documents for a large project.

Every project plan should reflect the company's effort to ensure the work will be performed in a timely manner. It should include:

- Resources required for compliance with the schedules.
- Permit acquisition.
- Material staging.
- Coordination with other trades on the project.
- Scheduled meetings.
- Overall job schedule.
- Security and safety plan.
- Materials list.
- Tools list.
- Task list and description.
- Job requirements.
- Labor estimates.
- Acceptance plan.
- Inspection schedules.

- Special circumstances.
- Labeling system.

Routing for horizontal and backbone cables and their respective pathways should be identified. All labeling schemes should be discussed. Telecommunications closet layouts should be reviewed to ensure all items can be installed in their respective places both on the floor or on the wall spaces. Cables and equipment to be removed should be identified. Grounding and bonding methods should be discussed to ensure compliance with the scope of work, codes, and standards. Safety plans should be reviewed for applicability on the project. Appropriate personal protective equipment should be itemized.

Prior to beginning work on the project, the job plan should be reviewed by the installation team as part of the initial construction meeting. If changes are required, they should be made at this time.

Initial construction meeting

After completing the site survey, verify the existing infrastructure, and formulate the job plan. The telecommunications project manager should then hold an initial meeting with the entire installation team. At this meeting the project manager and team leader should detail the responsibilities of everyone involved. This ensures that everyone is informed regarding every aspect of the project. The job plan can be reviewed and updated as necessary. Questions can be asked and answered. Communications between all personnel involved in the project are critical to its success.

Minutes of this meeting should be taken and a printed copy provided to each person attending and any additional persons having a position of responsibility on the project.

Materials

Materials should be ordered for the project, allowing time for any back orders and shipping delays. As the materials arrive, they must be inventoried, inspected for damage, and checked to ensure that the contents match the labeling on the boxes.

The material list should contain the description of each item, the catalog number, and quantity ordered. It should also have a column for materials received and materials dispensed. These two columns are used by field personnel to manage inventory during the actual installation work.

Receiving materials is one of the most important tasks on a project. It can affect the profitability of the project just as much as the labor employed. Determining where the materials are to be stored, how they are to be dispensed, and how unused materials are to be disposed of is critical to the success of the project.

The materials need to be organized by job site and kept separated from other projects. Failure to do so may result in the delay of the project because another crew may use the materials. If the materials are to be delivered to the site, an on-site representative will be responsible for receiving, inventorying, and storing them in a secure location.

Materials can be stored at the job site, company location, or with the distributor but be concerned with their safety until the installation is completed.

Depending on the size of the project, all of these options should be considered. The best alternative for the particular project may be a combination of all options.

Only designated persons should be allowed to distribute materials on the job site. Control of access to job materials will determine who is allowed to distribute them. When materials are distributed, some record of accountability should be made to ensure the materials are not lost or misplaced. Excess materials should always be accounted for at the end of each work day and stored for use later or returned to the company storage area for use on another project.

Plan a storage area on site regardless of where the bulk of the materials and tools are being stored.

Development of a project schedule

Once all of the items associated with planning an installation have been identified, develop a project schedule. If the job is at a building under construction, the general contractor's construction schedule should be obtained and referenced. It includes all of the trades working on the project and indicates their specific time frames for accomplishing work on the project. Of particular concern are the schedules for completion of the supporting structure inside the building. Backbone cables or horizontal cables cannot be installed until the electrical contractor or the general contractor has completed installation of the pathways and spaces that will be used to house these cables.

With new building construction, identify the finishing schedules for other trades. For example, faceplates cannot be installed until the wall covering is completed. Racks may not be installed until the floor covering is installed.

The project schedule should begin with the award of the contract and is complete upon acceptance by the customer. The detail required is directly proportional to its importance in completing items that precede or follow it. The project schedule should indicate the planned time required for each item, and provide space for inserting the actual time needed to perform the job task.

Project scheduling software is now available to simplify this task and provide information relating to the status of a project. Most types of project scheduling software track the resources, materials, tools, and expenditures, as well as the actual schedule of progress of the work.

Upon completion of the project schedule, copies of the schedule should be provided to all concerned parties. This schedule should be updated daily, indicating the progress of the day's work and whether the installation is on schedule, ahead of schedule, or behind schedule. Any supporting documentation that will lend credibility to delays should also be referenced in the project schedule updates.

Preinstallation meeting

Once the project schedule is compiled, an internal meeting, called a preinstallation meeting, should be convened by the project manager, the telecommunications installation team, and appropriate contractors. All aspects relating to the project should be addressed, discussed, and adjustments made to the project schedule based on the results of the meeting. The project checklist should be reviewed and everyone should be completely aware of their individual responsibilities. To ensure the work is performed in a timely and professional manner, the project should be reviewed in detail so that each team member can work in concert.

Meetings

Periodic meetings should be held for every project. These meetings can occur as often as once a week or as necessary to ensure that everyone knows what is going on and what is expected of them. Work progress, as well as roadblocks and ways of overcoming them, should be discussed and agreed on by all concerned parties. Each team can contribute to the success of the project by thoughtful participation in these meetings.

On a project where the building is under construction, the general contractor and subcontractors should be requested to attend and participate in these meetings. The project manager should also attend the general contractor's construction meetings to ensure that all concerns are addressed. This will ensure proper coordination between the telecommunications cabling installation team and other contractors working on the project.

Project safety plan

Safety is the first item of importance on a project. The safety of workers, customer personnel, and the subcontractors is of paramount importance. Workplace accidents can disrupt the best-planned job and cause costly delays.

The installation company should already have an approved safety plan. Before beginning any work operation, the contents of that safety plan should be reviewed with each employee working on the project. Each employee should fully understand how the rules of safety should be implemented as each installation task is performed. Time should be taken to ensure that each employee is equipped with the proper safety equipment and has the knowledge to use them safely. As a contractor or subcontractor, the installation team may be contractually required to attend periodic safety meetings.

> **Note.** It is better to halt a work operation if questions of safety occur rather than risk an on-the-job accident.

Strategy

When unforeseen circumstances occur, plan how to overcome them and implement a backup strategy in case the initial plan meets with obstacles. Proper project planning is essential for the successful completion of a project.

System Cutover

Overview

With the system acceptance testing complete, the installation team is ready to put the new system into service (a process known as *system cutover*).

The cutover is handled differently for every installation. Some are as simple as installing a new system, turning it on, and using it. Other retrofit systems are installed next to old systems but are completely separate, so that both can operate at the same time until the final cutover.

Situations in which the new system can be installed while the old system is left in service are the safest way to handle the cutover. The new system is installed, tested, and then put into service. The users are asked to start using the new equipment. The old system remains in operation as a backup. Any users who experience problems can go back onto the old system until the new system can be corrected or the users can be properly trained in the use of the new system.

The old system can be disconnected and removed after the users are comfortable with the new system, and the system has had a chance to burn in. The term *burn in* refers to the electronic circuits getting warm when they are turned on. A new system should have a low percentage of its circuit boards fail after they have been initially turned on and allowed to heat up.

Each job may require different amounts of burn-in time. It can range from a week to six months. It may be phrased as "The new system shall be considered accepted after a 45-day burn-in and 31 days without a failure." This protects the customer from getting a new system that is losing a circuit board every other day.

> **Note.** There have been systems installed that have not been accepted for over a year due to continuous equipment failures.

Ask the customer the best time for cutover and how long the system can be out of service. Check for any circuits that must remain operational throughout the entire project.

Documentation and installation manuals for all the equipment being installed must be on site during the cutover. Also, a set of manuals for the equipment being removed should be on site. They can be useful if it is necessary to return to the old system.

There are several phone numbers that can be helpful if complications occur during a cutover. The numbers should be researched in advance so time is not wasted when the customer is without communications. Some of these phone numbers include:

- Internet access to download files from the manufacturer.
- Manufacturer's technical assistance.
- Local central office.
- Long distance providers.
- Special circuits providers, T-1s, optical fiber, WAN.

Make a list of every circuit number entering the facility so they may be given to service providers to provide assistance.

Most installers think of a cutover taking place on the weekend or in the middle of the night. These may not be the best times for the customer or the local service providers. The customer may download files from around the world between the hours of midnight and 4:00 a.m. each morning. Technical assistance from the service providers after hours may also be a problem. Most service providers use a reduced staff after normal working hours.

If a problem is experienced with the new system, it may be difficult to receive technical assistance from the manufacturer after hours. Contact the manufacturer ahead of time and ask about their after-hours assistance policy. If required, establishing an account with the vendor for technical assistance may be beneficial.

Note. Many manufacturers now charge for technical assistance. Most also require a technician's certification number prior to giving assistance.

Performing a cutover on the weekend requires a great deal of coordination. Notify the users of the pending cutover and provide initial training on how to use the new system.

When installing a PBX, the cabling installer should:

- Deactivate and remove the existing system.
- Install the new system.
- Cross-connect the backbone and horizontal cables to the new system.
- Transfer the entrance trunks and tie lines to the new system.
- Install, test, and label new user instruments.
- Perform acceptance testing of the whole system.

When installing a new data network, the cabling installer should:

- Install the new cabling system.
- Install and configure the new system electronics.
- Install and configure the new network interface card (NIC) and existing NIC card in each terminal.
- Cross-connect the backbone and horizontal cables to the new system.
- Test each terminal's ability to log onto the network.
- Install test and label peripheral devices (e.g., printers, modems).
- Transfer any entrance tie lines to the new system.
- Perform acceptance testing of the whole system.
- Deactivate and remove the existing system.

The new system should be ready for the users when they arrive. No matter how thoroughly the system is tested and put through its paces, be prepared for problems. When the users start loading the system with full-scale usage, trouble reports will be received.

A crew should be on site and a notice placed on every user's telephone, explaining how to use the system and listing one telephone number to report troubles. The users should be informed that the cabling installers are on site ready to fix troubles and provide training. This should encourage users to report troubles and ask questions right away.

Steps—Cutover

Step	Cutover
1	Method of implementation - Confirm whether work will be performed after or during normal business hours. - Identify cables to be removed prior to the completion of the project. - Remove all cables and cross-connects that are not supporting active circuits and are scheduled for removal. - Develop cutover sheets. - Install temporary cabling, if required. - Float backboards, racks, and equipment, as required. - Install new cabling. - Install and configure new equipment. - Confirm the cutover time with the customer. - Notify the users of the cutover and provide training on the new system. - Verify who is responsible for the interconnection of voice and data systems to the new cable plant. - Decide if the system will be cutover under a flash-cut scenario or phased in over a period of time in preplanned stages. - Install, label, and configure user's telephone instruments as required. - Install and configure the NIC cards at each terminal, as required. - Test each component within each system. - Provide on-site support both during and after cutover. - Firestop all penetrations not being used to remove abandoned cables.
2	Testing - Perform copper testing of all new cables for either link or channel tests, as determined by the customer to TIA/EIA TSB-67 and ANSI/TIA/EIA-568-A-1 specifications. - Perform fiber testing, using either a power meter or OTDR, as determined by the customer.
3	Documentation - Label all equipment, pathways, and cabling as appropriate according to ANSI/TIA/EIA-606. - Update blueprints for "as-built" documentation. - Provide updated cross-connect records. - Provide interconnection drawings. - Provide rack layout drawings.

- Compile all test results in both hard copy and electronic medium for the customer and the installation company's records.

Remove abandoned cable and equipment

Cables may be removed only after they have been checked to verify that they do not support any active circuits. Active cables must not be removed until the cutover is complete and the system has been accepted.

The removal of abandoned cable requires patience and an organized method of removal. It is very easy to cut the wrong cable or to damage the good cables.

Cables in conduits can be difficult, if not impossible, to remove without damage to the other cables in the conduit. Cables often twist around themselves as they are installed in conduits. If pulled too hard, the abandoned cable will damage the good cables and become cinched even tighter around the other cables. Cables may be freed by:

- Gently pulling as a helper pushes from the other end.
- Gently pulling all the cables back and forth a few inches while gently pulling the abandoned cable out.
- Pulling all the cables out together with a pull string attached. Discard the abandoned cables and pull the good cables back and reterminate.
- Using a lubricant.

Cables above a suspended ceiling can offer some problems. When pulling a cable at one end, an assistant at the other end might mistake a moving cable as the correct one. Just as in the conduits, cables get wrapped around each other in cable trays and J-hooks. As it is pulled, the cable may actually be moving three separate cables heading in three different directions.

Cables in open ceilings are often bundled by using tie wraps. They are also tie wrapped to cable ladders and other supports. These fasteners must be removed prior to pulling out the cables.

Remove any equipment or cross-connects from both ends of a cable prior to cutting the cable. When the cable is cut, all the conductors are momentarily shorted together. This can cause severe damage to electronic equipment.

Remove any cable-supporting hardware that will not be used. Abandoned hardware clutters the spaces and could possibly be reused on another job. Recycle hardware that is in good condition wherever possible.

Remove all trash and debris from above the ceiling tiles, attics, and under the crawl spaces. For safety reasons, piles of tie wraps, chunks of cable, cored holes, and the cabling installer's tools should be removed.

Dispose of trash in a responsible manner. Scrap cable can be recycled. On large projects, a salvage yard will leave a dumpster for usable scrap materials. At the end of the project, the salvage yard will pick up the scrap and pay the company for the materials.

Recycle the cardboard containers that come with the new systems, user instruments, and horizontal cable reels.

Steps—Postcutover removal

Step	Postcutover cable removal
1	Identify cables to be removed. ■ Confirm cable is not in use by testing with a wand and toner. ■ Check identification labels on both ends of the cable. ■ Never start by cutting in the middle and working towards the ends.
2	Remove cable support hardware. ■ Disconnect all fasteners from cable bundles to be removed (i.e., D-rings, grommets, etc.). ■ Dismantle all tie wraps from cables to be removed.
3	Remove cable terminations. ■ Disconnect all cables from work area outlets being removed. ■ Remove all terminations in wiring closet.
4	Remove cables.
5	Carefully retrieve cables for disposal.
6	Firestop ■ Firestop all new cable penetrations, all penetrations that cables were removed from, and any penetrations for which the installation company is responsible.
7	Remove all trash. ■ Remove all trash and debris from above the ceiling tiles.
8	Recycle appropriate materials.

Glossary

Numbers

10BASE-2 An IEEE specification for 802.3 network using a thin 50 Ω coaxial cable.

10BASE-5 An IEEE specification for 802.3 network using thick 50 Ω coaxial cable.

10BASE-T An IEEE specification for 802.3 network using UTP cable.

10BROAD-36 An IEEE specification for 802.3 network using 75 Ω cable.

802.3 An IEEE designation for a networking scheme which uses a collision detection access method. Also called Ethernet.

802.4 An IEEE designation for a networking scheme which uses a token-passing access method on a bus topology.

802.5 An IEEE specification for a networking scheme which uses a token-passing access method on a ring topology.

A

ablative Material that develops a hard char that resists erosion from fire and flame. Typical of firestops utilizing a silicone-based technology.

accelerator A chemical agent used to cause a chemical reaction for setting permanent bonds on epoxy glues.

acceptance angle The half angle of the cone wherein incident light is totally and internally reflected by the fiber core. The light within this cone is coupled into reflected modes of the fiber.

access floor A system of flooring that has removable and interchangeable floor panels. The floor panels are supported on adjustable pedestals and, sometimes, stringers to allow access to the area beneath.

active circuit A voice/data/video channel currently in use.

aerial cable Telecommunications cable installed on supporting structures such as poles, sides of buildings, and other structures.

aerial plant Wires and cables installed on poles with the assistance of guys, anchors, and pole attachment hardware.

air bottle Compressed air source used to propel a foam ball or other object through conduit for the purpose of attaching a pull string.

allthreaded rods A straight section of round rod stock which has threads installed over its entire length. Also known as threaded rod.

alternating current (ac) A current which changes polarity at a uniformly repetitious rate (e.g., 60 Hz).

alternating current equipment ground AC equipment ground conductor installed from the equipment grounding bus inside the electrical panel to a telecommunications grounding busbar or telecommunications main grounding busbar.

American Institute of Architects (AIA) An organization which provides resources, continuing education, and networking for architects.

American National Standards Institute (ANSI) The ANSI Federation is a private, nonprofit membership organization focused on meeting the standards and conformity assessment requirements of its diverse constituency. It provides a neutral forum for the development of consensus agreements on issues relevant to voluntary standardization. The U.S. representative to ISO, and through the U.S. National Committee, to IEC.

American Wire Gauge (AWG) An American copper wire sizing system.

ampere Unit of electric current; one ampere is equal to the current produced by one volt acting through a resistance of one ohm.

analog Transmission utilizing continuously varying electronic signals which directly follow the changes in loudness and frequency of the input signal.

anchor A device used to insert into a prepared hole in which a screw or bolt is inserted. It is so secure in the hole that it affixes itself to the penetrated item and usually does not come out without removal of the screw or bolt.

annular space The ring outside of the pipe or cable being protected and inside the hole or sleeve.

ANSI/TIA/EIA-568-A The telecommunications standard entitled *Commercial Building Telecommunications Cabling Standard*.

ANSI/TIA/EIA-568-A-1 The telecommunications standard entitled *Propagation Delay and Delay Skew Specifications for 100 Ω 4-pair Cable*.

ANSI/TIA/EIA-569-A The telecommunications standard entitled *Commercial Building Standard for Telecommunications Pathways and Spaces*.

ANSI/EIA/TIA-570 The telecommunications standard entitled *Residential and Light Commercial Telecommunications Wiring Standard*.

ANSI/TIA/EIA-606 The telecommunications standard entitled *Administration Standard for the Telecommunications Infrastructure of Commercial Buildings*.

ANSI/TIA/EIA-607 The telecommunications standard entitled *Commercial Building Grounding and Bonding Requirements for Telecommunications*.

aramid A liquid crystal polymer material with exceptional tensile strength and coefficient of thermal expansion near that of glass. Widely used as a strength member in optical fiber cables.

aramid yarn A strength element used in cable to provide support and additional protection of the cable bundles.

Architectural, Mechanical, Electrical, Structural Drawings or blueprints that include architectural, mechanical, electrical, and structural designs.

as built
1. Documentation that indicates cable routing, connections and blueprint attributes upon job completion that reflects changes from the planned to the finished state.
2. A drawing that details how something was built or how field conditions were found.

Asynchronous Transfer Mode (ATM) A high-speed transmission technology with packet-like switching and multiplexing.

attenuation The decrease in signal power between two points.

attenuation-to-crosstalk ratio (ACR) A ratio comparing the received signal with the near-end crosstalk of a cable pair.

aught Conductors classified larger than a 1 AWG and smaller than conductors classified in circular mill.

authority having jurisdiction The building official, electrical inspector, fire marshal or other individuals or entities responsible for interpretation and enforcement of local building and electrical codes.

autorange Automatic ranging by an optical time-domain reflectometer (OTDR).

autotest A feature which allows for sequential testing without operator intervention.

B

backboard A wood or metal panel used for mounting equipment.

backbone cable Cabling from the entrance facility to the equipment room, telecommunications closets, or between buildings.

backbone pathway Pathway from the entrance facility to the equipment room, telecommunications closet, or between buildings.

backscatter The scattering of light into a direction opposite to the original direction.

backscatter coefficient The ratio of backscattered light to transmitted light. The backscatter is a fixed percentage of the transmitted light.

balun A balanced-to-unbalanced circuit coupling device.

bandwidth A range of frequencies, usually the difference between the upper and lower limits of the range, expressed in hertz (Hz). It is used to describe the information-carrying capacity of a medium, such as an optical fiber. In copper and optical fibers, the bandwidth decreases with increasing length and is specified in MHz-km. It is not the same as data rate.

baseband signaling Transmission of an analog or digital signal at its original frequency.

basic link test configuration Horizontal cable of up to 90 m (295 ft) plus up to 2 m (6.5 ft) of test equipment cord from the main unit of the tester to the local connection and up to 2 m (6.5 ft) of test equipment cord from the remote connection to the remote unit of the tester. Maximum length is 94 m (308 ft).

baud A measure of signaling speed equal to the number of signal transitions per second, which may or may not be equal to the data rate in bits per second (b/s); the latter is the preferred specification.

Bayonet Neil-Concelman (BNC) The BNC connector is named for the designers of these coaxial connectors. It is a bayonet locking connector used with 10BASE-2 thin coax cable segments. These connectors, used throughout the cable length, attach to T-connectors, which in turn connect to network devices. Also, Bayonet Navel Connector.

beam clamp Device attached to a beam above the ceiling to hold cable supports, such as bridle rings.

bel A measure of analog signal strength named in honor of telephone pioneer Alexander Graham Bell.

bend radius Maximum radius which a cable can be bent to avoid physical or electrical damage or cause adverse transmission performance.

bending radius 1. For fibers, the bend radius that causes excess attenuation due to light leaking from the core.
2. The smallest permitted bend in a cable; determined by the construction. Influences the design of cable pathways and installation practices.

biconic A type of optical fiber connector.

BICSI® A telecommunications association, formerly known as Building Industry Consulting Service International.

bill of materials A list of quantity and specific types of materials to be utilized on a project. This list should also include a consideration for exempt materials (screws, bolts, etc.).

binary Indicates a state or condition, such as current flow or no current, on or off, a logical one or a logical zero.

binary digital system A system that determines the true rate of data transfer based upon baud and bit rates. The numbers will vary depending on such items as encoding schemes.

binder group A group of wire pairs found in a large cable. Groups can be distinguished from one another through the use of colored threads. Standard color coding provides for 25 pairs per binder group.

bit A binary digit, the smallest element of information in binary systems. A one (1) or zero (0) of binary data.

bit error rate (BER) The ratio of incorrectly transmitted bits to correctly transmitted bits. A primary specification for all transmission systems. Usually expressed as a power of 10.

bits per second (b/s) The measure of digital data rate, usually abbreviated b/s. Not equal to baud or hertz.

block (connecting) Device used to connect one group of wires to another.

blueprint A reproduction of an architectural plan and/or technical drawing that provides details of a construction project or an existing structure. These drawings are printed on special paper which allows the graphics and text to appear as blue on a white background.

bonding The permanent joining of metallic parts to form an electrically conductive path that will assure electrical continuity the capacity to conduct safely any current likely to be imposed, and the ability to limit differences in potentials between the joined parts.

bonding conductor for telecommunications Interconnects the building's service equipment (power) ground to the telecommunications grounding system.

branch splice A splice in which one cable is spliced to multiple smaller pair-count cables.

bridging clips Metal clips utilized to couple cable conductors on a 66-series block.

bridle ring A ring that is circular in shape but is open rather than closed. It has a pointed shaft at its apex that is threaded for installation into wood or prethreaded devices. It is available in various sizes.

broadband A general term for transmission of signals that have wide bandwidth (e.g., Broadband ISDN) or multiple modulated channels (e.g., 10BROAD-36).

broadband signaling This refers to multiple signals on a media at the same time. Each transmission channel has a bandwidth of 6 kHz and occupies a different frequency on the cable. Coax cable and multimode optical fiber cables are popular transmission media for use in this type of signaling.

buffer coating A protective thermoplastic material, which is applied to the acrylate to protect against environmental hazards. May be more than one layer.

buffer tube Loose-fitting cover over the optical fibers in loose-tube construction used for protection and isolation.

building entrance protector A device or devices used to terminate cables entering or leaving buildings. It provides housing for the voltage and current modules protecting the cable pairs from lightning and foreign voltage.

building grounding electrode system A network of grounded building components, i.e., metal underground water piping, metal building frame, concrete-encased electrode, a ground ring and rod and pipe electrodes.

bullwheel Large wheel used to maintain an arc when feeding large cables into a backbone pathway.

bundle 1. Many individual fibers contained within a single jacket or buffer tube. Also, a group of buffered fibers distinguished in some fashion from another group in the same cable core.
2. Also used to indicate time and common handling of multiple cable routed together.

bundled fiber 1. Many fibers contained within a single jacket or buffer tube.
2. A group of buffered fibers distinguished in some fashion from another group in the same manner.

buried cable A cable installed under the surface of the ground (not in conduit) in such a manner that it cannot be removed without disturbing the soil. See also: Trench; Underground Cable.

burn-in The time required for electronic circuits to get warm after they are turned on without being put in service.

bus topology A linear configuration where all networks devices are placed on a single length of cable.

butt set See telephone test set.

butt splice A splice in which cables enter the same endcap of the splice closure.

C

cabinet Cabinets are enclosed relay racks. They are normally equipped with two sides, a front door, and a rear door. They are available in various sizes with a wide variety of miscellaneous equipment mounted inside them (e.g., fans, power strips, etc.).

cable-end locator kit Set of numbered 8-pin modular plugs which can be identified by the cable tester (sometimes referred to as an office locator kit).

cable head The end of the cable attached to the pulling device.

cable labeling system Scheme adapted for labeling cables to identify them based on ANSI/TIA/EIA-606, *Administration Standard for the Telecommunications Infrastructure of Commercial Buildings*. The scheme employed when identifying cable or its associated hardware.

cable reel Spool which cable is wrapped around.

cable reel brake See reel brake.

cable run The route or path of a cable.

cable support system A combination of conduits, cable trays, support hooks, tie wraps, and any other hardware pieces used in a cabling installation to support cables. The cable support system keeps excess stress off cables and may provide some mechanical protection to the cables being supported.

cable termination 1. Item used for attaching the pairs of a cable to allow for connecting the cable to other cables or devices. Examples of cable termination hardware are: (1) patch panels, (2) connecting blocks, (3) patch blocks 66M-, 110- or BIX-type, and (4) modular jacks.
2. The connection of the wire or fiber to a device, such as equipment, panels or a wall outlet.

cable tray A metal tray with sides that allows cables to be placed inside of the sides over its entire length. It is usually supported by threaded rods suspended from the ceiling structure or from wall-mount brackets fastened to the wall structure.

cable tree Vertical rack with multiple arms for holding small reels of cable.

calibration Task of verifying test equipment against a reference to ensure proper operation.

campus The buildings and grounds of a complex, such as a university, college, industrial park, or military establishment.

CAN/CSA-T524 The Canadian Standards Association standard entitled *Residential Wiring for Telecommunications* (harmonized with ANSI/EIA/TIA-570).

CAN/CSA-T527 The Canadian Standards Association standard entitled *Bonding and Grounding for Telecommunications in Commercial Buildings* (harmonized with ANSI/TIA/EIA-607).

CAN/CSA-T528 The Canadian Standards Association standard entitled *Telecommunication Administration Standard for Commercial Buildings* (harmonized with ANSI/TIA/EIA-606).

CAN/CSA-T529-A The Canadian Standards Association standard entitled *Design Guidelines for Telecommunications Wiring Systems in Commercial Buildings* (harmonized with ANSI/TIA/EIA-568-A).

CAN/CSA-T530 The Canadian Standards Association standard entitled *Building Facilities, Design Guidelines for Telecommunications* (harmonized with ANSI/TIA/EIA-569-A).

capacitance The tendency of an electronic component to store electrical energy. Pairs of wire in a cable tend to act as a capacitor, which has two conductors or plates that are separated by a dielectric.

capstan A device for pulling cable.

carding brush A brush used for scuffing (abrading) the surface of a cable sheath.

category ANSI/TIA/EIA-568-A describes mechanical properties and transmission characteristics of unshielded twisted-pair (UTP) cables and screened twisted-pair (ScTP) and assigns a unique number classification (Category 3, Category 4, and Category 5).

Category 3 UTP 100-Ω unshielded twisted-pair copper cable which meets or exceeds specifications in ANSI/TIA/EIA-568-A, *Commercial Building Telecommunications Cabling Standard,* for transmissions up to 16 MHz.

Category 4 UTP 100-Ω unshielded twisted-pair copper cable which meets or exceeds specifications in ANSI/TIA/EIA-568-A, *Commercial Building Telecommunications Cabling Standard,* for transmissions up to 20 MHz.

Category 5 UTP 100-Ω unshielded twisted-pair copper cable which meets or exceeds specifications in ANSI/TIA/EIA-568-A, *Commercial Building Telecommunications Cabling Standard,* for transmissions up to 100 MHz.

cementitious Material that is mixed with water. Similar in appearance to lightweight concrete or mortar, it can be troweled to a smooth finish.

central office The location where the telecommunications service providers terminate customer circuits and locate switching equipment.

certification test set Cable test set designed specifically to measure the electrical properties of wire to determine whether the wire meets certification standards.

channel The end-to-end transmission path over one or more data links through which information may travel. A path between two telecommunications devices that includes the data patch cords at the device location and at the telecommunications closet. TIA/EIA TSB-67 defines a channel as up to 90 m (295 ft) of horizontal cable with connectors, plus up to 10 m (33 ft) of patch and equipment cords.

channel stock A metallic U-shaped bar with spaced holes utilized as a trapeze raceway or for support of raceway systems, such as conduits.

characteristic impedance The input impedance of a cable pair of infinite physical or electrical length.

chase nipple A conduit insert within an end connector, or a plastic ring that is threaded onto the sharp ends of a conduit (or fitting) that reduces cable sheath damage during pulling operations.

chromatic dispersion One of the effects which limits the transmission properties of optical fibers fibers by producing pulse spreading.

cladding The outer concentric glass layer that surrounds the fiber core and has a lower index of refraction. It provides total internal reflection and protects against scattering from contaminants at the core surface.

Class E standard An ISO/IEC proposed standard for transmission performance up to 200 MHz. This standard will harmonize with the proposed Category 6 cabling performance.

Class F standard An ISO/IEC proposed standard for transmission performance up to 600 MHz. This will harmonize with the proposed Category 7 cabling performance.

cleave The process of breaking an optical fiber by a controlled fracture of the glass to obtain a fiber end which is flat, smooth, and perpendicular to the fiber axis.

cleaver Device which square-cuts the end of glass fibers.

clove hitch A knot consisting of two half hitches made in opposite directions, forming a nonslip loop.

clustered star Also known as a bus star topology. It is like a tree topology except that there are clusters of devices at the end of each branch.

coating A material put on a fiber during the drawing process to protect it from the environment. See buffer coating.

coaxial cable A type of copper cable in which a central copper wire conductor is surrounded by an insulator, a metallic shield and, usually, an outer insulator.

coefficient of expansion This coefficient is used when determining the need for conduit/tubing expansion fittings as relating to exposure to extreme temperatures.

composite Made up of disparate or separate parts (i.e., copper and fiber cables).

conductance 1. A factor which represents the power loss in the conductor insulation.

2. The ability of an electrical circuit or component to pass (conduct) current.

conduit A round or circular raceway that comes in either 3.1 m (10 ft) or 6.1 m (20 ft) lengths (called sticks). It can be metallic or plastic.

conduit elbow A bend in a section of conduit usually at a specified radius and degree of turn.

conduit run Multiple sections of conduit.

conduit shoe A device placed in a conduit to assist in directing cable into a conduit during pulling operations. This device helps to prevent cable sheath damage.

conduit stub-out A short section of conduit that is installed from a receptacle box, usually in a wall, through a suspended ceiling space a short distance to an adjacent hallway.

conduit stub-up A short section of conduit that is installed from a receptacle box, usually in a wall, to a suspended ceiling space immediately above the receptacle box.

cone Safety marker that is used to designate a secure "off limits" area for nonworker.

connector A device used to terminate a cable. Connectors form reusable joints. Connectors may mate with other connectors in an adapter or mate with sources and detectors in a receptacle. See coupling.

connector insertion The attenuation associated with the physical attachment of two connectors.

consolidation point A location for interconnection between horizontal cables that extend from building pathways and horizontal cables that extend into work area pathways.

Construction Specifications Institute Creates and maintains a construction specifications book that is used by the American Institute of Architects (AIA).

continuity test Test that validates whether a material can conduct sound, current, light, or heat without significant interruption or degradation.

continuous noise A constant noise signal.

contractor An individual or company that has been employed to perform specific tasks or functions. In this document, contractors will be installing telecommunications cabling and hardware.

core The light-carrying part of an optical fiber. It has an index of refraction higher than that of the surrounding cladding.

corners Turns in a cable path.

coupled bonding conductor A bonding conductor placed (e.g., strapped) on the outside surface of a communications cable; used to reduce transient noise.

coupler Device for connecting two other devices, such as connectorized cables, together.

coupling A device that is used to connect two sections of conduit together.

crimp The act of clamping connectors to a cable.

crimp head A section of a splicing rig that fits over an assembled modular connector on the splicing head to provide the crimping of the module.

cross-connect A facility for interconnection or cross-connection between cable runs, subsystems, and equipment using patch cords or jumpers.

crossed pairs Error condition in twisted-pair wiring in which pairs are reversed.

crosstalk Unwanted coupling of signal from one or more circuits to other circuits.

cross threading When the thread pattern of a nut and bolt do not match, a stripping (cross threading) of the thread pattern can occur during installation.

curing A chemical process expected over time.

current Flow of electrons in a conductor measured in amperes.

customer An individual or company that has employed contractors to install their telecommunications system(s).

cutover The process of switching from old network components to new network components. This term is used when describing the switch of a piece of equipment, such as a computer terminal or telephone, from an existing channel to a newly installed channel. See flash cutover and hot cut.

cutsheet A listing of cable pair assignments used to specify desired circuit connections in a splice or cross-connect. This listing can also serve as the "as built" of a splice or cross-connect field.

cutsheets Cable documentation that shows the existing cable plant, the new cable plant, and the cross-connects that will be relocated during cutover.

D

D-subminiature connector See DB connector.

daisy chaining The practice of wiring devices in series.

dark fiber   Fiber which is not in use and has no light transmitted. Excess fiber installed in anticipation of system expansion. May or may not be terminated.

DB connector   Also known as a D-subminiature connector, there is a D-shaped metal skirt surrounding the connector's pins. This connector is widely used for connections between data equipment and is available in a variety of pin configurations.

dBm   Decibel referenced to one milliwatt, 0 dBm is equal to 1 milliwatt; 20 dBm is equal to 100 milliwatts.

dc loop resistance   Cable conductor resistance with the far end of the cabling shorted. This is the resistance for both conductors of a coax cable.

dead zone   A space on a fiber trace following a fresnel reflection in which no measurement can be made.

decibel   1. A standard logarithmic unit for the ratio of two values of power, calculated by 10 log (P_1/P_2).

2. The unit of measure for relative signal strength.

delay skew   The maximum difference in propagation delay between pairs in a cable.

demarc   Demarcation point; the point inside of a commercial building where the location service provider stops and the customer's cabling begins.

design specifications   Plans which identify with words and graphics a goal or set of goals. For example, a design specification might identify the installation of a system that is to transport information from one point to another using cables, connecting hardware, and associated electronics. It usually includes specific performance and design parameters that the customer desires.

designation strips   Colored labels placed on terminal blocks and used for identification (e.g., circuits).

detector   An optoelectronic transducer which converts optical power to electrical current. In optical fiber, usually a photodiode.

die   A steel block or plate with small conical holes through which wire is drawn.

dielectric   An electrically nonconductive material. Glass fibers are a dielectric. Being nonconductive, they are unaffected by many electromagnetic noise sources. Dielectric cables do not attract lightning.

digital   1. A nominally discontinuous signal which changes from one state to another in a limited number of discrete steps.

2. A data format that uses at least two distinct states to transmit information.

digital signal   A signal with a fixed number of discrete values. Commonly a binary signal with two values. The two values are used to transmit the two states (0, 1) used by digital computers. Most data transmission in optical fibers is by digital optical pulses.

direct connection   The act of connecting a cable to customer equipment without the use of a patch panel or terminal block.

direct current (dc) A current that does not vary (cycle), as does an alternating current (ac). A unidirectional current.

dispersion
1. In most materials, the optical propagation parameters depend on wavelength. This causes nonideal optical pulses to be broadened and have longer time durations. The three types of dispersion in optical fibers are modal, material, and waveguide.
2. The broadening of the input light pulses along the length of the fiber.

distribution frame
1. Wall- or floor-mounted frame with protectors or terminal blocks (or both) used to terminate cable pairs.
2. Structure with terminations for connecting the horizontal and backbone cabling in such a manner that interconnections and cross-connections may be made.

divestiture On January 1982, AT&T signed a Consent Decree with the U.S. Department of Justice, stipulating that on midnight December 30, 1983, AT&T would divest itself of its 22 telephone operating companies and reorganize into seven regional Bell operating companies (RBOCs).

dopant Placing of an impure material into another material.

download The transfer of a file from one network device to another. Often refers to the transfer of a file from a larger device such as a mainframe to a smaller device such as a personal computer.

downtime Term used to describe the amount of time required to transfer services from an existing cabling system to a newly installed infrastructure.

drain wire A conductor used to bond a cable shield to ground.

dressing Placing cables into a neat and symmetrical pattern for proper alignment and positioning for termination.

dressing block A plastic receptacle used to form and hold connectors in a connector for termination.

D-ring Wire management ring made of metal or plastic, shaped like the letter "D" for routing and supporting distribution cables or cross-connections on a backboard.

drop cable
1. A branch cable.
2. The cable allowing connection and access to and from the trunk cable of a network.

drywall An interior wall construction consisting of gypsum or plaster board.

dynamic range Determines length of fiber that can be measured.

E

Electronics Industries Alliance (EIA) A standards association which publishes telecommunications criteria.

elastomeric Made of one of several substances which resemble rubber (i.e., flexible).

Electrical Distribution Panel (EDP) Serves all of the electrical service contained within the walls of a closet. An ac equipment ground (ACEG) must be connected to this panel in each closet.

electromagnetic field tester This tester measures the presence of electromagnetic interference (EMI).

electromagnetic induction Current flow in telecommunications conductors produced by coupling of a magnetic field (i.e., by current in power lines, the cable shield, or other cable pairs).

electromagnetic interference (EMI) Any electrical or electromagnetic interference that causes undesirable signals in electronic equipment. Optical fibers neither emit nor receive EMI. A more general term than radio frequency interference (RFI).

electromagnetic pulse (EMP) A broadband, high-intensity, short-duration burst of electromagnetic energy.

endcap The endplate of a splice enclosure specifically arranged for the cable(s) entering and exiting the enclosure.

endplate See endcap.

endothermic Absorbing heat energy.

engineered judgment Decision by a firestopping professional for a slight variance in usage of an approved firestop system.

entrance facility An entrance to a building for both public and private network service cables (including antennae) including the entrance point at the building wall and continuing to the entrance room or space.

equipment grounding conductor The conductor used to connect the noncurrent-carrying metal parts of equipment raceways and other enclosures to the system grounded conductor, the grounding electrode conductor, or both, at the service equipment.

equipment room A centralized space for telecommunications equipment that serves the occupants of a building. Equipment housed therein is considered distinct from a telecommunications closet because of its nature or complexity.

event table The results of some OTDRs, for optical fibers, that list each event, its location, loss, and reflectance (if any).

exothermic A chemical change that is accomplished through the release of heat.

exposed When a circuit is in such a position that, in case of failure of supports or insulation, contact with another conductor may result.

F

F connector Used with coaxial cable. They do not have center pins as a result of having solid copper center conductors. They are usually crimp connectors and are widely known as CATV connectors, although they are also used for video transmission systems.

false ceiling A ceiling that creates an area or space between the ceiling material and the structure above the material.

false event When measuring long fibers with an OTDR, the far end of the trace may be very noisy, with individual data points fluctuating up and down. This may cause a false event to be listed.

fanned Separated cable conductors, strands, or pairs.

fan out
1. Used to describe the physical preparation of wire pairs exiting the jacketed cable to facilitate placement and termination in a splice or connecting block.
2. A device used to enable termination of optical fiber strands. See furcating harness.

far-end crosstalk (FEXT) Value of the signal transfer between pairs as measured at the cable end remote from the signal application point.

fastener A screw, bolt, or nail-like device that is used to secure an item to a wall, floor, or ceiling with an anchor.

FC A type of optical fiber connector identifiable by its round, screw-operated locking nut. It is usually metal. Its ruggedness leads it to be widely used in test equipment.

Federal Communications Commission (FCC) Regulatory body for the U.S. interstate telecommunications services as well as international service originating in the U.S.

ferrule A mechanical fixture, generally a rigid tube, used to protect and align the stripped end of a fiber.

fiber distributed data interface (FDDI) Operates at 100 Mb/s. Developed by the ANSI X3T9 committee.

fiber distribution unit An administrative housing used to terminate cables and connectors for the purpose of interconnections and cross-connection.

fiber optic cable A cable containing one or more optical fibers. The nonfiber components of the cable usually include the sheath, strength members, and buffer. The purpose of these components is to protect the fiber from mechanical and environmental damage.

fiber optic flashlight A small flashlight equipped with a special fitting which mates with a connector.

fiber optics Optical glass strands utilized for communications or signaling by means of light transmission.

fiber span Length of a fiber under test.

fiber strand identifier This clamp-on testing unit for optical fiber inserts a microbend into the cable and thereby is able to detect the light escaping from the fiber.

filled cable A cable with water-blocking material inside the sheath. The material helps prevent the penetration of moisture.

fire break A material, device, or assembly of parts installed in a cable system (but not at a cable penetration of a fire barrier) to prevent the spread of fire along a cable system.

fireproof A property of a material such as masonry, block, brick, concrete, and gypsum board that does not support combustion even under accelerated conditions. No material is entirely fireproof.

fire resistance The time in hours (or fractions of hours) wherein full-scale material designs or assemblies show an acceptable (fire-rated) resistance to fire. This resistance is compared to industry-recognized standards.

fire retardant Any substance added to delay the start of ignition or fire of any material.

fire shield A material, device, or assembly of parts used to prevent propagation of flames from one cable system to an adjacent cable system (i.e., between two parallel cable trays or between layers in vertically stacked trays).

firestop Materials, devices, or an assembly of parts installed in a penetration of a vertical or horizontal smoke- or fire-rated partition to restore the penetration to its original fire rating. This assembly will prevent the passage of fire or smoke from one side of the partition to the other.

firestopping The process of installing specialty materials into penetrations in fire-rated barriers to reestablish the integrity of the barrier.

fire wall A barrier that helps prevent the spread of fire or smoke from one area of a building to another. The barrier is usually constructed from the floor slab to the structural ceiling.

fire zone A contained area completely enclosed by fire-resistant rated walls, floors, and ceilings or floor/ceilings.

fishtape A temporary device that can be extended from the beginning of a pathway to the other end to assist in installing a pull line or to retrieve a cable.

flash cutover The process of cutting over from the old system to the new one. The cutover is done in one continuous process. Also known as a hot cut. Usually performed at night or on the weekend.

flashing Pieces of sheet metal or the like used to cover and protect certain joints and angles.

flexible conduit A type of conduit, usually made of flexible metal to allow it to be bent in different directions without distorting it. Flexible conduit is normally used to connect rigid pathways to other pathways where they may not join in exact alignment.

floor plan A scaled diagram of a building or other structure, shown as if seen from above. Floor plans are usually to scale and include all of the information associated with the type of architectural view it represents.

foldback splicing Process of folding back conductors in a splice for future maintenance or rearrangements.

footprint The area where a piece of equipment or furniture rests on the floor.

form and dress cable Line up cables side by side, shape into sweeping arcs, and join cable into bundles with tie wrap or other means to hold bundles together.

forming Placing cables into a neat and symmetrical pattern for proper alignment and positioning.

frequency The number of identical cycles per second of a periodic wave.

fresnel reflection Whenever light traveling in a material encounters a different density material, some of the light is reflected back to the light source while the rest continues.

furcate The process of covering a 250 μm coated fiber with a 900 μm buffer tube to facilitate field connectorization.

furcating harness An assembly used to increase the effective outer diameter of strands within an optical fiber cable to enable connector termination, typically provided in either 6 or 12 strand configurations. Sometimes referred to as a "fanout."

furcation tubing Flexible tubes used to increase the effective outer diameter of coated optical fiber strands (typically from 250 μm to 900 μm) to enable connector termination. Also provides physical protection of the coated strands.

fusion splice A permanent joint accomplished by applying localized heat sufficient to melt the ends of the optical fiber, forming a continuous single fiber.

G

GANTT chart A chart used to indicate a task associated with a job. It is generally used to check progress, or delay of a project.

gigabit Ethernet A standard developed by the IEEE 802.z task group operating at 1 Gb/s.

gigabits per second (Gb/s) One billion bits per second.

gigahertz (GHz) One billion Hertz.

gopher pole Telescoping pole for lifting and moving cable in open ceiling.

graded-index fiber An optical fiber where the core index of refraction decreases with distance from the axis. The purpose is to reduce modal dispersion and thereby increase fiber bandwidth.

ground A conducting connection, intentional or accidental, between a circuit (or equipment) and the earth (or to some conducting body that acts in place of the earth).

ground loop Interference in electrical communication links due to the ground at each end being at different potentials.

grounded Connected to earth or to some conducting body that serves in place of the earth.

grounded conductor A system or circuit conductor that is intentionally grounded.

grounding bushing A fitting for attaching a ground wire to a conduit.

grounding electrode A conducting item such as a metal water pipe, building steel, metal frame, bare copper conductor, rod, pipe, or plate in contact with the earth.

grounding electrode conductor The conductor used to connect the grounding electrode to either the equipment grounding conductor or the grounded conductor (or to both) of the circuit at the service equipment.

grounding system A system of hardware and wiring which provides an electrical path from a specified location to an earth ground point.

grunt sack A bag that is raised and lowered on a rope to provide a means of safely passing tools and small materials between individuals working in a construction environment.

H

half tap Splicing of a cable or individual pairs where the through cable/conductor is not cut.

hand trace Physically hand trace a cable from the workstation to its termination to determine the path of the cable.

hanger A device that is used to hold something in an elevated position.

heat shrink tubing Rubber tubing which shrinks upon the application of heat.

hermaphroditic connector A 4-contact connector that is neither male nor female and is designed for token ring applications.

hertz (Hz) A unit of frequency equal to one cycle per second.

hierarchial star An extension of the star topology utilizing a central hub. It is the required topology for structured cabling backbone systems in buildings and in campus environments.

high-pair-count cable Cables consisting of multipair conductors formed into binder groups of 25 pairs or less.

horizontal cable Cable which runs from the telecommunications closet to a device location. It may be installed in either horizontal or vertical plane.

horizontal cross-connect (HC) A cross-connect of horizontal cabling to other cabling and equipment.

hot cut Describes a cutover where the cables are unplugged from the old system and plugged into the new system. The customer is totally without communications during this process. See cutover and flash cutover.

hub A device which provides connections to and from multiple network devices. A hub enables communications between the multiple network devices, including servers and peripheral devices.

hybrid cable An assembly of two or more cables (of the same or different types or categories) bound to form a single unit.

I

impedance   A measurement of the opposition to the flow of alternating current (ac).

impulse noise   Discrete noise spikes which occur on a regular or irregular basis.

index-matching gel   A material used at optical fiber interconnections which has a refractive index close to that of the fiber core; used to reduce reflections from the residual air gap.

index of refraction   The ratio of the velocity of light in a vacuum to the velocity of light in a given material.

inductance   The opposition to change in current flow in an ac circuit.

inductive amplifier   Test device used to detect a signal placed on a cable for the purpose of tracing and identification. Sometimes referred to as a wand or probe.

infrared   The electromagnetic spectrum having wavelengths between 0.75–1 µm.

infrared conversion card   Allows a cabling installer to visually detect an infrared signal when that signal is directed at the card's phosphorus material.

infrastructure   Permanently installed cable plant.

in-line splice   A splice in which, for example, cable enters one endcap and, after splicing the cable, exits the other endcap of the closure.

innerduct   Conduit placed inside a larger diameter conduit.

input impedance   The ratio of the voltage at the sending end of the line to the current in the line at the sending end.

insertion loss   1. Signal loss when the transmission line is terminated in other than its characteristic impedance.
2. The loss of power that results from inserting a component, such as a connector or splice, into a previously continuous optical path.

insulation   The material that physically separates wires.

insulation displacement connection (IDC)   A wire connection device which penetrates the insulation of a copper wire when it is being inserted (puncheddown) into a metal contact, allowing the electrical connection to be made.

interbuilding backbone cable   Cable that runs between buildings in a campus environment. Outside plant cabling.

interbuilding (campus)   A backbone network providing communications between more than one building.

interconnection   A connection scheme that provides for the direct connection of a cable to another cable or to an equipment cable without a patch cord or jumper.

intermediate cross-connect The connection point between a backbone cable which extends from the main cross-connect (first-level backbone) and the backbone cable from the horizontal cross-connect (second-level backbone).

International Electrotechnical Commission Founded in 1906, the Commission sets international and electronics standards.

intrabuilding backbone cable Cable that runs between closets inside a commercial building. Can be vertical or horizontal in physical orientation but are backbone cables because they serve closets.

intumescent Enlarging or swelling (under the influence of heat, etc.).

ISO/IEC Defines international cabling standards.

ISO/IEC Standard 11801 International standard is titled *Information Technology—Generic Cabling for Customer Premises*. Accepts both 100 and 120 Ω UTP cabling, in addition to shielding options for the media.

J

jacket The outer layer of a cable. See sheath.

jackstand A device for holding a large cable reel off the floor so the cable can be removed from the reel.

J-hook A supporting device for horizontal cables that is shaped like a J. It is attached to some building structures. Horizontal cables are laid in the opening formed by the J to provide support for the cables.

job change order A written request from the customer, another contractor, a subcontractor, or other person or company on the project to add, delete, or change some work operation on the project. This document must be signed by the customer prior to the work described therein is started.

job plan A comprehensive outline of all aspects of the project. It includes all work operations and scheduling, how and when the work is to be performed, how each aspect of the work will affect the remaining areas, and how the work will fit into the general contractor's construction schedule.

job site The physical location where work is to be performed during the installation of a telecommunications system.

jumper An optical fiber cable with connectors installed on both ends. In the case of copper, no connectors are present.

K

key service unit An electromechanical telephone system in which telephones have multiple buttons permitting the user to manually select outgoing or incoming central office phone lines. It is not necessary to dial "9" for an outgoing line. The system uses 25-pair horizontal cable to every telephone.

key system An electronic telephone system in which telephones have multiple buttons permitting the user to manually select outgoing or incoming central office phone lines. It is not necessary to dial "9" for an outgoing line. The system can operate using UTP Category 3, 4, or 5 cabling.

keyed   A jack, outlet, or connector is considered keyed when it requires a specific orientation in order to prevent mismating.

L

labor list   A complete list of all major units of labor to be employed on the project.

ladder rack   A device similar to a cable tray but more closely resembles a single section of a ladder. It is constructed of metal with two sides affixed to horizontal cross members.

laser   (*L*ight *a*mplification by *s*timulated *e*mission of *r*adiation.) A light source producing coherent, highly directional, nearly monochromatic light through stimulated emission.

laser diode   Laser for optical fiber applications, usually made from a p–n junction in a semiconductor material. Laser diodes give higher performance at higher cost than light-emitting diodes. Laser diodes are commonly used with single-mode fiber.

lashing   Attachment of optical fiber cable to a supporting cable by wrapping thin steel or dielectric strands about them.

launch cable   Length of optical fiber cable used to condition the launch.

leg   The portion of the conduit elbow that is straight.

legend   A definition of symbology on a blueprint.

level   The power of a signal measured at a certain point in the circuit.

light-emitting diode (LED)   A semiconductor diode that spontaneously emits incoherent light from the p–n junction when forward current is applied. LEDs give moderate performance at lower prices than laser diodes. LEDs are commonly used with multimode fiber in data enterprise and industrial applications.

light source   A piece of test equipment used to create a light wavelength for testing optical fiber cable.

line width   The spread of wavelengths around the central wavelength of a laser source.

link   A transmission path between two points, not including terminal equipment, work area cables, and equipment cables.

Local Area Network (LAN)   A set of personal computers and peripheral devices, such as printers and CD-ROM drives connected together in a defined, limited geographic area.

logical topology   The actual method by which different nodes in a network communicates with one another as compared to the physical connections.

loop resistance   A measurement of the resistance of both conductors in a pair of conductors connected in series.

loose tube A type of optical fiber cable construction where one or more fibers are laid loosely in a larger tube. Its advantages are smaller size, wider temperature range, and greater tensile strength.

loose-tube fiber Optical cable constructed of fiber strands individually covered with a 250 μm acrylate coating, usually encased in bundles of 6 or 12 strands.

loss Attenuation of the optical signal, usually measured in decibels.

loss budget The total allowable loss between source and detector—allocated among fiber, connectors, splices, and safety margin.

loss level A change or sudden spike up from the backscatter which indicates fresnel reflections.

loss resolution Setting on an OTDR to determine data points.

low intensity laser Also known as a hot red light, this device operates in the visible light range. It is used to identify individual fibers and will glow red at the point of a fiber break.

LSA splice loss method A mathematical technique utilized in the measurement of optical fiber splice loss.

M

macrobends Cable bends with curvatures of 1–100 mm radius and from a fraction of a turn to many turns, usually due to cable installation. Results in greater stress, leading to shorter lifetime. At smaller radius, it may lead to extra attenuation.

main building ground electrode The designated point to which all utilities in a building are connected.

main cross-connect The cross-connect, normally located in the equipment room, is utilized for cross-connection and interconnection of entrance cables, backbone cables, and equipment cables.

main distribution frame A wiring arrangement which connects outside telephone lines to a building's interior wiring.

main distribution panel Electrical service entrance facility.

maintenance hole 1. Space used to access and maintain underground cable plant.
2. A hole through which a person may go to gain access to an underground or enclosed structure.

manchester encoding A digital encoding scheme where a voltage change occurs in the middle of each binary digit sent. A high-to-low change represents the binary digit zero and a low-to-high change represents the binary digit one.

mandrel A rod or a shaft.

manhole See maintenance hole.

material list A complete list of all materials to be ordered and received for the project. This includes all capital items and miscellaneous materials.

measured tape A calibrated tape used to measure and pull lengths of conduit.

measurement resolution Setting on the OTDR to determine spacing of data points.

mechanical splicing Joining two fibers together by mechanical means.

megabits per second (mb/s) A measurement of the data rate at which the LAN operates.

megahertz (MHz) A unit of frequency equal to one million Hertz.

megger A device which can be used to measure electrical resistance in a grounding system.

mesh grip A device attached to the end of a cable to facilitate pulling the cable.

Metropolitan Area Network (MAN) A network designed to provide regional connectivity.

microbends Bends with curvatures of a few micrometers and spaced by distances of a few millimeters, usually due to the cable construction. Results in increased attenuation.

micron One-millionth of a meter. Also denoted micrometer. Abbreviation is μm.

modal dispersion Dispersion resulting from the different optical path lengths in a multimode fiber. Also called modal distortion.

mode Loosely, a possible path followed by light rays. Strictly, a distribution of electromagnetic energy that satisfies Maxwell's equations and boundary conditions in guided wave propagation, such as through a waveguide or optical fiber.

mode field diameter The diameter of one mode of light propagating in a singlemode fiber. The mode field diameter replaces core diameter as a practical parameter in singlemode fiber.

modem Acronym for modulator/demodulator. Device which converts between analog signals and digital signals.

modular connector Modular multipair connectors consisting of a base, body, and cover.

modular furniture Groups of low-wall partitions, desks, and furniture assembled in the field in open spaces within an office.

modular jack A telecommunications female connector. A modular jack may be keyed (exclusion feature) or unkeyed and may have up to eight contacts.

modulation A process whereby certain characteristics of a wave, often called the carrier, are varied or selected in accordance with a modulating function. This includes amplitude, frequency or phase, and other modulation techniques.

multimeter Test equipment that can be set up to perform a variety of electrical property measurements, usually including resistance, voltage, and current.

multimode Transmits or emits more than one propagating mode.

multimode fiber An optical waveguide in which light travels in multiple propagation modes. Typical core/cladding sizes (measured in microns) are 50/125 μm and 62.5/125 μm.

multiplexer A device which combines two or more signals over a single communications channel. Examples include time-division multiplexing and wavelength-division multiplexing.

multiplexing Combining two or more signals into a composite signal from which the original signals may be individually recovered.

mushroom A plastic guide in the shape of a mushroom used for routing jumpers.

Multiuser Telecommunications Outlet Assembly (MUTOA) A group of telecommunications outlets that are arranged together in a single assembly housing. They may be located within the confines of a group of modular, low-wall partitioned furniture and serve only that group. A line cord is extended from the MUTOA to the work area contained by the modular partitions and is plugged directly into a device at that location.

mutual capacitance Effective capacitance between the two conductors of a pair.

N

N connector Used as a connector for RG-8, Thicknet, and RG-11U coaxial cables. N-type connectors have a center pin that must be installed over the cable's center conductor.

nanometer (nm) A unit of measurement equal to one billionth of a meter, abbreviated nm. The most common unit of measurement for optical fiber operating wavelengths.

National Electrical Code® (NEC) A safety code written and administered by the National Fire Protection Association (NFPA).®

network A group of three or more nodes that can communicate with each other, either directly through common wiring or indirectly through repeaters to separated wiring.

Near-end crosstalk (NEXT) Signal transfer between pairs at the end of a cable nearest the point of transmission.

National Fire Protection Association (NFPA) This association writes and administers the *National Electrical Code* (NEC).

node An intelligent communication device on a network. May be a modem, repeater, multiplexer, etc. Synonym for station.

noise Unwanted signal on a wire which provides a random or persistent disturbance which interferes with the clarity or quality of the expected signal.

nominal velocity of propagation (NVP) The speed of transmission along a cable relative to the speed of light in a vacuum.

nonreflective break When a fiber is cut or broken, the end may shatter so that the angle at where the light hitting the end may not reflect at all.

nonreturn to zero An encoding scheme.

numerical aperture (NA) The "light-gathering ability" of a fiber, defining the maximum angle to the fiber axis at which light will be accepted and propagated through the fiber. NA = sin Θ, where Θ is the acceptance angle. NA also is used to describe the angular spread of light from a central axis, as in exiting a fiber, emitting from a source, or entering a detector.

O

offset Degree to which the cable or conduit changes direction.

ohm Unit of measurement for the opposition to the flow of direct current (dc), called resistance, or opposition to the flow of alternating current (ac), called impedance. Abbreviation is Ω.

ohm's law The voltage in volts is equal to the current in amps multiplied by the resistance in ohms.

ohm-meter Device used to measure voltage and resistance.

open A lack of continuity.

optical fiber A thin filament of glass capable of carrying signals in the form of light.

optical fiber flashlight A device utilizing a LED to check for fiber continuity.

optical power meter Test equipment that measures in dBm the strength of a light wave over a fiber cable.

Occupational Safety and Health Administration (OSHA) This agency develops and enforces safety and health standards that apply to the work conditions, practices, means, methods, operations, installations, and processes performed at telecommunications locations and at telecommunications field installations.

optical time-domain reflectometer (OTDR) A device for measuring optical fibers based on detecting backscattered (reflected) light. Used to measure attenuation of fiber, splices, connectors, and locate faults. It can be used as a measure of splice and connector locations.

P

padding down A term used to describe the intentional attenuation of an optical fiber or copper circuit.

pair count Indicates how many pair of wires are in a cable or the pair identification serving a location.

pair scanner See wire map tester.

pair twist The uniform twist of an insulated copper pair that helps to improve the effects of capacitance imbalance and electromagnetic induction.

patch cord A connecting cable. In common usage, a reference to cross-connect equipment and work area cables.

patch panel Device containing multiple jacks and utilized for interconnecting circuits in order to provide flexibility.

pathway A device or group of devices that allow the installation of telecommunications cables and conductors between building spaces. The vertical and horizontal route of the cable.

payout box Cardboard container with hole for cable distribution directly from the box.

Private Branch Exchange A small telephone switching system that requires the user to dial an access code (dial "9") for an outgoing line.

pedestal An enclosure for access or termination of cable.

penetrations Openings made in fire-rated barriers (architectural structures or assemblies). There are two kinds of penetrations:

- Membrane penetrations pierce or interrupt the outside surface of only one side of a fire-rated barrier.
- Through penetrations completely transmit a fire-rated barrier, piercing both outside surfaces of the barrier.

peripheral device Equipment not integral to but working with a voice or data system.

personal protective equipment Any number of safety apparatuses worn or used, such as goggles, gloves, or clothing that shields against possible injury while performing tasks.

PERT chart A chart for viewing and altering dependencies between tasks, generally used by the project manager to see how one change in the project affects the remaining tasks.

phased cutover positions of the old system are transferred to the new system in groups.

photon A fundamental unit of light. Photons are to optical fiber what electrons are to copper wires.

physical topology Refers to the physical appearance of how devices are attached on the LAN.

picofarad One-trillionth of a farad. Used to designate capacitance unbalance between pairs and capacitance unbalance of the two wires of a pair to ground.

pigtail A short length of fiber cable with a connector on one end and bare fiber on the other.

pinout A wiring scheme for a jack or plug.

plenum In telecommunications, a space that is used to distribute environmental air.

plenum cable A cable with flammability and smoke characteristics that allow it to be routed in a plenum area without being enclosed in a conduit.

plenum rated Meeting the flammability requirements of UL as defined by the National Electrical Code.

plugger Stopper for the end of a syringe.

point of demarcation The point at which the service provider interfaces with the customer (demarcation point).

poke-thru A penetration through the fire-resistive floor structure to permit the installation of electrical or communications cables.

poststressed concrete Concrete poured into a mold with metal cables exposed at each end.

power A term which applies to the energy required to operate an electrical device.

power pole Correctly termed a utility column. It is a pathway used to house cables that run from above a suspended ceiling to the termination location in a work area.

prefusing The machine cleaning of the fiber endfaces prior to performing a fusion splice.

prestressed concrete Concrete poured into a mold containing tension rods or cable.

premises A generic term that includes interbuilding, intrabuilding, and horizontal cabling that is owned by a single tenant or landlord.

primary protector A device that limits voltage between telecommunications conductors and ground.

prime contractor The master contractor on a job site that may be serviced by several general contractors.

project log A written log of everything that happens on a project, hour by hour, day by day, item by item. It should contain any information relating to events that occur on the project that can affect the project. The project manager or job foreman usually maintains this log.

project schedule A chronological order of events that will be accomplished on a project and in the order that they must occur. This device can be manually generated on paper or can be developed using any of the software packages available on the market.

propagation delay The time interval required for a signal to travel from one end of the transmission path to the other end of the path.

proposal The act of offering or suggesting something for acceptance.

protector A device used to limit dangerous foreign voltages on metallic telecommunications conductors.

proton   An elementary particle that is a fundamental constituent of all atomic nuclei.

pull point   Location where it is possible to physically access the cables to pull them.

pull rope   Attached between pull string and cable to obtain an increased amount of strength for pulling or moving high-pair-count cable.

pull string   Line attached to a cable to pull it through conduit trays or open ceiling supports.

pulling eye   A factory-installed device on a length of cable to which a swivel eye and pull rope are attached.

pulling sheave   A pulley having a grooved rim for retaining a rope or cable.

pulling technique   Collectively refers to the methods and materials employed to install cables.

pulse repetition rate (PRR)   The rate at which an OTDR transmits a test pulse.

punch down   The termination of twisted-pair wire to connecting hardware.

punching down   The process of terminating cable conductors on IDC terminals by use of a handheld tool.

Q

quad cable   A 4-conductor nontwisted-pair cable with a red, green, black, and yellow conductor.

quote   A statement of the current or market price of a product or service.

R

raceway   Any enclosed channel designed expressly for holding wires, cables, or busbars.

radio frequency interference (RFI)   A disturbance in the reception of an electrical signal due to conflict with radio frequency signals.

rat   Lightweight object that can be sent into a conduit to aid in installing a pull string.

rayleigh scattering   Optical attenuation due to scattering from small fluctuations in the index of refraction. These may be due to the molecular structure and variations in that structure. In glass, these set an inherent lower limit to attenuation because the toughness and flexibility of glass is due to its amorphous, varying structure.

rebar   Reinforcing bar to add strength to concrete.

receiver (RX)   An optoelectronic circuit that converts an optical signal to an electrical serial logic signal. It contains a photo detector, amplifier, discriminator, and pulse-shaping electronics.

reel brake Device used to control the rate of removal of a cable from a cable reel.

reel dolly A jackstand with wheels used to assist in carrying out and paying out cable.

reflection The abrupt change in direction of light as it travels from one material into a dissimilar material. In optical fibers, some of the light at a core-air interface is reflected back to the source. A critical phenomenon for OTDR operation.

reflective break The amount of reflection at the location of a break. See fresnal reflection.

reflection coefficient The degree of reflection caused by a mismatch between the line and the load.

refraction The angular change in direction of a beam of light at an interface between two dissimilar media or a medium whose refractive index is a continuous function of position (graded index medium).

relay rack A vertical system of metallic structures that are equipped with threaded holes which will accept screws at a predefined spacing on the front of the rack (or on both the front and rear sides of the rack). They are used to mount termination hardware, electronic equipment, or a combination of both. They can be floor mounted (free standing) or wall mounted. They are generally available in 0.6 m (2 ft), 1 m (3 ft), 1.2 m (4 ft), 2 m (6 ft), and 2.1 m (7 ft) heights.

request for proposal (RFP) A detailed document of requested design services and equipment of a buyer and submitted to others for responses.

request for quotation (RFQ) A document that solicits quotes for telecommunications project or equipment and provides vendors with all the information necessary to prepare a quote.

resistance A measure of opposition to the flow of direct current (dc).

resistance unbalance The difference in resistance, expressed in ohms, between the conductors of a pair. May also be expressed in terms of percentage by the ratio of unbalance in ohms to the lowest conductor resistance.

retrofitting To modify systems which are already in service using parts made available after the time of original installation.

return loss 1. A ratio, expressed in decibels, of the power of the outgoing signal to the power of the reflected signal.
2. The amount of reflected power compared to the amount of incident power at an interface, expressed in dB. Return loss is a critical factor in singlemode links because reflected light can destabilize some lasers.

return to zero Term used to describe an encoding scheme.

ribbon An assembly of several fibers laid side by side in a plane and fastened together. Commonly used in outside plant telephone cables.

rigging A system of ropes or pulleys used to move material and equipment.

ring A means for identification of one conductor of a pair. Historically, associated with the wire connected to the "ring" portion of an operator's telephone plug.

ringing tool A device used to remove cable sheaths.

ring network A network topology in which nodes are connected in a point-to-point serial fashion in an unbroken circular configuration. Each node receives and retransmits the signal to the next node.

riser cable Cables intended for use in vertical shafts between floors in a building. Obsolete term, replaced by backbone cable.

roll bar Metal shaft placed inside the center hole of the reel head to pay off (remove) the cable.

rolling cut A cutover where the cross-connects are relocated one pair after another.

S

safety margin A power loss (dB) value used to assure optical fiber cable performance criteria will be satisfied over the life of the network. Includes expected losses in source power, splice losses due to repairs, wear and tear of connectors. Typically 2–3 dB.

SC A type of optical fiber connector identified by the square cross section of its plastic housing, colored blue for singlemode and beige for multimode.

SC Subcommittee as defined by standards' committees.

scanner Device which checks cables for opens, shorts, crossed pairs and, sometimes, cable length; however, they do not measure cable performance.

scattering A property of a fiber which causes light to deflect from the fiber and contribute to losses.

screened twisted-pair cable (ScTP) A cable with one or more pairs of twisted copper conductors covered with an overall metallic shield.

scribing tool Device used to remove cable sheaths.

sensitivity In optical fiber receivers, the minimum power required to achieve a specified bit error rate (BER). See bit error rate.

server Combines hardware and software to offer, or serve, network resources to other attached devices. Servers manage the shared resources on a LAN.

service loop Field-configured coil of cable arranged at the point of termination to facilitate future arrangements.

sheath The outer covering of a cable. See jacket.

shield 1. Metallic layer placed around a conductor or group of conductors to prevent electrostatic or electromagnetic coupling between the enclosed wires and external fields.

2. Housing, screen, or cover which substantially reduces the coupling of electric and magnetic fields into or out of circuits. Prevents the accidental contact of objects or persons with parts or components operating at hazardous voltage levels.

shielded twisted-pair (STP) A cable consisting of two or more pairs of which two are individually shielded and are covered with an overall shield. The shielded pairs are capable of supporting transmission to 20 MHz.

short Accidentally caused low-resistance contact between conductors of a circuit.

shorting bar Maintains continuity of the ring after removal of a hermaphroditic connector.

shorting plug A device to create a direct connect between two or more conductors at one end of a cable for test purposes.

signal generator Test equipment that generates a distinctive tone(s) that is placed on a cable pair for identification purposes. Sometimes referred to as a toner.

signal-to-noise ratio (SNR) The ratio between the detected signal power and noise in a receiver, expressed in dB. The prime determining factor in BER. See bit error rate.

sine wave The variation of a wave from zero to maximum (positive) back through zero to minimum and back to zero (negative).

singlemode fiber An optical fiber that supports only one mode of light propagation. This does not necessarily imply single-wavelength operation. This singlemode avoids modal dispersion.

sleeve A short section of metallic conduit that is placed in an opening between floors or in a wall where cables can be installed between two spaces.

slot A rectangular opening in a floor that allows cables to be installed vertically between floors.

Standard Network Interface (SNI) A device that serves as a demarcation point between the local exchange carrier (LEC) and the customer. Usually a modular jack with some type of mounting for the modular jack.

spatial resolution Setting on an OTDR to determine how close individual data points are spaced in time and distance.

spike An instantaneous surge of energy.

splice 1. The joining of two or more cables together by connecting the conductors pair to pair.
2. A permanent joint between two optical fibers.

splice bank Placement of 25-pair modules in a symmetrically spaced configuration within a splice enclosure.

splice closure A device for physical protection of a splice consisting of a solid or split-sleeve cover, endcaps, and steel hose clamps.

splice tray A container used to organize and protect splices and spliced fibers.

splicing head A section of a splicing rig that supports the crimp head. Can be either single or dual.

splicing rig A specific manufacturer's tool kit for terminating modular connectors.

split grip A wire mesh grip that is open on one side.

split pair Inadvertent transposition of two conductors of separate pairs.

spools Cylindrical containers of cable. See cable reel.

ST A type of optical fiber connector identified by its bayonet housing. The housing may be metallic or plastic.

standing wave ratio The ratio of the amplitude of a standing wave at an antinode to the amplitude at a node.

star topology Network devices are connected to a central hub like the points on a star.

star-wired ring Physical star configured as a ring. Also known as a collapsed ring.

station Telecommunications end-user location. Usually dedicated to a single-user location and function, such as a telephone or computer hook-up work area outlet.

station location Telephone or computer location.

step-index fiber An optical fiber, either multimode or singlemode, in which the core refractive index is uniform throughout so that a sharp step in refractive index occurs at the core-to-cladding interface.

stick Slang term used to describe a section of conduit.

STP-A A cable consisting of two individual shielded pairs capable of supporting transmission to 300 MHz.

strand identifier This clamp-on unit inserts a microbend into optical fiber cable and thereby is able to detect light escaping from the fiber.

strength member That part of an optical fiber cable composed of aramid yarn, steel strands, fiberglass filaments, or fiberglass-reinforced epoxy composite rod that increases the tensile strength of the cable.

structural return loss Measurement of the distance between the test signal amplitude and the amplitude of signal reflections returned by the cable.

stub-out Conduit installed from a wall outlet to a raceway. Provides physical and electrical protection (ground).

stub-up Conduit installed from a wall outlet into ceiling space. Used for physical protection only.

surface-mounted raceway Plastic or metallic raceway that is installed on the surface of a wall, floor, or ceiling.

surge arrestor Device used to prevent transient voltage surges from reaching electronic equipment.

sweep Bend that has a gentle arc rather than a sharp bend.

swing floor phasing The act of removing personnel and property from one location to another in order to facilitate renovation of the space vacated.

swingset See bullwheel.

T

TIA/EIA Trade associations involved in developing telecommunications industry standards.

TIA/EIA TSB-67 1. The Telecommunications Systems Bulletin entitled *Transmission Performance Specifications for Field Testing of Unshielded Twisted-Pair Cabling Systems.*
2. Defines field test parameters for Category 3, 4, and 5 UTP cabling.

TIA/EIA TSB-72 The Telecommunications Systems Bulletin entitled *Centralized Optical Fiber Cabling Guidelines.*

TIA/EIA TSB-75 The Telecommunications Systems Bulletin entitled *Additional Horizontal Cabling Practices for Open Offices.*

telecommunications In this manual, low voltage and optical systems which carry information inside a building.

telecommunications bonding backbone (TBB) Conductor extending from the telecommunications main grounding busbar to all telecommunications grounding busbars.

telecommunications bonding backbone interconnecting bonding conductor (TBBIBC) A conductor utilized to interconnect two or more telecommunications bonding backbones (TBBs).

telecommunications closet (TC) An enclosed space within the confines of a building used to house telecommunications cable terminations and associated hardware. This closet is the recognized cross-connect between the backbone cables and the horizontal cables. There may be several in a large installation.

telecommunications equipment bonding conductor (TEBC) Should be installed from each piece of equipment to the telecommunications grounding busbar (TGB) or telecommunications main grounding busbar (TMGB).

telecommunications grounding busbar (TGB) A busbar located in a telecommunications closet connected to the telecommunications main grounding busbar (TMGB) via a telecommunications bonding backbone (TBB).

Telecommunications Industry Association (TIA) A standards association which publishes telecommunications criteria.

telecommunications main grounding busbar (TMGB) A busbar located in the telecommunications closet which is connected to the service equipment ground.

telephone test set A voice circuit testing device used to identify circuits and perform diagnostics. Also butt set.

temporary cabling Cables and equipment that are installed to provide service on a temporary basis. (i.e., providing parallel system during a retrofit installation where the cables are removed after cutover).

terminal block A device that provides a cable pair point of termination utilizing insulation displacement connections.

terminate See cable termination.

termination point A cable connection point, such as a terminal block, wall plate, or UTP modular plug.

terminator An impedance matching device placed at the end of the telecommunications transmission line.

thicknet See 10BASE-5.

thinnet See 10BASE-2.

throughput The useful amount of information transferred during a specified time period.

tie wraps Plastic or hook and loop strips used for binding and dressing cable.

tight buffer A glass fiber is covered with a tightly bonded, protective thermoplastic coating to a diameter of 900 µm. High tensile strength gives durability, ease of handling, and ease of connectorization.

tight-buffered fiber Optical fiber strand covered by a 900 µm coating.

timber hitch A type of knot.

time-division multiplexing (TDM) A transmission technique whereby several low-speed channels are multiplexed into a high-speed channel for transmission. Each low-speed channel is allocated a specific time position in the bit stream.

time-domain reflectometer (TDR) A device which sends a signal down a cable, then measures the magnitude and amount of time required for the reflection of that signal to return. TDRs are used to measure the length of cables as well as locate cable faults.

tip A means for identification of one conductor of a pair. Historically, associated with the wire connected to the "tip" portion of an operator's telephone plug.

token ring A topology in which a token must be received by a terminal or workstation before that terminal or workstation can start transmitting.

toner Device used to apply an electrical signal to a circuit to assist in identification or fault location.

topology Physical routing of cable plant within a building either in a star, bus, or ring configuration—such as the physical topology of a LAN.

total internal reflection Confinement of light within a fiber because the angle of incidence is within the numerical aperture.

topology The configuration in which wires and cables are installed from the closet to the work area. A topology can be a star, a bus, or a ring configuration.

Twisted-Pair Physical Medium-Dependent Allows 100 Mb/s transmission over twisted-pair cable.

trade size Nominal name given to materials to identify a nominal size.

trailer string Line attached to the end of a cable(s) being pulled. To be used for future additional pulls.

transceiver A transmitter and receiver combined in one package.

transfer impedance A ratio of the resulting voltage induced on a shielded transmission line to the magnitude of the induced current on the shield. Low transfer impedance values correlate to higher shield effectiveness.

transition point Location of a change in facilities or means (e.g., where flat cable connects to round conventional telecommunications wires).

transmission media The physical carriers of electromagnetic energy (e.g., copper, fiber, and air) radiation.

transmitter (TX) An optoelectronic circuit which converts an electrical logic signal to an optical signal. In fiber optics, using a source such as an LED or laser.

trapeze A support device using threaded rod and channel stock.

trench A narrow furrow dug into the earth for the direct installation of buried cable or for the installation of troughs or ducts.

tugger Device that acts as an assist mechanism for advancing a cable or groups of cables during installation.

twisted-pair cable A multiconductor cable comprising two or more copper conductors twisted in a manner designed to cancel electrical interference.

two-point method A method used to measure optical fiber cable loss utilizing two closely spaced cursors on an OTDR.

U

underground cable A telecommunications cable installed in an underground trough or duct system separating the cable from direct contact with the soil. See also: Buried Cable; Trench.

Underwriters Laboratory (UL) A U.S.-based independent testing laboratory that sets tests and standards for electrical equipment for safety.

Underwriters Laboratory—Canada (ULC) A Canadian-based independent testing laboratory that sets tests and standards for electrical equipment for safety.

unshielded twisted-pair Cable containing one or more pairs of twisted copper without metallic shielding.

V

V-groove Position in fusion splicer in which fiber strand is placed.

vampire tap Device used to tap into thick Ethernet cable.

velocity of propagation The speed of transmission along a cable relative to the speed of light in a vacuum.

volt Unit of electromotive force or potential difference which will cause a current of one ampere to flow through a resistance of one ohm.

volt-ohmmeter (VOM) An instrument used to measure electrical characteristics.

W

wall-mount brackets Devices that are constructed at a right angle, having a diagonal brace between the vertical section and the horizontal section, and mount to a wall. They support sections of conduit, ladder rack, and cable trays.

wand Test device used to detect a signal placed on a cable for the purpose of identification.

wavelength The length of a wave measured from any point on one wave to the corresponding point on the next wave, such as from crest to crest.

wide area network (WAN) Computer networks in which devices are connected over extended distances using telecommunications links, such as telephone lines, satellites, and microwave connections, rather than a length of cable.

wire management Components placed on racks or walls to support the routing of cables.

wire map tester An instrument used to determine circuit opens, shorts, crossed pairs, improper wiring, and the determination of proper pin configuration; additionally, some units indicate cable length.

wireway An enclosed pathway for cables.

wirewrap Termination of conductors by wrapping around a pair.

work area The space in an office or open area of a building in which people perform specific tasks.

work area outlets Device placed at user workstations for termination of horizontal media.

working group (WG) Working group as defined by standards' committees.

workstation A telecommunications device used in communicating with another telecommunications device.

Z

Z-gap Spacing in fusion splicing.

Abbreviations

ac alternating current
ACEG alternating current equipment ground
ACR attenuation-to-crosstalk ratio
AIA American Institute of Architects
AMES architectural, mechanical, electrical, structural
ANSI American National Standards Institute
ANSI/TIA/EIA American National Standards Institute/Telecommunications Industry Association/Electronics Industries Alliance
ASTM American Society for Testing Materials
ATM asychronous transfer mode
ATR all threaded rod
AWG American wire gauge

BCT bonding conductor for telecommunications
BER bit error rate
BNC Bayonet Navel Connector
BNC Bayonet Neil-Concelman
b/s bits per second

CAD computer aided design
CATV community antenna television
CBC coupled bonding conductor
CLR circuit layout record
CMR communications media riser
CSA Canadian Standards Association
CSI Construction Specification Institute

dB decibel
dc direct current
demarc demarcation

EDP electrical distribution panel
EIA Electronics Industries Alliance
EKS electronic key system

EKS executive key system
EMI electromagnetic interference
EMI/RFI electromagnetic interference/radio frequency interference
EMP electromagnetic pulse
EMT electrical metallic tubing

FC fiber connector
FCC Federal Communications Commission
FDDI fiber distributed data interface
FEP fluorinated ethylene propylene
FEXT far-end crosstalk
FOTP fiber optic test procedure

Gb/s gigabits per second
GFI ground fault interrupter
GHz gigahertz
GRC galvanized rigid conduit

HC horizontal cross-connect
HVAC heating, ventilating, air conditioning
Hz hertz

IC intermediate cross-connect
ID inside diameter
IDC insulation displacement connection
IEEE Institute of Electrical and Electronic Engineers
IMC intermediate metallic conduit
ISO International Standards Organization
ISO/IEC International Standards Organization (ISO)/International Electrotechnical Commission (IEC)

JTC Joint Technical Committee

kHz kilohertz
KSU key service unit

LAN local area network
LEC local exchange carrier

LED light-emitting diode
LID local injection detection
LSA link state algorithm

mA milliampere
MAC move, add, or change
MAN metropolitan area network
mb/s megabits per second
MC main cross-connect
MDP main distribution panel
MHz megahertz
Modem modulator/demodulator
MSDS material safety data sheet
MTBF mean time between failures
MUTOA Multiuser Telecommunications Outlet Assembly

NEC *National Electrical Code*®
NEMA National Electrical Manufacturers Association
NEXT near-end crosstalk
NFPA National Fire Protection Agency
nm nanometer
NICAD nickel cadmium (commonly used in batteries)
NVP nominal velocity of propagation

OD outside diameter
OSHA Occupational Safety and Health Administration
OSP outside plant
OTDR optical time-domain reflectometer

PAS profile alignment system
PBX private branch exchange
PC personal computer
PO purchase order
POF polymeric optical fiber
POTS plain old telephone service (colloquial)
PPE personal protective equipment

PRR   pulse repetition rate

PTC   positive temperature coefficient

PVC   polyvinyl chloride

RBOC   regional bell operating company

RCDD®   Registered Communications Distribution Designer

REA   Rural Electrification Administration

RFI   radio frequency interference

RFP   request for proposal

RFQ   request for quotation

RUS   Rural Utilities Services

RX   receiver

SC   subcommittee

ScTP   screened twisted-pair

SNI   standard network interface

SNR   signal-to-noise ratio

SP   service provider

STP   shielded twisted-pair

TBB   telecommunications bonding backbone

TBBIBC   telecommunications bonding backbone interconnecting bonding conductor

TC   telecommunications closet

TDR   time-domain reflectometer

TEBC   telecommunications equipment bonding conductor

TEF   triethylene fluoride

TGB   telecommunications grounding busbar

TIA   Telecommunications Industry Association

TMGB   telecommunications main grounding busbar

TP   transition point

TP-PMD   twisted-pair physical medium dependent

UDC   universal data connector

UL   Underwriters Laboratory

ULC   Underwriters Laboratory—Canada

UPS uninterruptible power supply
USOC universal service order code
UTP unshielded twisted-pair
UV ultraviolet

VOM volt-ohmmeter

WAN wide area network
WG working group

Bibliography

ANSI/EIA/TIA-455-61. Currently Standard Proposal No. 2837-B, Proposed Revision of EIA/TIA-455-61, FOTP-61. *Measurement of Fiber or Cable Attenuation Using as OTDR* (if approved, to be published as TIA/EIA-455-61A).

ANSI/EIA/TIA-526-14A. Currently Standard Proposal No. 2981, Proposed Revision of EIA/TIA-526-14, OFSTP-14. *Optical Power Loss Measurements of Installed Multimode Fiber Cable Plant* (if approved, to be published as TIA/EIA-526-14A).

ANSI/EIA/TIA-570. *Residential and Light Commercial Telecommunications Wiring Standard,* June 1991.

ANSI/EIA/TIA-598-A. *Optical Fiber Cable Color Coding,* 1995.

ANSI/TIA/EIA-455-171A. Currently Standards Proposal No. 3017, Proposed Revision of EIA-455-171, FOTP-171. *Attenuation by Substitution Measurement for Short Length Multi-mode and Graded Index and Single-Mode Optical Fiber Cable Assemblies* (if approved, to be published as TIA/EIA-4550171-A).

ANSI/TIA/EIA-526-7. Currently Standard Proposal No. 2974-B, Proposed New OFSTP-7. *Measurement of Optical Power Loss of Installed Single-Mode Fiber Cable Plant* (if approved, to be published as TIA/EIA-526-7).

ANSI/TIA/EIA-568-A. *Commercial Building Telecommunications Cabling Standard,* October 25, 1995.

ANSI/TIA/EIA-569-A. *Commercial Building Standards for Telecommunications Pathways and Spaces,* February 1998.

ANSI/TIA/EIA-606. *Administrative Standard for the Telecommunications Infrastructure of Commercial Buildings,* February 1993.

ANSI/TIA/EIA-607. *Commercial Building Grounding and Bonding Requirements for Telecommunications,* August 24, 1994.

DSP-100/2000 LAN CableMeter®/Cable Analyzer Users Manual, Fluke Corporation, January 1997.

Earth Testing, Amprobe Instruments, December 1995.

Fundamentals of Field Cable Testing, Fluke Corporation, October 1997.

Getting Down to Earth, Biddle Instruments, March 1990.

Grounding of Industrial and Commercial Power System, IEEE Std 142-1991, Green Book, Institute of Electrical and Electronics Engineers, Inc., 1992.

Guide to Fiber Optic Measurements. Wavetek Corporation, March 1998.

Handbook of LAN Cable Testing. Wavetek Corporation, September 1992.

LAN Design Manual, 2nd edition, BICSI, 1997.

National Electrical Code Handbook, 7th edition, National Fire Protection Association, 1996.

Network Maintenance and Troubleshooting Guide, Fluke Corporation, August 1997.

Newton's Telecom Dictionary, Harry Newton, 1997.

Premises Cabling, Donald J. Sterling, Jr., Delmar Publishers, 1995.

Soares Book on Grounding, International Association of Electrical Inspectors, 5th edition, 1993.

Standard for Protectors for Paired Conductor Communication Circuits, UL-497, Underwriters Laboratories, 5th edition, September 24, 1992.

Standards-29 CFR, U.S. Department of Labor, Occupational Safety and Health Administration, 1998.

Telecommunication Project Management, James B. Pruitt, 1987.

Telecommunications Distribution Methods Manual, 8th edition, BICSI, 1998.

Understanding Fiber Optics, Jeff Hecht, 2nd edition, 1993.

Understanding OTDRs, John Lidh, GN Nettest-Laser Precision Division, 1997.

Index

66-block termination, 336–338
110-style hardware, 338, 339

A

Ablative firestops, 296
ac (*see* Alternating current)
Access flooring, 148
ACEG (alternating current equipment ground), 241
ACR (*see* Attenuation to crosstalk ratio)
Aerial entrances, structured cabling systems for, 13–14
AIA (American Institute of Architects), 49
Air propulsion, 253, 265
Alternating current (ac), 84–85
Alternating current equipment ground (ACEG), 241
American Institute of Architects (AIA), 49
American National Standards Institute (ANSI), 5, 7, 30, 189
American Society for Testing and Materials (ASTM), 30, 297
American Wire Gauge (AWG), 87–88, 103
Anaerobic termination, 354, 373–374
Analog transmission, 85–86
Anchors, 221–223
　for drywall installation, 221–222
　epoxy resin, 223
　expansion, 221
　for masonry and concrete-block installation, 222–223
　plastic, 221
ANSI (*see* American National Standards Institute)

Appearance, professional, 158–159
Architects and Builders Service, 6
Architectural drawings, 162
Asbestos, 152–153
ASTM (*see* American Society for Testing and Materials)
AT&T, 2–4, 6
Attenuation, 87
　copper cable, 90, 421–422
　optical fiber cable, 96
Attenuation to crosstalk ratio (ACR), 421–422, 492
AWG (*see* American Wire Gauge)

B

Backbone cable/cabling, 14–16
　ANSI/TIA/EIA-568-A on, 35
　bonding/grounding, 119–123
　　telecommunications bonding backbone (TBB), 121
　　unshielded backbone cable, 122
　copper cable splicing, 375
　definition of, 426
　main components of, 15
　pulling, 253–254
　　horizontal cable, 283–287
　　horizontal pathway, 283–287
　　vertical pathway, 269–283
　supporting structures for, 209–228
　　anchors for, 221–223
　　conduit formations, securing of, 220
　　elbows/bends, conduit, 216–217
　　electrical metallic tubing, 210–211
　　fasteners, 223–228

Backbone cable/cabling, supporting structures for (*Cont.*):
   grounding infrastructure, 240–241
   hangers, 217–218
   joint make-up, 215
   preparation, 220–221
   terminations, conduit, 218–220
 system design considerations, 15
 testing, 426–427
 unshielded backbone cable, 122
Backscatter, 452
Bandwidth, optical fiber cable, 95–96
Batteries, 152, 425
BCT (*see* Bonding Conductor for Telecommunications)
Bel, 86
Bell System, 2–4
Bends, conduit, 216–217
BICSI, 6
BIX hardware, 338–340
Blankets, firestop, 312
BNC coaxial connectors, 80
Bolts, 226–227
Bonding, 113–124
 conductors, bonding, 116
 connections, bonding, 113–115
 definition of, 103, 231
 exposed cables, 117
 inspections of, 116–117
 with optical fiber cable, 353
 system practices for, 117–124
   and backbone cable protection, 119–123
   coupled bonding conductors, 121–122
   sheath terminations, exposed, 123–124
   small systems, 117
   telecommunications bonding backbone (TBB), 121
   in telecommunications closets, 119
   unshielded backbone cable, 122
 (*See also* Grounding)
Bonding Conductor for Telecommunications (BCT), 233–234
Bradley, Woody, 6
Breathing protection, 141
Brick walls, firestopping/firestop systems for, 312–315
Building Industry Consulting Service, 6
Building Industry Consulting Service International, Inc. (BICSI), 6
Buried entrances, structured cabling systems for, 12, 13
Bus topology, 23–24

C

Cabinets:
 floor-mounted, 207
 wall-mounted, 208–209
Cable tracers, 483, 496
Cable trays, 195–202
 installation of, 200–202, 245
 multilevel, 198
 rod-stock, 197–200
 suspended, 196
 tubular, 195–197
Cable trees, 257–258
Cable-end locator kit, 412, 496
Cable(s), 52–68
 coaxial, 60–63
 optical fiber, 63–68
 screened twisted-pair copper, 58–59
 unshielded twisted-pair, 53–58
 (*See also specific headings*)
Canadian Electrical Code, Part 1 (CEC, Part 1), 33
Capacitance (copper cable), 89, 421, 492
Carbon blocks, 126
Carter, Tom, 2–3
"Carterfone," 3
CATV coaxial cable, 112
CATV industry, 5
Catwalks, 148–149
Caulks, firestop, 309
CBCs (*see* Coupled bonding conductors)
CEC, Part 1, 33
Ceilings:
 firestopping/firestop systems for, 317–319
 pulling cable in open, 266–269
Cementitious firestop materials, 310
Certification field testers, 412, 415–418, 422–425, 482–483
Certification test sets, 412–413
Change orders, 175–176, 183
Characteristic impedance, 89–90

Chemical hazards, 153–154
Circuit Layout Records (CLRs), 521–522
Circuit protectors, 124–128
   enhanced protection with, 127–128
   fuses/fuse links, 126–127
   installation practices for, 128
   primary protectors, 126, 128
   secondary protectors, 127
Cladding, optical fiber cable, 352
Clothing, work, 144–145
CLRs (*see* Circuit Layout Records)
Clustered star topology, 26, 27
CMR cable, 375–376
Coaxial cable, 60–63
   connectors for, 78–80
   terminations for, 349
   testing, 428–432
      50-Ohm, 429–430
      75-Ohm, 431–432
   troubleshooting, 504–505
Codes, 28–33
   Canadian Electrical Code, Part 1, 33
   grounding/bonding, 232–233
   National Electrical Code, 31–33
   National Fire Protection Association (NFPA), 31
   and pulling cable, 254
   for supporting structures, 235
   (*See also* Standards)
Collars, firestop, 308
Communication(s):
   data, 3
   and professionalism, 157–158
   and pulling cable, 254
   and safety, 132
   and troubleshooting, 485–486
   video, 3
   voice, 3
Communications Act of 1934, 2
Concrete and concrete block walls, firestopping/firestop systems for, 312–315
Conduit(s):
   and cable trays, 203–204
   metallic, 252
   pulling cable through, 252–253, 259–266
   stub-up/stub-out, 228
   support system, installation of, 246–248
Confined spaces, 149–151

Connector insertion loss, 87
Connectors, 69–83
   by cable type, 82
   coaxial, 78–80
   enhanced shielded twisted-pair, 78
   hardware for, 69–70
   optical fiber, 80–82
   screened twisted-pair, 77
   shielded twisted-pair, 77–78
   unshielded twisted-pair, 70–76
Construction Specifications Institute (CSI), 49
Continuity testing, 414–415
Contracts, 165, 526
Copper cable, 4, 87–94
   attenuation with, 90
   capacitance of, 89
   characteristic impedance, 89–90
   crosstalk with, 93–94
   digital transmission speed, 93
   hardware for, 94
   inductance of, 89
   insulation of, 91–92
   pair twist, 92
   resistance of, 87–89
   return loss with, 90–91
   shielding of, 92–93
   splicing, 375–408
      cable binders, 377
      color code, 376
      foldback splicing method, 378–379
      in-line splicing method, 378
      intrabuilding, 375
      labeling, 380
      listed cables, 377
      modular connectors for, 379
      safety guidelines for, 381
      steps in, 381–389
      wrapping, 380
   testing, 409–414
      backbone, 426–427
      coaxial cable, 428–432
      field testers, 411–414
      horizontal cabling, 414–426
   troubleshooting (*see under* Troubleshooting)
   (*See also specific types of cable, e.g.,* Coaxial cable)

Copper jumpers, 21
Cotton, Ross, 6
Coupled bonding conductors (CBCs), 121–122
Couplings, 109
   electrical metallic tubing, 211
   galvanized rigid conduit, 213–214
   intermediate metallic conduit, 212
Crawl spaces, 149
Crimp connectors, 328
Crimp-style fiber termination, 353, 368–373
Cross-connections, 20–21, 191
Crosstalk, 4, 87, 93–94, 421–422
CSI (Construction Specifications Institute), 49
Current, 84
Customer drawings, 162–163
Customer relations, 155–156

D

Data communications, 3
dc (see Direct current)
dc loop resistance, 420–421
Decibel, 86–87
Designated work areas, 133–134
Designer's drawings, 163–164, 177
Detection badges, 142–143
Digital transmission, 86
Digital transmission speed, 93
Direct current (dc), 85
Diversion, 109
Documentation (for troubleshooting), 485
Drawing set, 42–49
Drawings:
   customer, 162–163
   designer's, 163–164, 177
D-rings, 202–203, 244

E

Earplugs, 143
Earthing system, 97
EDP (electrical distribution panel), 241
EIA (see Electronic Industries Alliance)
Elbows, conduit, 216–217
Electric shock hazards, 98–100
Electrical distribution panel (EDP), 241
Electrical drawings, 163
Electrical exposure, 101–103
Electrical hazards, 145–148

Electrical metallic tubing (EMT), 210–211
Electrical power protection, 107–108
Electrical power systems, 105–106
Electrical service ground, 110–111
Electronic Industries Alliance (EIA), 5, 7, 30, 189
Emergency rescue, 131
EMT (see Electrical metallic tubing)
Endothermic firestops, 296
Enhanced shielded twisted pair, 78
Entrance facilities, structured cabling systems for, 9–14
   aerial entrances, 13–14
   buried entrances, 12, 13
   underground entrances, 10–11
Epoxy resin anchors, 223
Equalization, 109
Equipment grounding system, 97
Equipment rooms, 18–19, 190
Expansion anchors, 221
Exposure monitors, 142–143
Eye protection, 140–141

F

Fasteners, 223–228
   bolts, 226–227
   drywall screws, 226
   metal screws, 225–226
   nuts, 226–227
   screws (generally), 224
   specialty, 227–228
   washers, 227
   wood screws, 224–225
Federal Communications Commission (FCC), 2, 5, 30
Fiber optic cable (see Optical fiber cable)
Field testers, 411–414
Field-constructed patch cords, 344–345
Firestopping/firestop systems, 293–325
   ablative firestops, 296
   appropriate systems, 299
   for brick/concrete/concrete block walls, 312–315
   classifications, fire-rating, 300–301
   endothermic firestops, 296
   evaluation of, 302–305
   and fire protection, 297

Firestopping/firestop systems (*Cont.*):
   for floor/ceiling assemblies, 317–319
   general considerations with, 319
   for gypsum board walls, 315–317
   independent testing laboratories for, 296–297
   installation, 294–295
   intumescent firestops, 296
   materials for, 298, 299
   mechanical systems, 295, 305–307
   membrane penetrations, 294
   nonmechanical systems, 295, 307–312
     blankets, firestop, 312
     caulks, 309
     cementitious materials, 310
     collars, 308
     intumescent sheets, 310
     intumescent wrap strips, 310–311
     pillows, premanufactured, 311–312
     putties, firestop, 308, 319–321
     silicone foams, 311
     sprays, 312
   product types, 295
   qualified fire-rated assemblies, 299–300
   requirement setting for, 297
   seals, firestop, 298, 299, 318–319
   secondary functions of, 298
   selection of, 299, 319
   terminology related to, 296
   terms used, 296
   testing of, 294, 296–297, 300
   through penetrations, 294
   typical installations, 319–325
     floor penetration, sealing of, 319–321
     outlet box, sealing of, 322–323
     restoration of fire-rated penetrations, 323–325
   UL-Qualified Assemblies, 294
   walls
     brick/concrete/concrete block, 312–315
     gypsum board, 315–317
First aid, 129–130
Fishtape, 253, 259–266
Flooring/floor assemblies:
   access, 148
   firestopping/firestop systems for, 317–321
Foldback splicing, 378–379
Footwear, protective, 142

Frequency, 85
Fresnel reflection, 452
F-type coaxial connectors, 78–79
Fuses/fuse links, 126–127
Fusion splicing, 390, 398–404

G
Galvanized rigid conduit (GRC), 213–215
Gas tubes, 126
Global Engineering Documents, 31
Gloves, 142
Goggles, 140
GRC (*see* Galvanized rigid conduit)
Greene, Harold, 3
Grooming, 145
Ground wires, cable trays and, 204–205
Grounding, 97–124
   American Wire Gauge, 103–104
   bonding, definition of, 103
   with CATV coaxial cable, 112
   choices, grounding, 110
   electric shock hazards, 98–100
   and electrical exposure, 101–103
   electrical service ground, 110–111
   electrode system, 97, 111–112
   equipment, 97, 124
   exposed cables, 117
   ground, definition of, 103
   inspection of ground systems, 116–117
   and lightning exposure, 101–102
   with optical fiber cable, 353
   physical protection of, 112
   protection systems, 104–109
     electrical bonding and grounding, 107
     electrical power protection, 107–108
     electrical power systems, 105–106
     grounding electrode system, 106–107
     telecommunications bonding and grounding, 108–109
   receptacle outlet, 124
   references, 99–101
   and safety, 98
   single-point equipment, 124
   supporting structure, 231–242
     alternating current equipment ground, 241
     bonding, 231

Grounding, supporting structure (*Cont.*):
    and Bonding Conductor for Telecommunications, 233–234
    and building type, 232
    code requirements, 232, 235
    ground test, steps for, 241–242
    grounding conductor, definition of, 231
    hardware for, 236–237
    multistory infratructure, typical, 233, 234
    source, ground, 236
    telecommunications bonding backbone, 240
    telecommunications bonding backbone interconnecting bonding conductor, 240–241
    telecommunications equipment bonding conductor, 240
    telecommunications grounding busbar, 240
    telecommunications main grounding busbar, 237–240
  system practices for, 117–124
    and backbone cable protection, 119–123
    coupled bonding conductors, 121–122
    sheath terminations, exposed, 123–124
    telecommunications bonding backbone (TBB), 121
    in telecommunications closets, 119
    unshielded backbone cable, 122
  telecommunications bonding and grounding system, 97
  with water pipe, 113
  (*See also* Bonding)
Grounding conductor, 231
Grounding electrode systems, 97, 106–107, 111–112
Gypsum board walls, firestopping/firestop systems for, 315–317, 324–325

H

Hair length, 145
Handheld cable testers, 482
Hangers, conduit, 217–218
Hardhats, 139–140
Hardware:
  110-style, 338, 339
  BIX, 338–340

Hardware (*Cont.*):
  for connectors, 69–70
  for copper cable, 94
  for grounding infrastructure, 236–237
  LSA, 340–341
  for terminations, 332
Harnesses, safety, 144
Hazardous environments:
  indoor, 145–154
    access flooring, 148
    asbestos, 152–153
    batteries, 152
    catwalks, 148–149
    chemical hazards, 153–154
    confined spaces, 149–151
    crawl spaces, 149
    electrical hazards, 145–148
    lightning hazards, 148
    optical fiber hazards, 151–152
  outdoor, 154
HC (*see* Horizontal cross-connect)
Headgear, 139–140
Hearing protection, 143–144
Heat coils, 127
Heat-cured terminations, 353, 355–362
Hierarchical star topology, 26, 28
Horizontal cable/cabling, 16–17
  ANSI/TIA/EIA-568-A on, 35
  pulling, 252, 253
    backbone cable, 283–287
    in conduit, 259–266
    in open ceiling, 266–269
  splices in, 375
  supporting structures for, 228–231
    stub-up/stub-out conduits, 228
    surface-mounted raceway, 228–231
  testing, 414–427
    attenuation to crosstalk ratio, 421–422
    cable impedance, 421
    cable testers, 415–418, 422–425
    capacitance, 421
    continuity testing, 414–415
    dc loop resistance, 420–421
    measurement problems, 425–426
    tester performance, 422–425
    TIA/EIA TSB-67/ANSI/TIA/EIA-568-A/addenda requirements, 418–420
Horizontal cross-connect (HC), 14, 21

Housekeeping, general, 248–249
Hybrid topologies, 25–28
   clustered star topology, 26, 27
   hierarchical star topology, 26, 28
   star-wired ring topology, 26, 27
   tree topology, 26

I
IC (*see* Intermediate cross-connect)
IDC termination (*see* Insulation Displacement Connection termination)
IEC (*see* International Electrotechnical Commission)
IEEE (*see* Institute of Electrical and Electronics Engineers, Inc.)
IMC (*see* Intermediate metallic conduit)
Impedance, 89–90, 421, 495
Inductance, 89
Induction amplifier/tone generator, 411
Infrared conversion card, 413
Initial construction meeting, 169, 542
In-line splicing, 378
Institute of Electrical and Electronics Engineers, Inc. (IEEE), 30, 41–42
Insulation (copper cable), 91–92
Insulation Displacement Connection (IDC) termination, 327, 333–348
   66-block, 336–338
   110-style hardware, 338, 339
   BIX hardware, 338–340
   direct connections, 344
   field-constructed patch cords, 344–345
   LSA hardware, 340–341
   patch panels, 341–342
   screened twisted-pair, 345–346
   steps in, 334–336
   tools for, 333, 334
   work area outlets, 343–344
Intermediate cross-connect (IC), 14, 21
Intermediate metallic conduit (IMC), 211–212, 214–215
International Electrotechnical Commission (IEC), 42
International Organization for Standardization (ISO), 30, 42
Intumescent firestops, 296, 310–311
Intumescent wrap strips, 310–311

ISO (*see* International Organization for Standardization)

J
J-hooks, 244–245, 266
Job change orders, 175–176, 183

L
Ladder racks, 245
Ladder safety, 135–138
LANs (*see* Local area networks)
Laser, low-intensity, 414, 483
Lifting belts, 141–142
Lightning hazards, 101–102, 148
Lightning protection systems, 104–105
Local area networks (LANs), 3, 4, 8, 22
Log, project, 166, 174, 527
Low-intensity laser, 414, 483
LSA hardware, 340–341

M
Main cross-connect (MC), 14, 21
MANs, 8
Material list, 164, 178–179
Material Safety Data Sheets (MSDS), 153–154, 309
Materials:
   distribution of, 173
   for firestops, 298, 299, 310
   for installation of cable support systems, 244
   ordering, 170
   receiving, 170–171
   for retrofit installations, 542–543
   storage of, 171–172
MC (*see* Main cross-connect)
Mechanical drawings, 162–163
Meetings:
   initial construction, 169, 542
   periodic, 174–175, 544
   preinstallation, 174, 544
Megahertz, 86
Membrane penetration (firestopping), 294
Metallic conduit, 252
MSDS (*see* Material Safety Data Sheets)

Multilevel cable trays, 198
Multimeter, volt-ohm-milliampere, 411
Multimode optical fiber cable, 64–65, 351–352
Multipair cable, 58
Mushrooms, 202–203

N

National Electrical Code (NEC), 31–33
National Electrical Manufacturers Association (NEMA), 30
National Fire Protection Association (NFPA), 30, 31
Near-end crosstalk (NEXT), 493
NEC (*see National Electrical Code*)
NEMA (National Electrical Manufacturers Association), 30
Network Interface Device (NID), 7
Networks, 3–4
NEXT (near-end crosstalk), 493
NFPA (*see* National Fire Protection Association)
NID (Network Interface Device), 7
Noise, 495–496
N-type coaxial connectors, 79–80
Nuts, 226–227

O

Occupational Safety and Health Administration (OSHA), 30, 129
Ohm's law, 88
Optical fiber cable/cabling, 4, 63–68, 95–96
  ANSI/TIA/EIA-568-A on, 36
  attenuation with, 96
  bandwidth of, 95–96
  connectors for, 80–82
  multimode, 64–65, 351–352
  pulling, 287–291
  singlemode, 65–67, 352
  splicing, 389–408
    fusion splicing, 390
    length determination, 391–392
    mechanical splicing, 390
    for multimode/singlemode fibers, 390
    for outside plant locations, 390
    planning for, 390–391
    safety guidelines for, 391

Optical fiber cable/cabling, splicing (*Cont.*):
    single-fiber automatic fusion splicers, 398–404
    single-fiber mechanical splicing, 390, 393–398
    single-fiber semi-automatic splicers, 404–408
    stripping, 392–393
  terminations for, 328, 351–374
    anaerobic termination, 354, 373–374
    cladding, 352
    connectors for, 353
    core, 352
    crimp-style connectors, 353, 368–372
    grounding/bonding, 353
    heat-cured terminations, 353, 355–362
    multimode, 351–352
    precautions with, 351
    singlemode, 352
    UV-cured terminations, 353, 362–368
  testing, 432–479, 510–513
    acceptance testing, 433–434
    channel or link testing, 435–449
    with light source and power meter, 435–450
    with optical time domain reflectometer, 450–479
    patch cable testing, 449–450
    preinstallation testing, 432–433
    preventive maintenance testing, 434
  troubleshooting. (*see under* Troubleshooting)
Optical fiber flashlight, 413
Optical fiber hazards, 151–152
Optical light source, 414
Optical time-domain reflectometers (OTDRs), 414, 450–474, 483
  accuracy and linearity, 458–460
  automatic measurements, 472–474
  components of, 452–455
  dead zone, 456–458
  dynamic range of, 455–456
  measurement problems with, 467–471
  operation, 461–467
  resolution, 458
  testing procedures, 471–472
  theory of operation, 451–452
  uses of, 450–451
  wavelength, 460–462

OSHA (*see* Occupational Safety and Health Administration)
OTDRs (*see* Optical time-domain reflectometers)
Outlet boxes, sealing of, 322–323
Outlets:
   receptacle, 124
   work area, 343–344

P

Patch cable testing, 449–450
Patch cords, 21, 344–345
Patch panels, 341–342
Penetrations (firestopping), 245–246, 294, 323–325
Periodic meetings, 174–175, 544
Personal protective equipment, 139–145
   breathing protection, 141
   clothing, 144–145
   detection badges, 142–143
   exposure monitors, 142–143
   eye protection, 140–141
   footwear, protective, 142
   gloves, 142
   and grooming, 145
   headgear, 139–140
   hearing protection, 143–144
   lifting belts, 141–142
   safety harnesses, 144
Personnel lifts, 138
PIC cable, 376
Pillows, firestop, 311–312
Planning, 161–187
   and contract, 165
   and customer's drawings, 162–163
   and designer's drawings, 163–165, 177
   and initial construction meeting, 169
   and job change orders, 175–176, 183
   and material list, 164, 178–179
   and materials distribution, 173
   and materials ordering, 170
   and materials receiving, 170–171
   and materials storage, 171–172
   and periodic meetings, 174–175
   and preinstallation meeting, 174
   and project log, 166, 174
   and project plan, 168–169

Planning (*Cont.*):
   and project safety plan, 175
   and project schedule, 165, 173–174, 180–181
   for retrofit installations, 523–526
   and scope of work, 164–165, 184–187
   and site survey, 166–168, 182
Plans, 42–49
Plastic anchors, 221
Plumbing, 163
Plywood backboards (telecommunications closets), 191–195
Power, 84
Power meters, 414, 436
Preinstallation meeting, 174, 544
Preinstallation testing (optical fiber cable), 432–433
Pretermination functions, 329–332
Preventive maintenance testing, 434
Primary protectors, 126, 128
Professionalism, 154–159
   and appearance, 158–159
   and communication skills, 157–158
   and customer relations, 155–156
   and project team, 156–157
Project log, 166, 174, 527
Project plan, 168–169
Project schedule, 165, 173–174, 180–181, 527
Project team, 156–157
Protection systems, 104–109
   electrical bonding and grounding, 107
   electrical power protection, 107–108
   electrical power system, 105–106
   grounding electrode system, 106–107
   lightning protection systems, 104–105
   telecommunications bonding and grounding, 108–109
PTC resistors, 127
Pulling cable, 251–291
   air-propelled methods for, 253
   backbone cabling, 253–254
      horizontal pathway, 283–287
      vertical pathway, 269–283
   and building codes, 254
   and communication with coworkers, 254
   in conduit, 252–253, 259–266
   with fishtape, 253, 259–266

Pulling cable (*Cont.*):
  cable, 252, 253
    backbone, 283–287
    in conduit, 259–266
    in open ceiling, 266–269
  in open ceiling, 266–269
  optical fiber cable, 287–291
  precable pulling tasks, 251–252
  setup for, 254–259
  in vertical pathway
    from bottom up, 276–283
    from top down, 269–276
Putties, firestop, 308, 319–321

Q
Qualified fire-rated assemblies, 299–300

R
Raceway, surface-mounted, 228–231
Radios, two-way, 438
Rayleigh scattering, 451–452
RBOCs (*see* Regional Bell Operating Companies)
REA (*see* Rural Electrification Administration)
REA Cooperatives, 2
Reference adapters, 437
Regional Bell Operating Companies (RBOCs), 3, 5
Relay racks, 205–207
Requests for quote (RFQs), 51
Rescue, emergency, 131
Resistance (copper cable), 87–89
Respirators, 141
Retrofit installations, 519–550
  backup strategy for, 545
  construction meeting, initial, 542
  contracts for, 526
  and evaluation of existing infrastructure, 537–540
  job plan for, 541–542
  materials for, 542–543
  periodic meetings, 544
  planning for, 523–526
  preinstallation meeting, 544
  project log for, 527
  safety plan for, 544–545
  scheduling considerations with, 527, 543–544
  scope of work for, 526
  site survey for, 527–532
  and verification of existing infrastructure, 532–537
Return loss, 90–91
RFQs (requests for quote), 51
Ring topology, 24–25
Rod-stock cable trays, 197–200
Rural Electrification Administration (REA), 2, 5
Rural Utilities Service (RUS), 2, 5

S
Safety, 128–159
  and communication, 132, 133
  and designation of work areas, 133–134
  and emergency rescue, 131
  and first aid, 129–131
  grounding and, 98
  in hazardous environments, 145–154
    indoors, 145–154
    outdoors, 154
  ladder, 135–138
  and OSHA, 129
  and personal protective equipment, 139–145
    breathing protection, 141
    clothing, 144–145
    detection badges, 142–143
    exposure monitors, 142–143
    eye protection, 140–141
    footwear, protective, 142
    gloves, 142
    and grooming, 145
    headgear, 139–140
    hearing protection, 143–144
    lifting belts, 141–142
    safety harnesses, 144
  and personnel lifts, 138
  project safety plan, 175
  and pulling cable, 254–255
  and testing, 441–442
  and tools/equipment, 134–135
Safety ground, 97

Safety harnesses, 144
Safety plan (retrofit installations), 544–545
Schedule (*see* Project schedule)
Scope of work, 164–165, 184–187, 526
Screened twisted-pair (ScTP), 4, 58–59, 77, 345–346
Screws, 224
  drywall, 226
  metal, 225–226
  wood, 224–225
ScTP (*see* Screened twisted-pair)
Seals, firestop, 298, 299, 318–319
Secondary circuit protectors, 127
Service provider (SP), 7
"Shall," 29
Shielded twisted-pair (STP), 59–60
  ANSI/TIA/EIA-568-A on, 36
  connectors for, 77–78
Shielding, copper cable, 92–93
"Should," 29
Silicone foams, 311
Singlemode optical fiber cable, 65–67, 352
Site drawings, 163
Site survey, 166–168, 182, 527–532
Sneak current fuse, 127
SP (service provider), 7
Specialty fasteners, 227–228
Specifications, 42, 49–52
Splicing, 375–408
  copper cable, 375–389
    cable binders, 377
    color code, 376
    foldback splicing method, 378–379
    in-line splicing method, 378
    intrabuilding, 375
    labeling, 380
    listed cables, 377
    modular connectors for, 379
    safety guidelines for, 381
    steps in, 381–389
    wrapping, 380
  optical fiber cable, 389–408
    fusion splicing, 390
    length determination, 391–392
    mechanical splicing, 390
    for multimode/singlemode fibers, 390
    for outside plant locations, 390
    planning for, 390–391

Splicing, optical fiber cable (*Cont.*):
    safety guidelines for, 391
    single-fiber automatic fusion splicers, 398–404
    single-fiber mechanical splicing, 390, 393–397
    single-fiber semi-automatic splicers, 404–408
    stripping, 392–393
Sprays, firestop, 312
Standards, 7, 28–31, 34–42
  ANSI/EIA/TIA-570, 38–39
  ANSI/EIA/TIA-606, 39–40
  ANSI/EIA/TIA-607, 40–41
  ANSI/NFPA-70, 32
  ANSI/TIA/EIA-568-A, 16–19, 21, 34–37
  ANSI/TIA/EIA-569-A, 9, 14, 37–38
  IEEE, 41–42
  ISO/IEC, 42
  and pulling cable, 254
  (*See also* Codes)
Star topology, 22–23
Star-wired ring topology, 26, 27
STP (*see* Shielded twisted-pair)
Strand identifier, 414
Stripping (optical fiber cable), 392–393
Structural drawings, 163
Structural return loss, 87
Structured cabling systems, 6–28
  backbone cables, 14–16
  cross-connects, 20–21
  entrance facilities, 9–14
    aerial entrances, 13–14
    buried entrances, 12, 13
    underground entrances, 10–11
  equipment rooms, 18–19
  generic installation, 8–9
  horizontal cables, 16–17
  topologies, 21–28
    bus topology, 23–24
    hybrid topologies, 25–28
    ring topology, 24–25
    star topology, 22–23
  uniqueness of, 7
  work areas, 17–18
Supporting structures, 189–249
  backbone pathways, 209–228
    anchors for, 221–223

Supporting structures, backbone pathways (*Cont.*):
    elbows/bends, conduit, 216–217
    electrical metallic tubing, 210–211
    fasteners for, 223–228
    hangers for support of, 217–218
    joint make-up for, 215
    preparation of, 220–221
    and securing of conduit formations, 220
    terminations, conduit, 218–220
  equipment rooms, 190
  grounding infrastructure, 231–242
    alternating current equipment ground, 241
    bonding, 231
    and Bonding Conductor for Telecommunications, 233–234
    and building type, 232
    code requirements, 232, 235
    ground test, steps for, 241–242
    grounding conductor, definition of, 231
    hardware for, 236–237
    multistory infrastructure, typical, 233, 234
    source, ground, 236
    telecommunications bonding backbone, 240
    telecommunications bonding backbone interconnecting bonding conductor, 240–241
    telecommunications equipment bonding conductor, 240
    telecommunications grounding busbar, 240
    telecommunications main grounding busbar, 237–240
  horizontal pathways, 228–231
    stub-up/stub-out conduits, 228
    surface-mounted raceway, 228–231
  installation of, 242–249
    and accessibility, 243
    blueprints/specifications for, 243
    cable trays, 245
    conduits, 246–248
    D-rings, 244
    existing support structures, verifying adequacy of, 243
    and general housekeeping, 248–249
    J-hooks, 244–245

Supporting structures, installation of (*Cont.*):
    ladder racks, 245
    load capacity, verification of, 243
    materials/tools, availability of, 244
    and pathway documentation, 248
    penetrations, firestopping, 245–246
    and separation of cable, 243
    wire ducts, 248
  telecommunications closets, 19–20, 190–209
    cable trays, 195–202
    clearances, minimum, 192
    conduits, 203–204
    cross-connects, 191
    floor-mounted cabinets in, 207
    galvanized rigid conduit, 213–215
    ground wires/bus, 204–205
    installation checklist for, 209
    installation of, 202–203
    intermediate metallic conduit, 211–212, 214–215
    plan/design of, 191
    plywood backboards, 191–195
    relay racks, 205–207
    and type of cabling facility, 190–191
    wall-mounted racks/cabinets in, 208–209
Surface-mount boxes, 230–231
Surface-mounted raceway, 228–231
Suspended cable trays, 196
System cutover, 545–550

T

TBB (*see* Telecommunications Bonding Backbone)
TBBIBC (*see* Telecommunications Bonding Backbone Interconnecting Bonding Conductor)
TDR (time domain reflectometer), 413
Team, project, 156–157
TEBC (telecommunications equipment bonding conductor), 240
Telecommunications bonding and grounding system, 97
Telecommunications Bonding Backbone (TBB), 119–122, 240
Telecommunications Bonding Backbone Interconnecting Bonding Conductor (TBBIBC), 119, 120, 240–241

Telecommunications closets, 19–20, 190–209
　cable trays in, 195–202
　clearances, minimum, 192
　conduits, 203–204
　cross-connects, 191
　floor-mounted cabinets in, 207
　galvanized rigid conduit in, 213–215
　ground wires/bus, 204–205
　grounding/bonding in, 119
　installation checklist for, 209
　installation of, 202–203
　intermediate metallic conduit in, 211–212, 214–215
　plan/design of, 191
　plywood backboards, 191–195
　relay racks in, 205–207
　and type of cabling facility, 190–191
　wall-mounted racks/cabinets in, 208–209
Telecommunications equipment bonding conductor (TEBC), 240
Telecommunications Grounding Busbar (TGB), 119, 240
Telecommunications industry, 1–2
Telecommunications Industry Association (TIA), 5, 7, 31, 189
Telecommunications Main Grounding Busbar (TMGB), 118–119, 237–240
Telecommunications Reform Act of 1996, 5
Telephone service, 1–6
Telephone test set (butt set), 414
Termination(s), 327–374
　anaerobic-style fiber termination, 373–374
　backbone conduits, 218–220
　cable management hardware for, 332
　copper cable, 327–328
　Copper IDC Termination, 333–348
　　66-block, 336–338
　　110-style hardware, 338, 339
　　BIX hardware, 338–340
　　direct connections, 344
　　field-constructed patch cords, 344–345
　　LSA hardware, 340–341
　　patch panels, 341–342
　　screened twisted-pair, 345–346
　　steps in, 334–336
　　tools for, 333, 334
　　work area outlets, 343–344
　crimp connectors, 328

Termination(s) (Cont.):
　crimp-style fiber termination, 353, 368–372
　forming and dressing cables, 329–331
　heat-cured fiber termination, 353, 355–362
　IDC, 327
　length and slack determination, 331–332
　optical fiber cable, 328, 351–374
　　anaerobic termination, 354, 373–374
　　cladding, 352
　　connectors for, 353
　　core, 352
　　crimp-style connectors, 353, 368–372
　　grounding/bonding, 353
　　heat-cured terminations, 353, 355–362
　　multimode, 351–352
　　precautions with, 351
　　singlemode, 352
　　UV-cured terminations, 353, 362–368
　organizing by destination, 329
　pretermination functions, 329–332
　tools for
　　coaxial cable, 349
　　copper IDC, 333, 334
　for undercarpet flat cable, 328
　UV-cured fiber termination, 353, 362–368
　wirewraps, 327–328
Test jumpers, 437
Testing, 409–479
　backbone cable, 426–427
　coaxial cable, 428–432
　　50-ohm, 429–430
　　75-ohm, 431–432
　copper cable, 409–414
　　backbone, 426–427
　　coaxial cable, 428–432
　　field testers, 411–414
　　horizontal cabling, 414–426
　equipment selection, 474–477
　fiber link attenuation record, 479
　fiber performance calculations worksheet, 478
　of firestops, 294, 300
　ground, 241–242
　horizontal cabling, 414–426
　　attenuation to crosstalk ratio, 421–422
　　cable impedance, 421
　　cable testers, 415–418, 422–425
　　capacitance, 421

Testing, horizontal cabling (*Cont.*):
    continuity testing, 414–415
    dc loop resistance, 420–421
    measurement problems, 425–426
    tester performance, 422–425
    TIA/EIA TSB-67/ANSI/TIA/EIA-568-A/addenda requirements, 418–420
    optical fiber cable, 432–479, 510–513
      acceptance testing, 433–434
      channel or link testing, 435–449
      with light source and power meter, 435–450
      with optical time domain reflectometer, 450–479
      patch cable testing, 449–450
      preinstallation testing, 432–433
      preventive maintenance testing, 434
TGB (*see* Telecommunications Grounding Busbar)
Thicknet, 62–63
Thinnet, 63
Through penetration (firestopping), 294
TIA (*see* Telecommunications Industry Association)
Time domain reflectometer (TDR), 413
TMGB (*see* Telecommunications Main Grounding Busbar)
Tools:
   for installation of cable support systems, 244
   safety practices with, 134–135
   for terminations
     coaxial cable, 349
     copper IDC, 333–336
Topologies, structured cabling system, 21–28
   bus topology, 23–24
   hybrid topologies, 25–28
     clustered star topology, 26, 27
     hierarchical star topology, 26, 28
     star-wired ring topology, 26, 27
     tree topology, 26
   ring topology, 24–25
   star topology, 22–23
Transmission, 83–87
   ac vs. dc, 84–85
   analog, 85–86
   and current, 84
   digital, 86
   and frequency, 85

Transmission (*Cont.*):
   and power, 84
   and voltage, 84
Tree topology, 26
Troubleshooting, 481–518
   copper cable, 489–505
     10BASE-2 coaxial cable, 504–505
     10BASE-2 (thin ethernet) coaxial cable, 501–502
     attenuation-to-crosstalk ratio, 492
     cable length, incorrect, 493–494
     cables, locating, 496
     capacitance, incorrect, 492
     connections, incorrect, 494
     general guidelines, 489–491
     high attenuation, 491–492
     impedance mismatches, 495
     intermittent faults, 499–500
     interpreting test results, 491
     loop resistance, excessive, 494
     NEXT, excessive, 493
     noise, 495–496
     remote end cannot be located, 499
     test set selection, 499
     twisted pair, 496–500
     UTP cable, 502–503
   documentation for, 485
   equipment needed for, 481–485
   and isolation of problems, 486–489
   optical fiber cable, 505–518
     fault isolation, steps for, 513–515
     isolation of problem, 506–509
     OTDR fault presentations, typical, 516–518
     repair of problem, 509–510
     testing, 510–513
     verification, telephonic, 506
   skills/knowledge required for, 485–486
Tubular cable trays, 195–197
Two-way radios, 438

U

UDCs (universal data connectors), 347
UL (Underwriters Laboratories, Inc.), 31
Undercarpet cable, 56–57, 328
Underground entrances, structured cabling systems for, 10–11

Underwriters Laboratories, Inc. (UL), 31
Universal data connectors (UDCs), 347
Unshielded twisted-pair (UTP), 4, 53–58
   ANSI/TIA/EIA-568-A on, 36
   connectors for, 70–76
   troubleshooting, 502–503
UV-cured terminations, 353, 362–368

V

Vacuum hoses, 265–266
Vertical pathway, pulling cable in:
   backbone cabling, 269–283
   from bottom up, 276–283
   from top down, 269–276
Video communications, 3
Voice communications, 3

Voltage, 84
Volt-ohm-milliampere multimeter, 411

W

Walls, firestopping/firestop systems for:
   brick/concrete/concrete block walls, 312–315
   gypsum board walls, 315–317
WANs (wide area networks), 8
Washers, 227
Wide area networks (WANs), 8
Wire map testers, 411–412, 482
Wirewraps, 327–328
Work area outlets, 343–344
Work areas, 17–18, 133–134
Work clothing, 144–145

We wrote the book!

The BICSI Approach

Recognized throughout the world as a leader in telecommunications education, BICSI specializes in low-voltage cabling. Focusing on commercial distribution design and installation, BICSI will soon add residential installation to its telecommunications mix. BICSI offers a full line of educational products and services, including registration programs, courses, technical publications, and conferences, including the annual BICSI Cabling Workshop.

Professional Registrations

Earn international credentials by demonstrating your proficiency in a specific area. Currently, BICSI offers professional registration programs in cabling installation (BICSI Registered Installer, Level 1; Installer, Level 2; and Technician), distribution design (Registered Communications Distribution Designer—RCDD®), and LAN design (RCDD/LAN Specialist). An RCDD/Outside Plant Specialty and a Registered Residential Installer designation is planned for the future. Many leaders in the premises cabling components industry now endorse the BICSI Telecommunications Cabling Installation Training and Registration Program. This recognition may save you time and money by allowing you to participate in one vendor-neutral training program!

Vendor-Neutral Training

BICSI offers a variety of learning opportunities, from structured classroom instruction to flexible Internet home study. With instruction now available via the Web and at 100 licensed training centers, BICSI presents leading-edge technical training in many parts of the world. BICSI's virtual campus features courses such as *The Residential Telecommunications Cabling Standard—Understanding ANSI/TIA/EIA-570-A* course. BICSI's *Educational Resource Catalog* outlines BICSI's 25 courses, and is available upon request.

BICSI®
A Telecommunications Association

Technical Publications

Since 1974, BICSI has been developing a library of technical publications, several of which have become industry standards. These reference books span the subjects of cabling installation, outside plant design, LAN design, and more. Many of these valuable reference tools double as study guides for BICSI courses and exams. BICSI members receive a significant discount on manuals, and may enjoy even greater savings if the previous edition is owned. *Note: All manuals also available on CD-ROM at manual price.*

- BICSI's *Telecommunications Distribution Methods Manual (TDMM)*, now the accepted guideline of the industry, is a valuable reference tool for those who design the telecommunications infrastructure. The *TDMM* also serves as a detailed study guide for those preparing to take the Registered Communications Distribution Designer (RCDD®) exam. US$179 BICSI member; US$329 non-member. ISBN 1-928886-04-3 (manual); ISBN 1-928886-05-1 (CD).

- In the *Telecommunications Cabling Installation Manual,* you'll find step-by-step procedures for installing telecommunications cable, and useful information included in the BICSI Installer and Technician exams. Also available as a hard cover, 3-ring binder. US$99 BICSI member; US$99 non-member (3-ring binder version).

- The *LAN and Internetworking Design Manual* describes all aspects of networking—LANs, local internetworks (backbone/campus), and wide area internetworks and associated telecommunications links. US$179 BICSI member; US$329 non-member. ISBN 1-928886-01-9 (manual); ISBN 1-928886-02-7 (CD).

- Learn design guidelines for the integration of emerging applications into existing LANs in the *LAN and Internetworking Applications Guide.* US$29 BICSI member; US$29 non-member. ISBN 1-928886-00-0.

- Ideal for those with previous outside plant design experience, our *Customer-Owned Outside Plant Design Manual* offers an overview of outside plant design, including pathways and spaces, CO-OSP cabling infrastructure, and more. US$99 BICSI member; US$179 non-member.

- BICSI's newest release, an *Introduction to Commercial Voice/Data Cabling Systems* Video and Workbook, provides a visual tour of voice/data cabling for the modern commercial building, as well as the spaces and systems that comprise its infrastructure. US$249 BICSI member; US$349 non-member.

Introduction to Commercial Voice/Data Cabling Systems Video and Workbook

- The *BICSI Telecommunications Dictionary* is a compilation of glossaries from all of BICSI's publications. Acronyms, abbreviations, symbols, and international telecommunications standards can all be found in this convenient book. US$19 BICSI member; US$19 non-member. ISBN 1-928886-03-5.

- BICSI also offers the *On-the-Job Training Booklet,* which provides a performance checklist of key cabling installation tasks—perfect for candidates studying for certification as a BICSI Installer or Technician, or for the supervisor looking to evaluate employee performance. US$9 BICSI member; US$9 non-member.

- And finally, BICSI developed the *Telecommunications Quick Reference Guide for Code Officials: Summary and Excerpts from the NEC® 1999,* which outlines portions of the National Fire Protection Association (NFPA 70), *National Electrical Code®* *(NEC), 1999 Edition.* Free to both BICSI members and non-members.

New Member Instant Reference Library Deal

Save over US$250! For a limited time only, BICSI is offering a New Member Instant Reference Library Deal, allowing new members to purchase all of the above publications (with their CD-ROM counterpart) at a significant savings—20% off of the discounted member price! New members who purchase the entire library within 30 days of becoming a BICSI member pay only US$995 plus shipping!

Check out BICSI's publications in more detail. Visit the BICSI Web site (www.bicsi.org) for additional information on any of BICSI's reference books. There you'll find sample chapters of all BICSI manuals, a complete price listing, and an online order form. You may also call BICSI's Customer Service Department at 800-242-7405 or 813-979-1991 to request a publications brochure (also available on the BICSI Web site). For your convenience, an information request form may be found on the following page.

Membership

To remain competitive in our changing environment, you need to stay abreast of telecommunications issues, standards, and technology—locally and around the globe. Fortunately, BICSI is here to keep you informed and knowledgeable in all aspects of the telecommunications profession. With more than 22,000 members anticipated by the end of the year 2001, BICSI offers members substantial discounts on our quality technical publications, design courses, and conference fees. In fact, the cost of membership can more than pay for itself with the purchase of just one BICSI manual.

We recognize that you may have questions about BICSI, so we encourage you to contact BICSI for a BICSI Information Packet. BICSI membership is your key to a successful career in telecommunications, and we want you to appreciate all of the member benefits.

BICSI Information Packet Available

To request a BICSI Information Packet, complete the form below and fax it to 813-971-4311. You may also request a packet by calling 800-242-7405 or 813-979-1991 or e-mailing bicsi@bicsi.org.

Yes, I want to find out more about BICSI. Please send me a BICSI Information Packet:

_____
name

_____
address

_____
city   state/province   zip/postal code   country

_____
phone   fax

_____
e-mail

Would you like to receive information about BICSI's annual Cabling Workshop?
❑ yes   ❑ no

For more information or to request a BICSI
Information Packet, contact BICSI today!

BICSI, 8610 Hidden River Parkway, Tampa, FL 33637-1000 USA
800-242-7405 or 813-979-1991; fax: 813-971-4311
e-mail: bicsi@bicsi.org; Web site: www.bicsi.org